Multi-Processor System-on-Chip 1

To my parents, sisters and husband, the loves and pillars of my life.

Liliana ANDRADE

I express my profound gratitude to Karine, my parents and all my family, for their help and support throughout all these years.

Frédéric ROUSSEAU

SCIENCES

Electronics Engineering, Field Director – Francis Balestra

Design Methodologies and Architecture,
Subject Head – Ahmed Jerraya

Multi-Processor System-on-Chip 1

Architectures

Coordinated by
Liliana Andrade
Frédéric Rousseau

iSTE

WILEY

First published 2020 in Great Britain and the United States by ISTE Ltd and John Wiley & Sons, Inc.

ISTE Ltd
27-37 St George's Road
London SW19 4EU
UK

www.iste.co.uk

John Wiley & Sons, Inc.
111 River Street
Hoboken, NJ 07030
USA

www.wiley.com

Library of Congress Control Number: 2020940076

British Library Cataloguing-in-Publication Data
A CIP record for this book is available from the British Library
ISBN 978-1-78945-021-7

ERC code:
PE6 Computer Science and Informatics
PE6_1 Computer architecture, pervasive computing, ubiquitous computing
PE6_10 Web and information systems, database systems, information retrieval and digital libraries, data fusion
PE7 Systems and Communication Engineering
PE7_2 Electrical engineering: power components and/or systems

MIX
Paper from responsible sources
FSC www.fsc.org FSC® C013604

Contents

Foreword

Ahmed JERRAYA

Cyber Physical Systems Programs, CEATech, Grenoble, France

Multi-core and multi-processor SoC (MPSoC) concepts started in the late 1990s, mainly to mitigate the complexity of application-specific integrated circuits (ASICs) and to bring some flexibility. The integration of instruction-set processors into ASIC design aimed both to structure the architecture and to allow for programmability. The concept was adopted for general-purpose CPU and GPU in the second phase. Among the pioneers of MPSoC design, we can list the MPA architecture from ST that used eight specific cores to implement MPEG4 in 1998. This evolved 10 years later to give rise to MPPA, the Kalray's general-purpose MPSoC architecture. Another pioneer is the emotion engine from Sony that used five cores (two DSP and three RISC) to build the application processor for the PlayStation (PS2). This also evolved and later converged to bring the CELL architecture (developed jointly by Sony, IBM and Toshiba) in 2005. In 2000, Lucent announced Daytona (quad SPARC V8), and in 2001, Philips designed the famous Viper architecture that combined a MIPS architecture and a DSP (Trimedia). In 2004, TI introduced the OMAP architecture that combined an ARM and a DSP. Using MPSoC to build specific architectures is continuing, and almost every SoC produced today is a multi (or many) core architecture. An important evolution took place in 2005 with the ARM MPCore, the first general-purpose quad core. This was followed by several commercial, general-purpose multi-cores, including Intel Core Duo Pentium, AMD Opteron, Niagra Spark, the Cell processor (8 Cell cores + PowerPC, ring network).

MPSoC started a new computing era, but brought a twofold challenge: building multi-core HW that can be used easily by SW designers, and building distributed SW that fully exploits HW capabilities. To deal with these challenges, the design communities from Academia and Industry began a series of conferences and workshops to rethink classical distributed computing. The study of new methods, models and tools to deal with these new distributed HW and SW architectures generated new concepts, such as the interconnect architectures called network-on-chip (NoC). The MPSoC Forum, created in 2001, was the first interdisciplinary forum that brought together the leading thinkers from the different fields to design multi-core and multi-processor SoC. Over the last 20 years, MPSoC has been a unique opportunity for me to meet so many of the world's top researchers and to communicate with them in person, in addition to enjoying the high-quality conference programs. The confluence of academic and industrial perspectives, and hardware and software, makes MPSoC not "yet another conference". I have learned how emerging SW and HW design technologies and architectures can benefit from advanced semiconductor manufacturing technologies to build energy-efficient multi-core architectures that can serve advanced computing (image, vision and cloud) and distributed networked systems. This book, in two volumes (Architectures and Applications), was published to celebrate the 20th anniversary of MPSoC with outstanding contributions from previous MPSoC events.

This first volume on architectures covers the key components of MPSoC: processors, memory, interconnect and interfaces.

Acknowledgments

Liliana ANDRADE and Frédéric ROUSSEAU

Université Grenoble Alpes, CNRS, Grenoble INP, TIMA, 38000 Grenoble, France

The editors are indebted to the MPSoC community who made this book possible. First of all, they acknowledge the societies that supported this project. EDAA and IEEE/CAS partially funded the organization of the first two events. Since its creation, IEEE/CEDA has sponsored the event. Industrial sponsors played a vital role in keeping MPSoC alive for the last 20 years; special thanks to Synopsys, Arteris, ARM, XILINX and Socionext. The event was created by a nucleus of several people who now form the steering committee (Ahmed Jerraya, Hannu Tenhunen, Marilyn Wolf, Masaharu Imai and Hiroto Yasuura). A larger group has, for the last 20 years, been working to form the community (Nicolas Ventroux, Jishen Zhao, Tsuyoshi Isshiki, Frédéric Rousseau, Anca Molnos, Gabriela Nicolescu, Hiroyuki Tomiyama, Masaaki Kondo, Hiroki Matsutani, Tohru Ishihara, Pierre-Emmanuel Gaillardon, Yoshinori Takeuchi, Tom Becnel, Frédéric Pétrot, Yuan Xie, Koji Inoue, Masaaki Kondo, Hideki Takase and Raphael David). The editors would like to acknowledge the outstanding contribution of the MPSoC speakers, and especially those who contributed to the chapters of this book. Finally, the editors would like to thank the people who participated in the careful reading of this book (Breytner Fernandez and Bruno Ferres).

PART 1

Processors

1

Processors for the Internet of Things

Pieter VAN DER WOLF[1] and Yankin TANURHAN[2]

[1] Solutions Group, Synopsys, Inc., Eindhoven, The Netherlands

[2] Solutions Group, Synopsys, Inc., Mountain View, USA

The Internet of Things (IoT) enables a "smart world" in which many billions of devices communicate to provide advanced functionalities. More specifically, a broad variety of IoT edge devices that can "sense", "listen" and "see" is emerging to capture data for further processing and communication. Such devices have a broad range of compute requirements for implementing a mixture of control processing, digital signal processing (DSP), machine learning, security, etc. In this chapter, we analyze the compute requirements of IoT edge devices and discuss processor capabilities that support the efficient implementation of such devices. More specifically, we focus on IoT edge devices that demand low power consumption. We present concrete examples of versatile, configurable and extensible processors that provide capabilities for control processing, DSP (e.g. voice/audio processing, communications) and machine learning. The processor examples are complemented with benchmark data for illustrative IoT edge application functions.

1.1. Introduction

In recent years, computing paradigms have evolved significantly. One such paradigm is cloud computing, a centralized paradigm that aims to offer computing

---

For a color version of all figures in this book, see www.iste.co.uk/andrade/multi1.zip.

Multi-Processor System-on-Chip 1 – Architectures,
coordinated by Liliana ANDRADE and Frédéric ROUSSEAU. © ISTE Ltd 2020.

as a utility. Another complementary paradigm is edge computing, a decentralized paradigm that aims to offer smart compute capabilities at the edge of the network. A related term is the Internet of Things (IoT), which refers to large numbers of interconnected computing devices aimed at offering services to a great variety of applications.

In this chapter, we specifically focus on *IoT edge devices*: smart devices at the edge of the network that interact with the "real world". These devices acquire data from the environment using sensors. This data is subsequently processed locally on the IoT edge device and/or on computing devices in the network. For each application, a proper trade-off must be made about which functions to perform where based on the requirements for computing, bandwidth, latency, connectivity, security, reliability, etc.

The number of IoT edge devices is predicted to grow to tens of billions over the coming years. Some example IoT edge devices are:

– smartphones and tablets;

– smart doorbells with cameras, performing face detection for triggering an alert, accompanied by an image or video, on the owner's smartphone;

– smart speakers with voice control, employing local speech recognition for a limited vocabulary of voice commands while relaying other speech data into the cloud for more advanced analysis;

– smart sensing devices used in agriculture to monitor and control, for example, soil quality, crop yield and livestock, while sporadically communicating data over cellular connections using, for example, NB-IoT protocols for low power consumption.

Many IoT edge devices are battery-operated and demand an optimized implementation in order to enable a long battery life. Therefore, we must target low power consumption for functions that need to be performed in software locally on the IoT edge device. This, in turn, requires programmable processors that are optimized for executing these software functions efficiently, which is the topic of this chapter.

1.2. Versatile processors for low-power IoT edge devices

1.2.1. *Control processing, DSP and machine learning*

Low-power IoT edge devices typically perform a range of different functions locally on the device. They run a local application that controls the device, its sensors and other interfaces, such as a communications interface to the network and a user interface. For this purpose, a processor must have capabilities for efficient processing of control code, including low branch overheads, efficient interrupt handling, timers, efficient integration with peripherals, support for real-time kernels, etc.

Furthermore, IoT edge devices typically perform some processing on the data acquired through their sensors. These can be sensors to monitor physical phenomena, such as thermometers, gyroscopes, accelerometers and magnetometers. Let us consider, for example, a personal health device or the smart sensing devices used in agriculture, mentioned above. Data rates for this type of sensors are typically low. Microphones are another type of sensors that have higher data rates. For example, a 16 kHz sample rate is often used for voice data. Even higher data rates can be observed in IoT edge devices that use a camera, such as a smart doorbell performing face detection. Data rates for image and video data can vary largely, based on resolution and frame rates. Data rates of hundreds of MB/s are not unusual in high-end devices, but for more power-sensitive camera-based applications, much lower data rates can be observed.

The processing of sensor data typically involves *digital signal processing* (DSP) with functions such as filtering (e.g. FIR, correlation, biquad), transforms (e.g. FFT, DCT), and vector and matrix operations. Voice data can be processed by various DSP functions, including noise reduction and echo cancellation. In addition, the IoT edge device can perform encoding and/or decoding of voice or audio data. For example, consider an audio playback function on the device.

Communicating data involves further DSP functions. For example, some key functions in an NB-IoT protocol stack involve FFT, auto- and cross-correlations, and complex multiplications and convolutions. Furthermore, trigonometric functions such as sine and cosine must be performed. In addition, such protocol stacks perform convolutional coding, for example, Viterbi.

We conclude that the efficient processing of sensor data on an IoT edge device requires processors equipped with DSP capabilities. The relevant DSP capabilities are:

– support for fixed-point data types and arithmetic, including fixed-point multiply-accumulate (MAC) instructions, wide accumulators, and efficient saturation and rounding;

– support for floating-point data types and instructions, including fused multiply-add instructions;

– advanced address generation for efficient memory access, including circular and bit-reversed addressing for DSP kernels such as FIR filters and FFTs;

– zero-overhead loops;

– support for complex data types and arithmetic, including complex multiply and MAC instructions;

– support for vector or SIMD processing to enable increased efficiency by exploiting data parallelism;

– efficient divide and square root operations;

– high load/store bandwidth, as DSP functions can be memory-access intensive.

In addition to control processing and DSP, *machine learning* has recently emerged in various application areas as a technology for building IoT edge devices with advanced functionalities. Some illustrative examples are smart speakers, wearable activity trackers and smart doorbells. These devices apply machine learning technology that has been trained to recognize certain complex patterns (e.g. voice commands, human activity, faces) from data captured by one or more sensors (e.g. a microphone, a gyroscope, a camera). When such a pattern is recognized, the device can perform an appropriate action. For example, when the voice command "play music" is recognized, a smart speaker can initiate the playback of a song. In the following sections, we dig deeper into the requirements and processor capabilities for efficient machine learning in low-power IoT devices.

Integrated circuits for low-power IoT edge devices may use one or more processors for implementing the different types of processing. Multiple processors are required if a single processor cannot handle the complete software workload. A further reason for using multiple processors is that specialized processors can be used for the different types of processing. More specifically, different processors can be used for control processing, DSP and machine learning.

However, there are also good reasons to aim to reduce the number of processors. Lower cost is a key benefit, which is particularly relevant for low-cost IoT edge devices that are produced in high volumes. The use of fewer processors also reduces design complexity, as it simplifies the interconnect and memory subsystem required to integrate the processors. Furthermore, if multiple interacting functions are combined to be executed on a single processor, then this will limit data movements and reduce the software overhead for communication. An additional benefit for software developers is that a single tool chain can be used. To enable the flexible combination of functions, we need versatile **processors** that can efficiently execute different types of workloads, including control tasks, DSP and machine learning. Such processors are also referred to as DSP-enhanced RISC cores. They add a broad set of instructions for DSP and machine learning to a RISC core. If done well, the hardware overhead of these additions is small, for example, by sharing the register file and having unified functional units (e.g. a multiplier) for control processing, DSP and machine learning. Today, optimized DSP-enhanced RISC cores are available from IP vendors.

1.2.2. *Configurability and extensibility*

Integrated circuits for low-power IoT edge devices are often built using off-the-shelf processor IP that can be licensed from IP vendors. Since such licensable processors are multi-purpose by nature, to enable reuse across different customers and applications, they may not be optimal for efficiently implementing a specific set of application functions. However, some of these licensable processors offer support for customization by chip designers, in order to allow the processors to be tailored to the functions they need to perform for a specific application (Dutt and Choi 2003). More specifically, two mechanisms can be used to provide such customization capabilities:

– *Configurability*: the processor IP is delivered as a parameterized processor that can be configured by the chip designer for the targeted application. More specifically, unnecessary features can be deconfigured and optimal parameters can be selected for various architectural features. This may involve optimization of the compute capabilities, memory organization, external interfaces, etc. For example, the chip designer may configure the memory subsystem with closely coupled memories and/or caches. Configurability allows performance to be optimized for the application at hand, while reducing area and power consumption.

– *Extensibility*: the processor can be extended with custom instructions to enhance the performance for specific application functions. For the application at hand, the performance may be dominated by specific functions that execute critical code segments. The execution of such code segments may be accelerated dramatically by adding a few custom instructions. A further benefit of using these custom instructions is that the code size is reduced.

Both configurability and extensibility need to be used at design time. This must be supported by a tool chain (i.e. compiler, simulator, debugger) that is automatically enhanced to support the selected configuration and the added custom instructions. For example, the compiler must generate optimal code for the selected configuration while supporting programmers in using the custom instructions. Similarly, simulation models must support the selected configuration and include the custom instructions. If done properly, large performance gains can be achieved while optimizing area, power and code size, with a minimal impact on design time.

As an example of extensibility, we consider Viterbi decoding, which is a prominent function in an NB-IoT protocol stack for performing forward error correction (FEC) in the receiver. When using a straightforward software implementation on an off-the-shelf processor, this kernel becomes one of the most computationally intensive parts of an NB-IoT modem. Viterbi or similar FEC schemes are used in many communication technologies, especially in the IoT field, and often are a bottleneck in modem design.

In (Petrov-Savchenko and van der Wolf 2018), a processor extension for Viterbi decoding is presented using four custom instructions, which enhance the performance to just a few cycles per decoded bit. The instructions include a reset instruction, two instructions to calculate the path metrics and one instruction for the traceback. The instructions can be conveniently used as intrinsic instructions in the C source code. The resulting implementation reduces the worst-case MHz requirements for the Viterbi decoding function in an NB-IoT protocol stack to less than 1 MHz.

We note that the ability to extend the processor with custom instructions is radically different from adding an external hardware accelerator on a system bus. Using an external bus-based hardware accelerator requires data to be moved over a bus, with additional memory and synchronization requirements (e.g. through interrupts), thereby impacting area, cycles, power consumption and code size. When using custom instructions, these can be used directly in the software thread on the processor, accessing data that is available locally on the processor, in local registers or in local memory. Hence, there are no overheads for moving data to/from an accelerator or for performing explicit synchronization. It also greatly simplifies software development.

1.3. Machine learning inference

In this section, we investigate in detail the requirements and processor capabilities for efficient machine learning in low-power IoT edge devices. The common theme in machine learning is that algorithms that have the ability to learn without being explicitly programmed are used (Samuel 1959). As shown in Figure 1.1, in machine learning, we distinguish between *training* and *inference*.

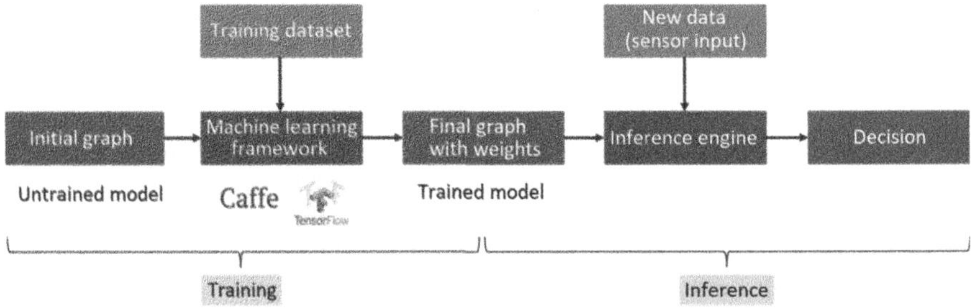

Figure 1.1. *Training and inference in machine learning*

Training starts with an untrained model, for example a multi-layered *neural network* with a chosen graph structure. In these neural networks, each layer transforms input data into output data while applying sets of coefficients or *weights*. Using a machine learning framework such as Caffe or TensorFlow, the model is trained using a large training dataset. The result of the training is a trained model, for example, a neural network with its weights tuned for classifying input data into certain categories. Such categories may, for example, be the different types of human activity in the activity tracker device mentioned above.

Inference uses the trained model for processing input data captured by sensors to infer the complex patterns it has been trained to recognize. For example, it can check whether the input data matches one of the categories that a neural network has been trained for, such as "walking" or "sitting" in the activity tracker device. Therefore, upon inference, the trained model is applied to new data. Inference is typically performed in the field. In this chapter, we focus on inference rather than on training. More specifically, we address the efficient implementation of machine learning inference on a programmable processor.

The processing requirements of machine learning inference can vary wildly for different applications. Some key factors impacting the processing requirements are:

– input data rate: this is the rate at which data samples are captured by the sensor(s). These samples can, for example, be pixels coming from a camera or pulse-code modulation (PCM) samples coming from a microphone. The input data rate can range from tens of samples per second, for example, for human activity recognition with a small number of sensors, to hundreds of millions of samples per second, for advanced computer vision with a high-resolution camera capturing images at a high frame rate;

– complexity of the trained model: this defines the number of operations to be performed for a set of samples (e.g. an input image) upon inference. For example, in the case of neural networks, the complexity depends on the number of layers in a graph, the sizes of the (multidimensional) input and output maps for each layer, and the number of weights to be applied in the calculation of the output maps. A low-complexity neural network has less than 10 layers, while a high-complexity neural network can have tens of layers (Szegedy *et al.* 2015).

Table 1.1 shows input data rates and model complexities for several example machine learning applications.

Machine learning application	Input data rate	Complexity of trained model
Human activity recognition	10s Hz (few sensors)	Low to medium
Voice control	10s kHz (e.g. 16 kHz)	Low to medium
Face detection	100s kHz (low resolution and frame rate)	Low to medium
Advanced computer vision	100s MHz (high resolution and frame rate)	High

Table 1.1. *Input data rates and model complexities for example machine learning applications*

As can be seen from Table 1.1, input data rates and model complexities can vary significantly. As a result, the compute requirements for machine learning inference can differ by several orders of magnitude. In this chapter, we focus on machine learning inference with low-to-medium compute requirements. More specifically, we target machine learning inference that can be used to build smart low-power IoT edge devices.

In the next section, we further detail the requirements for the efficient implementation of low/mid-end machine learning inference. Using the DSP-enhanced DesignWare® ARC® EM9D processor as an example, we discuss the features and capabilities of a programmable processor that enable efficient execution of computations often used in machine learning inference. We further present an extensive library of software functions that can be used to quickly implement inference engines for a wide array of machine learning applications. Benchmarks are presented to demonstrate the efficiency of machine learning inference implemented using this software library that executes on the ARC EM9D processor.

1.3.1. *Requirements for low/mid-end machine learning inference*

IoT edge devices that use machine learning inference typically perform different types of processing, as shown in Figure 1.2.

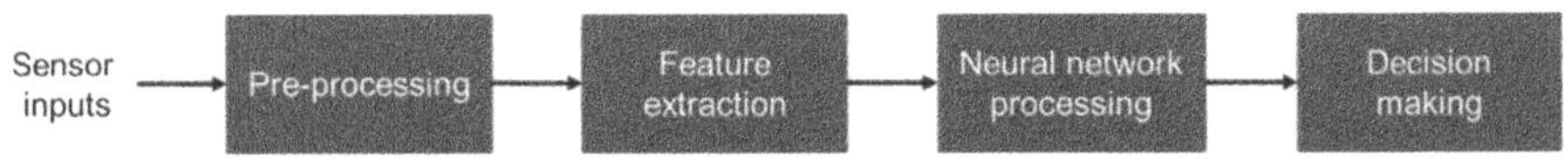

Figure 1.2. *Different types of processing in machine learning inference applications*

These devices typically perform some pre-processing and feature extraction on the sensor input data before performing the actual neural network processing for the trained model. For example, a smart speaker with voice control capabilities may first pre-process the voice signal by performing acoustic echo cancellation and multi-microphone beam-forming. It may then apply FFTs to extract the spectral features for use in the neural network processing, which has been trained to recognize a vocabulary of voice commands.

1.3.1.1. *Neural network processing*

For each layer in a neural network, the input data must be transformed into output data. An often used transformation is the *convolution*, which convolves, or, more precisely, correlates, the input data with a set of trained weights. This transformation is used in *convolutional neural networks* (CNNs), which are often applied in image or video recognition.

Figure 1.3 shows a 2D convolution, which performs a *dot-product* operation using the weights of a 2D weight kernel and a selected 2D region of the input data with the same width and height as the weight kernel. The dot product yields a value (M23) in the output map. In this example, no padding is applied on the borders of the input data, hence the coordinate (2, 3) for the output value. For computing the full output map, the weight kernel is "moved" over the input map and dot-product operations are performed for the selected 2D regions, producing an output value with each dot product. For example, M24 can be calculated by moving one step to the right and performing a dot product for the region with input samples A24–A26, A34–A36 and A44–A46.

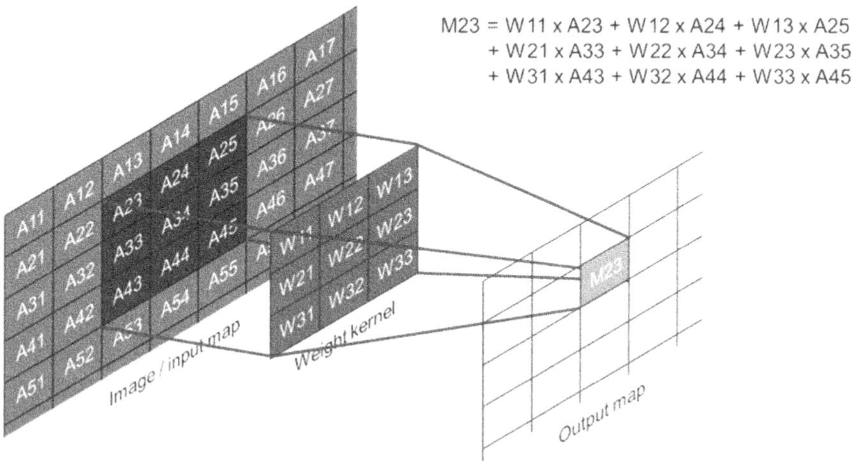

Figure 1.3. *2D convolution applying a weight kernel to input data to calculate a value in the output map*

Input and output maps are often three-dimensional. That is, they have a width, a height and a depth, with different planes in the depth dimension typically referred to as *channels*. For input maps with a depth > 1, an output value can be calculated using a dot-product operation on input data from multiple input channels. For output maps with a depth > 1, a convolution must be performed for each output channel, using different weight kernels for different output channels. *Depthwise convolution* is a special convolution layer type for which the number of input and output channels is the same, with each output channel being calculated from the one input channel with the same depth value as the output channel. Yet another layer type is the *fully connected* layer, which performs a dot-product operation for each output value using the same number of weights as the number of samples in the input map.

The key operation in the layer types described above is the dot-product operation on input samples and weights. It is therefore a requirement for a processor to implement such dot-product operations efficiently. This involves efficient computation, for example, using MAC instructions, as well as efficient access to input data, weight kernels and output data.

CNNs are feed-forward neural networks. When a layer processes an input map, it maintains no state that impacts the processing of the next input map. *Recurrent neural networks* (RNNs) are a different kind of neural network that maintain the state while processing sequences of inputs. As a result, RNNs also have the ability to recognize patterns across time, and are often applied in text and speech recognition applications.

There are many different types of RNN cells from which a network can be built. In its basic form, an RNN cell calculates an output as shown in equation [1.1]:

$$h_t = f(x_t \cdot W_x + h_{t-1} \cdot W_h + b) \qquad [1.1]$$

where x_t is the frame t in the input sequence, h_t is the output for x_t, W_x and W_h are weight sets, b is a bias, and $f()$ is an output activation function. Thus, the calculation of an output involves a dot product of one set of weights with new input data and another dot product of another set of weights with the previous output data. Therefore, also for RNNs, the dot product is a key operation that must be implemented efficiently. The long short-term memory (LSTM) cell is another well-known RNN cell. The LSTM cell has a more complicated structure than the basic RNN cell that we discussed above, but the dot product is again a dominant operation.

Activation functions are used in neural networks to transform data by performing some nonlinear mapping. Examples are rectified linear units (ReLU), sigmoid and hyperbolic tangent (TanH). The activation functions operate on a single data value and produce a single result. Hence, for an activation layer, the size of the output map is equal to the size of the input map.

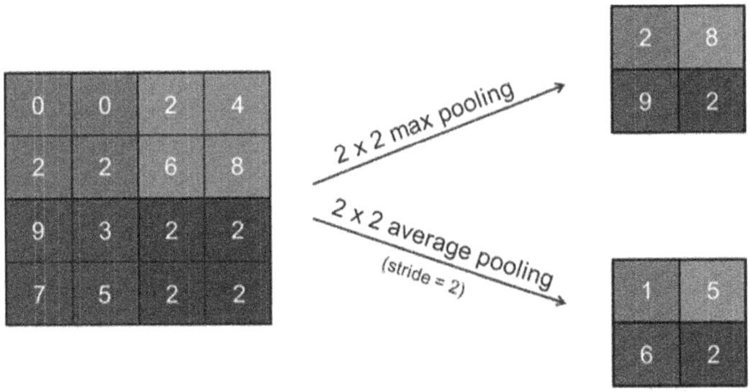

Figure 1.4. *Example pooling operations: max pooling and average pooling*

Neural networks may also have *pooling layers* that transform an input map into a smaller output map by calculating single output values for (small) regions of the input data. Figure 1.4 shows two examples: max pooling and average pooling. Effectively, the pooling layers downsample the data in the width and height dimensions. The depth of the output map is the same as the depth of the input map.

1.3.1.2. *Implementation requirements*

To obtain sufficiently accurate results when implementing machine learning inferences, appropriate data types must be used. During the training phase, data and weights are typically represented as 32-bit floating-point data. The common practice for deploying models for inference in embedded systems is to work with integer or fixed-point representations of quantized data and weights (Jacob *et al.* 2017). The potential impact of quantization errors can be taken into account in the training phase to avoid a notable effect on the model performance. Elements of input maps, output maps and weight kernels can typically be represented using 16-bit or smaller data types. For example, in a voice application, data samples are typically represented by 16-bit data types. In image or video applications, a 12-bit or smaller data type is often sufficient for representing the input samples. Precision requirements can differ per layer of the neural network. Special attention should be paid to the data types of the accumulators that are used to sum up the partial results when performing dot-product operations. These accumulators should be wide enough to avoid overflow of the (intermediate) results of the dot-product operations performed on the (quantized) weights and input samples.

Memory requirements for low/mid-end machine learning inference are typically modest, thanks to limited input data rates and the use of neural networks with limited complexity. Input and output maps are often of limited size, i.e. a few tens of kBs or less, and the number and size of the weight kernels are also relatively small. The use of the smallest possible data types for input maps, output maps and weight kernels helps us to reduce memory requirements.

In summary, low/mid-end machine learning inference applications require the following types of processing:

various types of pre-processing and feature extraction, often with DSP-intensive computations;

– neural network processing, with the dot-product operation as a dominant computation and regular access patterns on multidimensional data. Additional requirements come from the use of scalar activation functions and pooling operations working on 2D data;

– decision-making, which is performed after the neural network processing, is more control-oriented.

The different types of processing may be implemented using a heterogeneous multi-processor architecture, with different types of processors to satisfy the different processing requirements. However, for low/mid-end machine learning inference, the total compute requirements are often limited and can be handled by a single processor running at a reasonable frequency, provided it has the right capabilities. As we discussed above, the use of a single processor eliminates the area and communication overhead associated with multi-processor architectures. It also simplifies software development, as a single tool chain can be used for the complete application. However, it requires that the processor performs DSP, neural network processing and control processing with excellent cycle efficiency. In the next section, we will discuss the capabilities of a programmable processor that enables such cycle efficiency in more detail.

For many IoT edge devices, low cost is a key requirement. Therefore, making IoT edge devices smarter by adding machine learning inference must be cost-effective. The main contributor to cost is silicon area, in particular, for high-volume products, so it is important that the processor implementing the machine learning inference minimizes the logic area and uses small memories. In addition, small code size is key to limiting the area of the instruction memory.

Many IoT edge devices are battery-operated and have a tight power budget. This demands a power-efficient processor, measured in uW/MHz, as well as an excellent cycle efficiency so that the processor can be run at a low frequency. Low power consumption is particularly important for IoT edge devices that perform always-on functions such as:

– smart speakers, smartphones, etc. with always-on voice command functions that are “always listening”;

– camera-based devices, performing, for example, face detection or gesture recognition that are “always watching”;

– health and fitness monitoring devices that are “always sensing”.

Such devices typically apply smart techniques to reduce power consumption. For example, an “always listening” device may sample the microphone signal and use simple voice detection techniques to check whether anyone is speaking at all. It then applies the more compute-intensive machine learning inference for recognizing voice commands only when voice activity is detected. A processor must limit power consumption in each of these different states, i.e. voice detection and voice command recognition. For this purpose, it must offer various power management features, including effective sleep modes and power-down modes.

1.3.2. *Processor capabilities for low-power machine learning inference*

Selecting the right processor is key to achieving high efficiency for the implementation of low/mid-end machine learning inference. In this section, we will describe a number of key capabilities of the DSP-enhanced ARC EM9D processor and illustrate how they can be used to implement neural network processing efficiently.

As described earlier, the dot-product operation on input samples and weights is a dominant computation. The key primitive for implementing the dot product is the *multiply-accumulate* (MAC) operation, which can be used to incrementally sum up the products of input samples and weights. *Vectorization* of the MAC operations is an important way to increase the efficiency of neural network processing. Figure 1.5 illustrates two types of vector MAC instructions of the ARC EM9D processor.

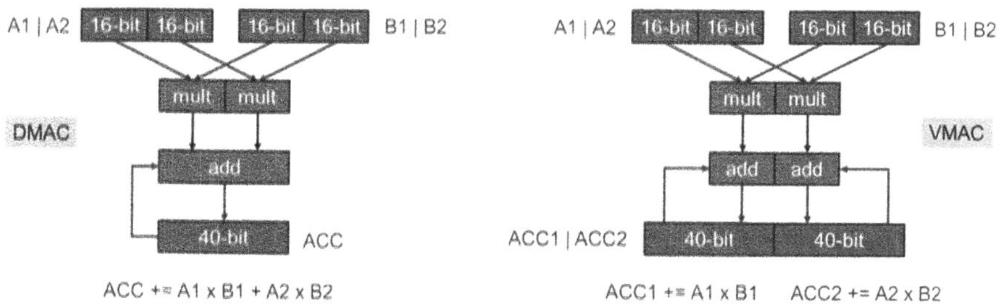

Figure 1.5. *Two types of vector MAC instructions of the ARC EM9D processor*

Both of these vector MAC instructions operate on 2x16-bit vector operands. The DMAC instruction on the left is a dual-MAC that can be used to implement a dot product, with A1 and A2 being two neighboring samples from the input map and B1 and B2 being two neighboring weights from the weight kernel. The ARC EM9D processor supports 32-bit accumulators for which an additional eight guard bits can be enabled to avoid overflow. The DMAC operation can effectively be used for weight kernels with an even width, reducing the number of MAC instructions by a factor of two compared to a scalar implementation. However, for weight kernels with an odd width, this instruction is less effective. In such cases, the VMAC instruction, shown on the right in Figure 1.5, can be used to perform two dot-product operations in parallel, accumulating intermediate results into two accumulators. In case the weight kernel "moves" over the input map with a stride of one, A1 and A2 are two neighboring samples from the input map and the value of B1 and B2 is the same weight that is applied to both A1 and A2.

Efficient execution of the dot-product operations requires not only proper vector MAC instructions, but also sufficient memory bandwidth to feed operands to these MAC instructions, as well as ways to avoid overhead for performing address updates, data size conversions, etc. For these purposes, the ARC EM9D processor provides *XY memory* with advanced address generation. Simply, the XY architecture provides up to three logical memories that the processor can access concurrently, as illustrated in Figure 1.6. The processor can access memory through a regular load, store instruction or enable a functional unit to perform memory accesses through *address generation units* (AGUs). An AGU can be set up with an address pointer to data in one of the memories and a prescription, or modifier, to update this address pointer in a particular way when a data access is performed through the AGU. After the setup, the AGUs can be used in instructions for directly accessing operands and storing results from/to memory. No explicit load or store instructions need to be executed for these operands and results. Typically, an AGU is set up before a software loop and then used repeatedly as data is traversed inside the loop.

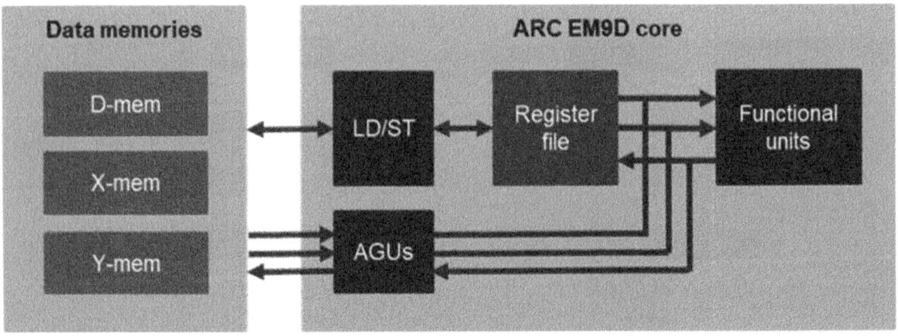

Figure 1.6. *ARC EM9D processor with XY memory and address generation units*

The AGUs support the following features relevant to machine learning inference:

– multiple modifiers per address pointer, which allow different schemes for address pointer updates to be prescribed and used. For example, a 2D access pattern can be supported by having one modifier prescribing a small horizontal stride within a row in the input map and another modifier prescribing a large stride to move the pointer to the next row in the input map;

– data size conversions, which allow, for example, 2x8-bit data to be expanded on the fly for use as a 2x16-bit vector operand. No extra instructions for unpacking and sign extension are required;

– replications, which allow data values to be replicated on the fly into vectors. For example, a single weight value may be replicated into a 2x16 vector for use in the VMAC instruction as discussed above.

In summary, the use of XY memory and AGUs enables very efficient code as no instructions are needed to load and store data, perform pointer math, or convert and rearrange data. All of these are performed implicitly while accessing data through the AGUs, with up to three memory accesses per cycle. In the next section, we present code examples that illustrate the use of the processor's XY memory and AGUs for machine learning inference.

Most other embedded processors have to issue explicit load and store instructions to perform accesses to memory. In a single-issue processor, the execution of these instructions may consume a significant portion of the available cycles, effectively reducing the throughput in MACs/cycle. Multi-issue processors, such as VLIW processors, aim to perform the load and store operations in parallel to compute operations (such as the MACs) to increase throughput. However, since wide instructions have to be used, this comes at the price of larger code size and higher power consumption in the instruction memory.

1.3.3. *A software library for machine learning inference*

After selecting the right processor, the next question is how to arrive at an efficient software implementation of the targeted machine learning inference application. For this purpose, we present a library of reusable software modules to show how these are implemented efficiently on the ARC EM9D processor.

Group	Kernels	Description
Convolution	2D convolution Depthwise 2D convolution	Convolve input features with a set of trained weights
Pooling	Average pooling Max pooling	Pool input features with a function
Recurrent and Core	Basic RNN Long-Short Term Memory (LSTM) Fully connected	Recurrent and fully connected kernels
Transform (Activation)	ReLU (Relu1, Relu6, Leaky ReLU, ..) Sigmoid and TanH SoftMax	Transform each element of input according to a particular non-linear function
Element-wise	Addition, subtraction, multiplication maximum, minimum	Apply multi-operand function element-wise to several inputs
Data Manipulation	Permute Concatenation Padding 2D	Move or extend input data by a specified pattern

Table 1.2. *Supported kernels in the embARC MLI library*

The *embARC machine learning inference* (MLI) library (embARC Open Software Platform 2019) is a set of C functions and associated data structures for building neural networks. Library functions, also called *kernels*, implement the processing associated with a complete layer in a neural network, with multidimensional input/output maps represented as tensor structures. Thus, a neural network graph can be implemented as a series of MLI function calls. Table 1.2 lists the currently supported MLI kernels, organized into six groups. For each kernel, there can be multiple functions in the library, including functions specifically optimized to support particular data types, weight kernel sizes and strides. In addition, helper functions are provided; these include operations such as data type conversion, data pointing and other operations not directly involved in the kernel functionality.

A notable feature of the embARC MLI library is the explicit support for recurrent layers. Sequential data series from sensors such as microphones or accelerometers are frequently used in the IoT domain. As explained above, RNNs maintain the state while processing sequences of inputs, thereby having the ability to recognize patterns across time. They have proven their effectiveness and are widely used in state-of-the-art solutions, such as speech applications (Amodei *et al.* 2016).

Data representation is an important part of the MLI definition. Despite the large variety of kernels, the data directly involved in the calculations has common properties. Deep learning frameworks therefore typically employ a unified data representation. For example, TensorFlow works with tensors, while Caffe uses similar objects called "blobs". Such objects represent multidimensional arrays of the proper sizes and may include additional data, such as links to related objects, synchronization primitives, etc. Similarly, mli_tensor is a universal data container in MLI. It is a lightweight tensor that contains the necessary elements for describing the data: a pointer to the data buffer, its capacity, shape, rank and data format-specific values. A kernel may take multiple tensors as inputs for producing an output tensor. For example, the 2D convolution kernel has three input tensors: the input map, the weights and the biases. All kernel-specific parameters, such as stride and padding for a convolution kernel, or the new order of dimensions for a permute kernel, are grouped into configuration structures. This data representation has several advantages:

– it provides a simple and clear interface for the functions;

– it allows the same interface to be used for several versions of one kernel;

– it matches well with layered neural networks, enabling ease of use and natural library extensibility.

The embARC MLI library uses signed fixed-point data types based on the Q-notation (*Q Number Format* 2019). Tensor structures with 8-bit and 16-bit data types are supported, both for input/output data and weights. When building an application, we should select the data types that provide sufficiently accurate results for each layer in the neural network. In addition to supporting kernels with the same data type (either 16-bit or 8-bit) for both data and weights, the MLI library also provides kernels with 16-bit data and 8-bit weights, in order to provide more flexibility for accuracy-vs-memory trade-offs.

The AGUs of the ARC EM9D processor support complex memory access patterns without spending cycles on load and store instructions. We illustrate the benefits of the AGUs with a code example of a fully connected layer. Each output value is calculated using a dot-product operation. We consider the case where inputs are vectors of 16-bit values and weights are vectors of 8-bit values. Typically, this implies the extension of the weight operands to 16-bit values, which takes additional process cycles inside a loop. The ARCv2DSP instruction set architecture (ISA) has several dual-MAC instructions, which allow two multiplications with accumulation in a single cycle (see the left diagram in Figure 1.5). These include 16x8 dual-MAC instructions where one operand is a 2x16-bit vector and the other operand is a 2x8-bit vector, which allows direct use of the 8-bit data. The assembly code in Figure 1.7, generated from high-level C-code, shows that this yields a zero-overhead loop with a loop body of just one DMAC instruction and a throughput of two 16x8 MACs/cycle. In addition to the dual-MAC, the DMAC instruction also performs two memory reads and two pointer updates through the two AGUs.

```
    ...        ; AGU 0 is used for input data
               ; AGU 1 is used for weights
    SR ...     ; setup AGU 0 for loading next 2x16-bit vector
    SR ...     ; setup AGU 1 for loading next 2x8-bit vector
    ...
    LP         lpend
    DMACHBL 0, %agu_u0, %agu_u1 ;
lpend: ...
```

Figure 1.7. *Assembly code generated from MLI C-code for a fully connected layer with 16-bit input data and 8-bit weights*

Similar optimizations using dual-MAC instructions are used for other kernels, including RNN kernels and some convolution kernels. However, this approach is not convenient for all cases. As an example, we consider a depthwise convolution for a channel–height–width (CHW) data layout with a 3x3 weight kernel size. This type of layer applies a 2D convolution to each input channel separately. Dual-MAC instructions cannot be used optimally here, due to the odd weight kernel size and the short MAC series for calculating an output value. If the convolution stride parameter is equal to 1, then neighboring input data elements are used for the calculation of neighboring output values. This implies that we can calculate two output values simultaneously using VMAC instructions that use two accumulators (see the right diagram in Figure 1.5), as shown in the generated assembly code in Figure 1.8. The VMAC instructions each perform two 16x16 MACs as well as two memory reads, sign extension and replication of the 8-bit value accessed through AGU 2 and two pointer updates.

This example demonstrates the flexibility of AGUs for complex data addressing patterns, including 2D accesses using two modifiers for the input data as well as sign extension and replication of weights. A typical approach for calculating convolution layers, for example, as popularized by Caffe, is to use additional image-to-column (im2col) transformations. Although such transformations are helpful on some processors as they simplify subsequent calculations for performing the convolutions, this comes at a price of a significant overhead for performing these transformations. The advanced AGUs, as used in Figure 1.8, make these transformations obsolete, thereby supporting efficient embedded implementations.

```
    ...         ; AGU 0 and AGU 1 are used for input data
                ; (one data pointer with two modifiers)
                ; AGU 2 is used for weights
    SR ...      ; setup AGU 0 for loading next two 16-bit values
    SR ...      ; setup AGU 1 for loading two 16-bit input values
                ; at next row of input data
    SR ...      ; setup AGU 2 for loading next 8-bit value with
                ; sign extension & replication
    ...
    LP          lpend
    VMAC2HNFR   0, %agu_u0, %agu_u2 ;
    VMAC2HNFR   0, %agu_u0, %agu_u2 ;
    VMAC2HNFR   0, %agu_u1, %agu_u2 ;
  lpend: ...
```

Figure 1.8. *Assembly code generated from MLI C-code for 2D convolution of 16-bit input data and 8-bit weights*

From the user's point of view, the embARC MLI library provides ease of use, allowing the construction of efficient machine learning inference engines without requiring in-depth knowledge of the processor architecture and software optimization details. The embARC MLI library provides a broad set of optimized functions, so that the user can concentrate on the application and write embedded code using familiar high-level constructs for machine learning inference.

1.3.4. *Example machine learning applications and benchmarks*

The embARC MLI library is available from embarc.org (embARC Open Software Platform 2019), together with a number of example applications that demonstrate the usage of the library, such as:

– CIFAR-10 low-resolution object classifier: CNN graph;

– face detection: CNN graph;

– human activity recognition (HAR): LSTM-based network;

– keyword spotting: graph with CNN and LSTM layers trained on the Google speech command dataset.

The CIFAR-10 (Krizhevsky 2009) example application is based on the Caffe (Jia *et al.* 2014) tutorial. The CIFAR-10 dataset is a set of 60,000 low-resolution RGB images (32x32 pixels) of objects in 10 classes, such as "cat", "dog" and "ship". This dataset is widely used as a "Hello World" example in machine learning and computer vision. The objective is to train the classifier using 50,000 of these images, so that the other 10,000 images of the dataset can be classified with high accuracy. We used the CIFAR-10 CNN graph in Figure 1.9 for training and inference. This graph matches the CIFAR-10 graph from the Caffe tutorial, including the two fully connected layers towards the end of the graph.

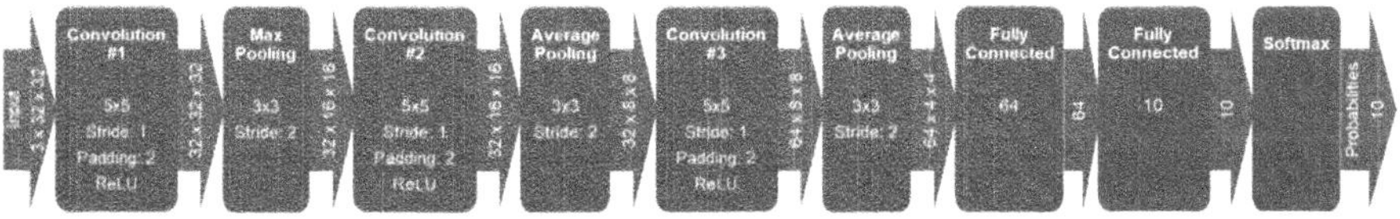

Figure 1.9. *CNN graph of the CIFAR-10 example application*

We used the CIFAR-10 example application with 8-bit for both feature data and weights to benchmark the performance of machine learning inference on the ARC EM9D processor. The code of this CIFAR-10 application, built using the embARC MLI library, is illustrated in Figure 1.10.

```
mli_krn_permute_fx8(&input, &permute_hwc2chw_cfg, &ir_Y);

ir_X.el_params.fx.frac_bits = CONV1_OUT_FRAQ;
mli_krn_conv2d_chw_fx8_k5x5_str1_krnpad(&ir_Y, &L1_conv_wt,
  &L1_conv_b, &conv_cfg, &ir_X);
mli_krn_maxpool_chw_fx8_k3x3(&ir_X, &pool_cfg, &ir_Y);

ir_X.el_params.fx.frac_bits = CONV2_OUT_FRAQ;
mli_krn_conv2d_chw_fx8_k5x5_str1_krnpad(&ir_Y, &L2_conv_wt,
  &L2_conv_b, &conv_cfg, &ir_X);
mli_krn_avepool_chw_fx8_k3x3_krnpad(&ir_X, &pool_cfg, &ir_Y);

ir_X.el_params.fx.frac_bits = CONV3_OUT_FRAQ;
mli_krn_conv2d_chw_fx8_k5x5_str1_krnpad(&ir_Y, &L3_conv_wt,
  &L3_conv_b, &conv_cfg, &ir_X);
mli_krn_avepool_chw_fx8_k3x3_krnpad(&ir_X, &pool_cfg, &ir_Y);

ir_X.el_params.fx.frac_bits = FC4_OUT_FRAQ;
mli_krn_fully_connected_fx8(&ir_Y, &L4_fc_wt, &L4_fc_b, &ir_X);

ir_Y.el_params.fx.frac_bits = FC5_OUT_FRAQ;
mli_krn_fully_connected_fx8(&ir_X, &L5_fc_wt, &L5_fc_b, &ir_Y);
mli_krn_softmax_fx8(&ir_Y, &output);
```

Figure 1.10. *MLI code of the CIFAR-10 inference application*

As the code in Figure 1.10 shows, each layer in the graph is implemented by calling a function from the embARC MLI library. Before executing the first convolution layer, we call a permute function from the embARC MLI library to transform the RGB image into CHW format so that neighboring data elements are from the same color plane. The code further shows that a ping-pong scheme with two buffers, `ir_X` and `ir_Y`, is used for buffering input and output maps.

A very similar CIFAR-10 CNN graph has been used by others for benchmarking machine learning inference on their embedded processors, with performance numbers published in (Lai *et al.* 2018) and (Croome 2018). Table 1.3 presents the model parameters of the CIFAR-10 CNN graph that we used, with performance data for the ARC EM9D processor and two other embedded processors presented in Table 1.4.

#	Layer type	Weights tensor shape	Output tensor shape	Coefficients
0	Permute	–	3 × 32 × 32	0
1	Convolution	32 × 3 × 5 × 5	32 × 32 × 32 (32K)	2400
2	Max Pooling	–	32 × 16 × 16 (8K)	0
3	Convolution	32 × 32 × 5 × 5	32 × 16 × 16 (8K)	25600
4	Avg Pooling	-	32 × 8 × 8 (2K)	0
5	Convolution	64 × 32 × 5 × 5	64 × 8 × 8 (4K)	51200
6	Avg Pooling	–	64 × 4 × 4 (1K)	0
7	Fully-connected	64 × 1024	64	65536
8	Fully-connected	10 × 64	10	640

Table 1.3. *Model parameters of the CIFAR-10 CNN graph*

The performance data for processor A is published in (Lai *et al.* 2018) in terms of milliseconds for a processor running at a clock frequency of 216 MHz. The cycle counts for processor A in Table 1.4 have been calculated by multiplying the published millisecond numbers with this clock frequency. The CIFAR-10 CNN graph reported in (Lai *et al.* 2018) has the same convolution and pooling layers as listed in Table 1.3, but uses a single fully connected layer with a 4x4x64x10 filter shape to directly transform the 64x4x4 input map into 10 output values. This modification of the Caffe CNN graph reduces the size of the weight data considerably, but requires retraining of the graph. The impact on the total cycle count is marginal.

The performance data for the RISC-V processor published in (Croome 2018) reports a total of 1.5 Mcycles for executing the CIFAR-10 graph on a highly parallel 8-core RISC-V architecture. For calculating the total number of cycles on a single RISC-V core, we consider that the performance is highly dominated by the cycles spent on 5x5 convolutions, which constitute more than 98% of the compute operations in this graph. For these 5x5 convolutions, (Croome 2018) reports a speed-up from a 1-core system to an 8-core system of 18.5/2.2 = 8.2. Hence, a reasonable estimate for the total number of cycles on a single RISC-V core is 1.5x8.2 = 12.3 Mcycles.

#	Layer type	ARC EM9D [*Mcycles*]	Processor A [*Mcycles*]	Processor B (RISC-V ISA) [*Mcycles*]
0	Permute	0.01	–	–
1	Convolution	1.63	6.78	–
2	Max Pooling	0.14	0.34	–
3	Convolution	3.46	9.25	–
4	Avg Pooling	0.09	0.09	–
5	Convolution	1.76	4.88	–
6	Avg Pooling	0.07	0.04	–
7	Fully-connected	0.03	0.02	–
8	Fully-connected	0.001		–
Total		**7.2**	**21.4**	**12.3**

Table 1.4. *Performance data for the CIFAR-10 CNN graph*

From Table 1.4, we conclude that the ARC EM9D processor spends 3x fewer cycles than processor A and 1.7x fewer cycles than the RISC-V core (processor B) for the same machine learning inference task, without using any specific accelerators. Thanks to the good cycle efficiency, the ARC EM9D processor can be clocked at a low frequency, which helps to save power in a smart IoT edge device.

1.4. Conclusion

Smart IoT edge devices that interact intelligently with their users are appearing in many application areas. These devices have diverse compute requirements, including a mixture of control processing, DSP and machine learning. Versatile processors are required to efficiently execute these different types of workloads. Furthermore, these processors must allow for easy customization to improve their efficiency for a specific application. Configurability and extensibility are two key mechanisms that provide such customization. Increasingly, IoT edge devices apply machine learning technology for processing captured sensor data, so that smart actions can be taken based on recognized patterns. We presented key processor features and a software library for the efficient implementation of low/mid-end machine learning inference. More specifically, we highlighted several processor capabilities, such as vector MAC instructions and XY memory with advanced AGUs, that are key to the efficient implementation of machine learning inference. The ARC EM9D processor is a universal processor for low-power IoT applications which is both configurable and extensible. The complete and highly optimized embARC MLI library makes effective use of the ARC EM9D processor to efficiently support a wide range of low/mid-end machine learning applications. We demonstrated this efficiency with excellent results for the CIFAR-10 benchmark.

1.5. References

Amodei, D., Ananthanarayanan, S., Anubhai, R., Bai, J., Battenberg, E., Case, C., Casper, J., Catanzaro, B., Cheng, Q., Chen, G., Chen, J., Chen, J., Chen, Z., Chrzanowski, M., Coates, A., Diamos, G., Ding, K., Du, N., Elsen, E., Engel, J., Fang, W., Fan, L., Fougner, C., Gao, L., Gong, C., Hannun, A., Han, T., Johannes, L.V., Jiang, B., Ju, C., Jun, B., LeGresley, P., Lin, L., Liu, J., Liu, Y., Li, W., Li, X., Ma, D., Narang, S., Ng, A., Ozair, S., Peng, Y., Prenger, R., Qian, S., Quan, Z., Raiman, J., Rao, V., Satheesh, S., Seetapun, D., Sengupta, S., Srinet, K., Sriram, A., Tang, H., Tang, L., Wang, C., Wang, J., Wang, K., Wang, Y., Wang, Z., Wang, Z., Wu, S., Wei, L., Xiao, B., Xie, W., Xie, Y., Yogatama, D., Yuan, B., Zhan, J., Zhu, Z. (2016). Deep speech 2: End-to-end speech recognition in English and Mandarin. *Proceedings of the 33rd International Conference on Machine Learning – Volume 48, ICML-16*, 173–182.

Croome, M. (2018). Using RISC-V in high computing, ultra-low power, programmable circuits for inference on battery operated edge devices [Online]. Available at: https://content.riscv.org/wp-content/uploads/2018/07/Shanghai-1325_GreenWaves_Shanghai-2018-MC-V2.pdf.

Dutt, N. and Choi, K. (2003). Configurable processors for embedded computing. *IEEE Computer*, 36(1), 120–123.

embARC Open Software Platform (2019). Available at: https://embarc.org/.

Jacob, B., Kligys, S., Chen, B., Zhu, M., Tang, M., Howard, A.G., Adam, H., and Kalenichenko, D. (2017). Quantization and training of neural networks for efficient integer-arithmetic-only inference. *Computing Research Repository*. Available at: http://arxiv.org/abs/1712.05877.

Jia, Y., Shelhamer, E., Donahue, J., Karayev, S., Long, J., Girshick, R.B., Guadarrama, S., and Darrell, T. (2014). Caffe: Convolutional architecture for fast feature embedding. *Computing Research Repository*. Available at: https://arxiv.org/abs/1408.5093.

Krizhevsky, A. (2009). Learning multiple layers of features from tiny images. Technical Report, University of Toronto, 2009.

Lai, L., Suda, N., and Chandra, V. (2018). CMSIS-NN: Efficient neural network kernels for arm cortex-M CPUs. *Computing Research Repository*. Available at: http://arxiv.org/abs/1801.06601.

Petrov-Savchenko, A. and van der Wolf, P. (2018). Get smart with NB-IoT: Efficient low-cost implementation of NB-IoT for smart applications. Technical paper, *Synopsys* [Online]. Available at: https://www.synopsys.com/dw/doc.php/wp/NB_IoT_for_Smart_Applications.pdf.

Q Number Format (2019). Available at: https://en.wikipedia.org/wiki/Q_(number_format).

Samuel, A.L. (1959). Some studies in machine learning using the game of checkers. *IBM Journal of Research and Development*, 3(3), 210–229.

Szegedy, C., Liu, W., Jia, Y., Sermanet, P., Reed, S., Anguelov, D., Erhan, D., Vanhoucke, V., and Rabinovich, A. (2015). Going deeper with convolutions. *2015 IEEE Conference on Computer Vision and Pattern Recognition (CVPR)*, 1–9.

2

A Qualitative Approach to Many-core Architecture

Benoît DUPONT DE DINECHIN

Kalray S.A., Grenoble, France

We present the design of the Kalray third-generation MPPA many-core processor, whose objectives are to combine the performance scalability of GPGPUs, the energy efficiency of DSP cores and the I/O capabilities of FPGA devices. These objectives are motivated by the consolidation of high-performance and high-integrity functions on a single computing platform for autonomous vehicles. High-performance computing functions, represented by deep learning inference and computer vision, need to execute under soft real-time constraints. High-integrity functions are developed under model-based design, and must meet hard real-time constraints. Finally, the third-generation MPPA processor integrates a hardware root of trust and implements a security architecture, in order to support trusted execution environments.

The MPPA software development tools and run-time environments conform to CPU standards, particularly the availability of C/C++/OpenMP programming environments, supported by POSIX operating systems and RTOSes. As these standards target multi-core shared memory architectures, there is an opportunity for higher-level application code generators to automate code and data distribution across the multiple compute units and the local memories of a many-core processor. For the MPPA3 processor, this opportunity is realized in cases of deep learning model inference, as well as of model-based software development using synchronous-reactive languages.

For a color version of all figures in this book, see www.iste.co.uk/andrade/multi1.zip.

Multi-Processor System-on-Chip 1 – Architectures,
coordinated by Liliana ANDRADE and Frédéric ROUSSEAU.

2.1. Introduction

Cyber-physical systems (CPSs) are characterized by software that interacts with the physical world, often with time-sensitive safety-critical physical sensing and actuation (Lee *et al.* 2017). Applications such as aircraft pilot support or automated driving systems require more than what classical CPSs provide. More specifically, application functionality increasingly relies on machine learning techniques, while cyber-security requirements have become significantly more stringent. We refer to CPSs enhanced with high-performance machine learning capabilities and strong cyber-security support as "intelligent systems". Given the state of CMOS computing technology (Kanduri *et al.* 2017), providing the processing performances required by intelligent systems while meeting the size, weight and power (SWaP) constraints of embedded systems can only be achieved by parallel computing and the specialization of processing elements. For example, automated driving systems targeting L3/L4 SAE J3016 levels of automation are estimated to require over 150 TOPS of deep learning inference in vehicle perception functions, while motion planning functions would require more than 50 FP32 TFLOPS (Figure 2.5).

In order to address the challenges of high-performance embedded computing with time predictability, Kalray has been refining a many-core architecture called the MPPA (Massively Parallel Processor Array) across three generations. The first-generation MPPA processor was primarily targeting accelerated computing (Dupont de Dinechin *et al.* 2013), but implemented the first key architectural features for time-critical computing (Dupont de Dinechin *et al.* 2014). Kalray further improved the second-generation MPPA processor for time predictability (Saidi *et al.* 2015), providing an excellent target for model-based code generation (Perret *et al.* 2016; Graillat *et al.* 2018, 2019). Accurate analysis of network-on-chip (NoC) service guarantees was achieved through a new deterministic network calculus formulation (Dupont de Dinechin and Graillat 2017). Unlike the first-generation MPPA processor that relied on cyclostatic dataflow programming (Bodin *et al.* 2013, 2016), the second-generation MPPA programming environment was able to support OpenCL and OpenVX applications (Hascoët *et al.* 2018).

In this chapter, we present the third-generation MPPA processor, manufactured in 16FFC CMOS technology, whose many-core architecture has significantly improved upon the previous ones in the areas of performance, programmability, functional safety and cyber-security. These features are motivated by application cases in defense, avionics and automotive where the high-performance, high-integrity and cyber-security functions can be consolidated onto a single or dual processor configuration. In section 2.2, we discuss many-core architectures and their limitations with regard to intelligent system requirements. In section 2.3, we present the main features of the third-generation MPPA architecture and processor. In section 2.4, we introduce the MPPA3 application software environments.

2.2. Motivations and context

2.2.1. *Many-core processors*

A multi-core processor refers to a computing device that contains multiple software-programmable processing units (cores with caches). Multi-core processors deployed in desktop computers or data centers have homogeneous cores and a memory hierarchy is composed of coherent caches (Figure 2.1). Conversely, a many-core processor can be characterized by the architecturally visible grouping of cores inside compute units: cache coherence may not extend beyond the compute unit; or the compute unit may provide scratch-pad memory and data movement engines. A multi-core processor scales by replicating its cores, while a many-core processor scales by replicating its compute units. A many-core architecture may thus be scaled to hundreds, if not thousands, of cores.

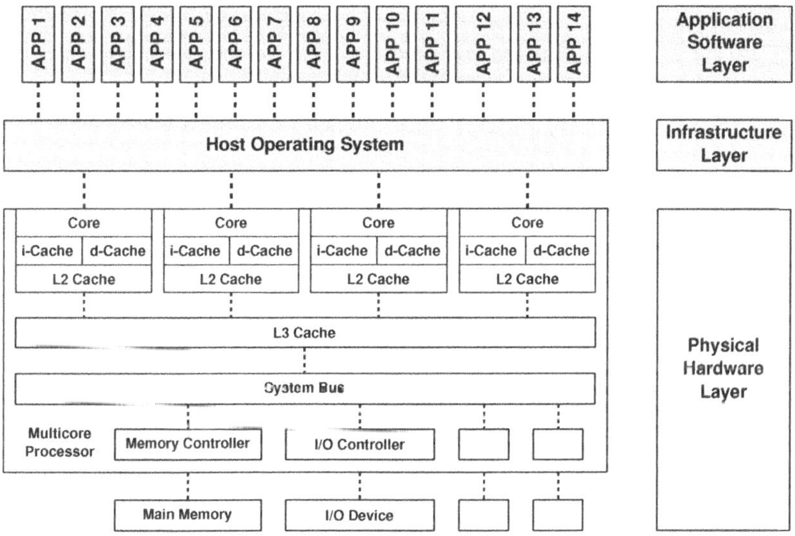

Figure 2.1. *Homogeneous multi-core processor (Firesmith 2017)*

The GPGPU architecture introduced by the NVIDIA Fermi (Figure 2.2) is a mainstream many-core architecture, whose compute units are called streaming multi-processors (SMs). Each SM comprises 32 streaming cores (SCs) that share a local memory, caches and a global memory system. Threads are scheduled and executed atomically by "warps", which are sets of 32 threads dispatched to SCs that execute the same instruction at any given time. Hardware multi-threading enables warp execution switching on each cycle, helping to cover the memory access latencies.

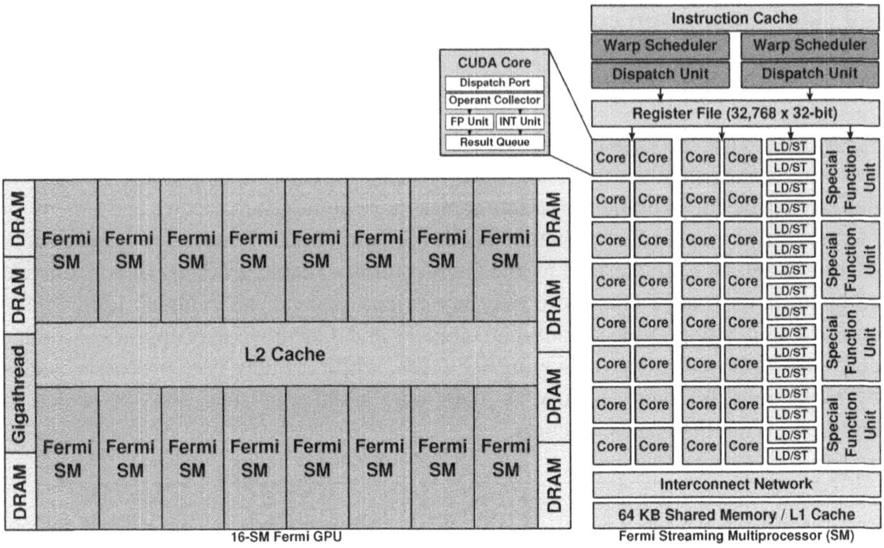

Figure 2.2. *NVIDIA fermi GPGPU architecture (Huang* et al. *2013)*

Although embedded GPGPU processors provide adequate performance and energy efficiency for accelerated computing, their architecture carry inherent limitations that hinder their use in intelligent systems:

– kernel programming environment lacks standard features of C/C++, such as recursion, standard multi-threading or accessing a (virtual) file system;

– performance of kernels is highly sensitive to run-time control flow (because of branch divergence) and data access patterns (because of memory coalescing);

– threads blocks are dynamically allocated to SMs, while warps are dynamically scheduled for execution inside an SM;

– coupling between the host CPU and the GPGPU relies on a software stack that results in long and hard-to-predict latencies (Cavicchioli *et al.* 2019).

2.2.2. *Machine learning inference*

The main uses of machine learning techniques in intelligent systems are inference of deep learning networks. When considering deep learning inference acceleration, several architectural approaches appear effective. These include loosely coupled accelerators that implement a systolic data path (Google TPU, NVIDIA NVDLA), coarse-grained reconfigurable arrays (Cerebras WSE) or a bulk-synchronous parallel graph processor (GraphCore IPU). Other approaches tightly couple general-purpose processing units with vector or tensor processing units that share the instruction stream and the memory hierarchy. In particular, the GPGPU architecture has further evolved with the NVIDIA Volta by integrating eight "tensor cores" per SM, in order to accelerate machine learning workloads (Jia *et al.* 2018). Each tensor core executes mixed-precision matrix multiply-accumulate operations on 4×4 matrices. Multiplication operand elements use the IEEE 754 binary 16 floating-point representation (FP16), while the accumulation and result operands use the IEEE 754 binary 16 or binary 32 (FP32) floating-point representation (Figure 2.3).

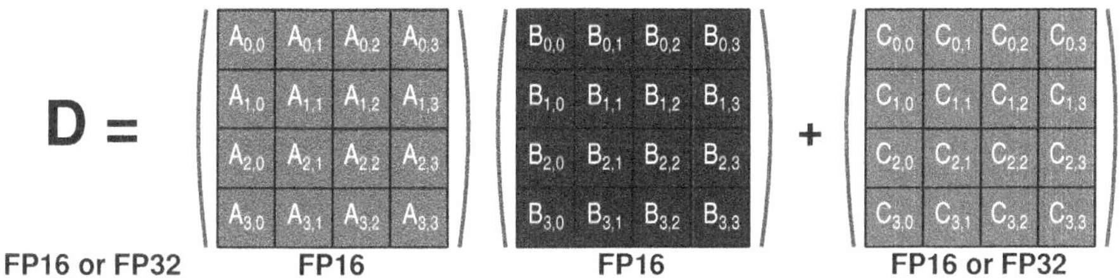

Figure 2.3. *Operation of a Volta tensor core (NVIDIA 2020)*

Machine learning computations normally rely on FP32 arithmetic; however, significant savings in memory footprint and increases in performance/efficiency can be achieved by using 16-bit representations for training and 8-bit representations for inference with acceptable precision loss. The main 16-bit formats are FP16 and BF16, which is FP32 with 16 mantissa bits truncated (Intel 2018), and INT16 that covers the 16-bit integer and fixed-point representations (Figure 2.4a). Those reduced bit-width formats are, in fact, used as multiplication operands in linear operations, whose results are still accumulated in FP32, INT32 or larger fixed-point representations.

While mainstream uses of 8-bit formats in convolutional network inference are signed or unsigned integers (Jacob *et al.* 2018; Krishnamoorthi 2018), floating-point formats smaller than 16-bit are also investigated. Their purpose is to eliminate the complexities associated with small integer quantization: fake quantization, where weights and activations are quantized and dequantized in succession during both the forward and backward passes of training; and post-training calibration, where the histogram of activations is collected on a representative dataset to adjust the saturation thresholds. Microsoft introduced the Msfp8 data format (Chung *et al.* 2018), which is FP16 truncated to 8 bits, with only 2 bits of mantissa left, along with its extension Msfp9. Among the reduced bit width floating-point formats, however, the Posit8 representations generate the most interest (Carmichael *et al.* 2019).

A Posit$n.es$ representation (Figure 2.4b) is parameterized by n, the total number of bits, and es, the number of exponent bits (Gustafson and Yonemoto 2017). The main difference with an IEEE 754 binary floating-point representation is the *regime* field, which has a dynamic width and encodes a power of $2^{2^{es}}$ in unary numerals. (de Dinechin *et al.* 2019) discuss the advantages and disadvantages of Posit representations. They advise the use of Posit as a storage-only format in order to benefit from the compact encoding, while still relying on standard IEEE binary floating-point arithmetic for numerical guarantees. Experimentally, Posit8 numbers provide an effective compressed representation of FP32 network weights and activations by rounding them to Posit8.1 or Posit8.2 numbers (Resmerita *et al.* 2020). Another approach is to use Posit8.1 on a log domain for the multiplicands, while converting to a linear domain for the accumulations (Johnson 2018). In both cases, however, the large dynamic range that motivates the use of the Posit representations in machine learning inference requires high-precision or exact accumulations.

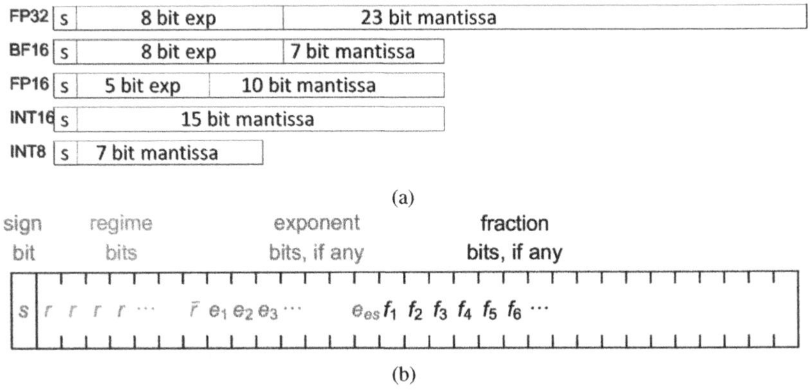

Figure 2.4. *Numerical formats used in deep learning inference (adapted from Gustafson (2017) and Rodriguez* et al. *(2018))*

2.2.3. *Application requirements*

In the case of automated driving applications (Figure 2.5), the perception and the path planning functions require programmability, high performances and energy efficiency, which leads to the use of multi-core or GPGPU many-core processors. Multi-core processing entails significant execution resource sharing on the memory hierarchy, which negatively impacts time predictability (Wilhelm and Reineke 2012). Even with a predictable execution model (Forsberg *et al.* 2017), the functional safety of perception and path planning functions may only reach ISO 26262 ASIL-B. Conversely, vehicle control algorithms, as well as sensor and actuator management, must be delegated to electronic control units that are specifically designed to host ASIL-D functions.

Similarly, unmanned aerial vehicle applications targeted by the MPPA processor are composed of two criticality domains, one being safety-critical and the other non-safety-critical (Figure 2.6). On the MPPA processor, these two domains can be segregated by physical isolation mechanisms, ensuring that no execution resources can be shared between them. The safety-critical domain hosts the trajectory control partition (DO-178C DAL-A/B). The non-critical domain hosts a secured communication partition (ED-202 SAL-3), a data management partition (DAL-E) running on Linux, machine learning and other embedded high-performance computing partitions running on a lightweight POSIX OS. Of interest is the fact that the secured partition is located in the non-safety-critical domain, as the availability requirements of functional safety are incompatible with the integrity requirements of cyber-security.

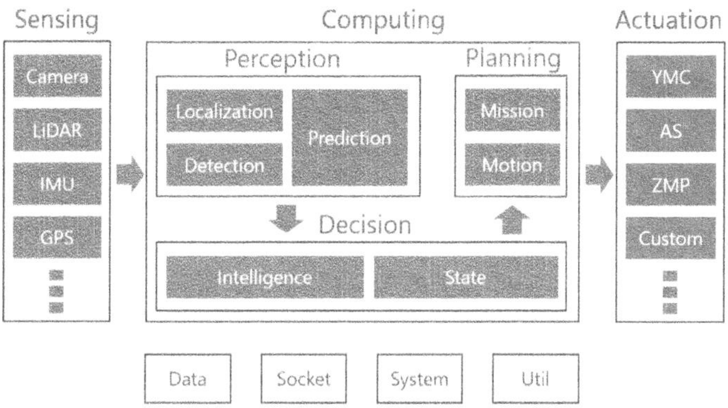

Figure 2.5. *Autoware automated driving system functions (CNX 2019)*

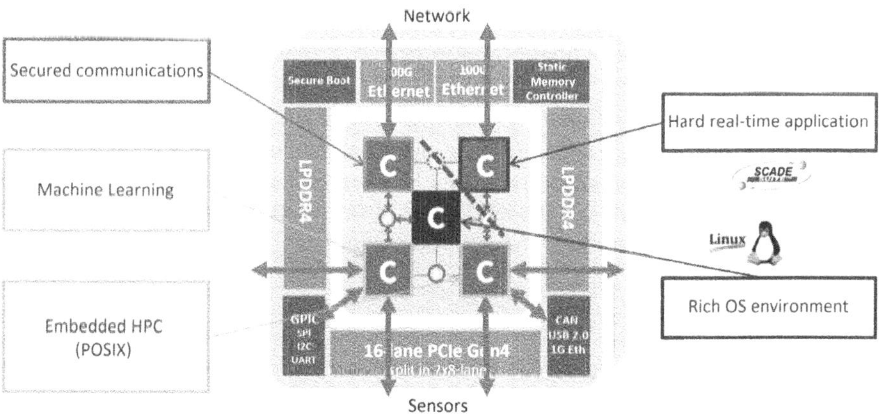

Figure 2.6. *Application domains and partitions on the MPPA3 processor*

Finally, embedded applications in the areas of defense, avionics and automotive have common requirements in the area of cyber-security (Table 2.1). The foundation is the availability of a hardware root of trust (RoT), i.e. a secured component that can be inherently trusted. Such RoT can be provided either as an external hardware security module (HSM), or integrated into the system-on-chip as a central security module (CSM). In both cases, this security module maintains the device's critical security parameters (CSP) such as public authentication keys, device identity and master encryption keys in a non-volatile secured memory. The security module embeds a TRNG, hashing, symmetric and public-key cryptographic accelerators, in order to support the chain of trust through digital signature verification of firmware and software.

	Defense	Avionics	Automotive
Hardware root of trust	✓	✓	✓
Physical attack protection	✓		✓
Software and firmware authentication	✓	✓	✓
Boot firmware confidentiality	✓	✓	
Application code confidentiality	✓	✓	✓
Event data record integrity		✓	✓

Table 2.1. *Cyber-security requirements by application area*

2.3. The MPPA3 many-core processor

2.3.1. *Global architecture*

The MPPA3 processor architecture (Figure 2.7) applies the defining principles of many-core architectures: processing elements (SCs on a GPGPU) are regrouped with a multi-banked local memory and a slice of the memory hierarchy into compute units (SMs on a GPGPU), which share a global interconnect and access to external memory. The distinguishing features of the MPPA many-core architecture compared to the GPGPU architecture are the integration of fully software-programmable cores for the processing elements, and the provision of an RDMA engine in each compute unit.

The structuring of the MPPA3 architecture into a collection of compute units, each comparable to an embedded multi-core processor, is the main feature that enables the consolidation of application partitions operating at different levels of functional safety and cyber-security, on a single processor. This feature requires provision of global interconnects with support for partition isolation. From experience with previous MPPA processors, it became apparent that chip global interconnects implemented as "network-on-chip" (NoC) may be specialized for two different purposes: generalization of busses and integration of macro-networks (Table 2.2).

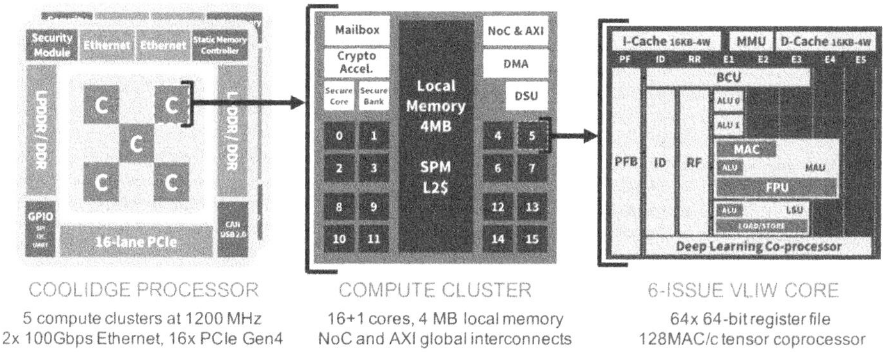

Figure 2.7. *Overview of the MPPA3 processor*

Generalized busses	Integrated macro-network
Connectionless	Connection-oriented
Address-based transactions	Stream-based transactions
Flit-level flow control	[End-to-end flow control]
Implicit packet routing	Explicit packet routing
Inside coherent address space	Across address spaces (RDMA)
Coherency protocol messages	Message multicasting
Reliable communication	[Packet loss or reordering]
QoS by priority and aging	QoS by traffic shaping
Coordination with the DDR controller	Termination of macro-networks

Table 2.2. *Types of network-on-chip interconnects*

Accordingly, the MPPA3 processor is fitted with two global interconnects, respectively identified as "RDMA NoC" and "AXI Fabric" (Figure 2.8). The RDMA NoC is a wormhole switching network-on-chip, designed to terminate two 100 Gbps Ethernet controllers, and to carry the remote DMA operations found in supercomputer interconnects or communication libraries such as SHMEM (Hascoët *et al.* 2017). The AXI Fabric is a crossbar of busses with round-robin arbiters, which connects the compute clusters, the external DDR memory controllers, the PCIe controllers and other I/O controllers. The main I/O interfaces of the MPPA3 processor are a PCI Express subsystem with 16 Gen1/Gen2/Gen3/Gen4 lanes for a peak throughput of 32 GB/s full-duplex, and an Ethernet subsystem composed of two controllers of four lanes each, for a total peak throughput of 200 Gbps full-duplex. Other high-speed I/O are supported by four CAN 2.0A/2.0B/FD controllers, and by two USB 2.0 OTG ULPI controllers.

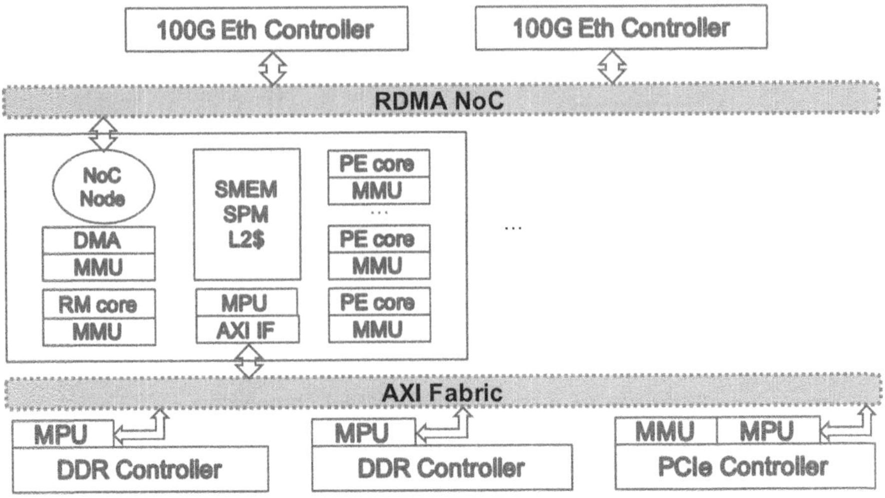

Figure 2.8. *Global interconnects of the MPPA3 processor*

Based on this global architecture, the consolidation of functions operating at different levels of functional safety and cyber-security is supported by two mechanisms:

– Cores and other bus initiators have their address translated from virtual to machine addresses by memory management units (MMUs). These MMUs actually implement a double translation: from virtual to physical, as directed by the operating system or the execution environment; from physical to machine, under the control of a partition monitor operating at the hypervisor privilege level. This first mechanism supports the requirements of isolating safety-critical application partitions in multi-core processors (CAST 2016).

– Memory protection units (MPUs) are provided on the AXI Fabric targets to filter transactions based on their machine addresses. Similarly, selected NoC router links can be disabled. This second mechanism has its parameters set at boot time, and then cannot be overridden without resetting the processor. Its purpose is to partition the processor and its peripherals into physically isolated domains, as in the unmanned aerial vehicle applications discussed in section 2.2.

2.3.2. *Compute cluster*

The compute unit of the MPPA3 processor, called the compute cluster, is structured around a local interconnect (Figure 2.9); it comprises a secure zone and a non-secure zone. The secure zone contains a security and safety management core (RM), a 256 KB dedicated memory bank and a cryptographic accelerator. The RM core of each compute cluster is also connected to the processor CSM and the internal boot ROM through a private bus. The purpose of the secure zone is to host a trusted execution environment (TEE), and a run-time system that performs on-demand code decryption for the applications that require it (Table 2.1). The non-secure zone contains 16 application cores (PE) that share a 16-banked local memory (SMEM), a DMA engine, a debug support unit (DSU), a cryptographic accelerator and cluster-local peripheral control registers. The SMEM can be configured to split its 4 MB capacity between scratch-pad memory (SPM) and level-2 cache (L2$). The non-secure zone of the MPPA3 compute cluster supports two types of execution environments:

– A symmetric multi-processing (SMP) environment, exposed through the standard POSIX multi-threading (supporting the OpenMP run-time of C/C++ compilers) and file system APIs. In this environment targeting high-performance computing under soft real-time constraints, all the core L1 data caches are kept coherent, and the local memory is interleaved across the banks at 64-byte granularity for the SPM and at 256-byte granularity for the L2$.

– An asymmetric multi-processing (AMP) environment, seen as a collection of 16 cores where each executes under an RTOS and is associated through linker maps with one particular bank (256KB) of the local memory. In this environment targeting high-integrity computing under hard real-time constraints, L1 cache coherence is disabled, and the local memory is configured as scratch-pad memory only.

By default, the L2 caches of the compute clusters are not kept coherent, although this can be enabled by running cache controller firmware in the RM cores and maintaining a distributed directory in the cluster SPMs. The third level of the memory hierarchy is composed of the external DDR memory (2x DDR4/LPDDR4 64-bit channels), and of the SPM of other compute clusters. The DDR memory channels can be interleaved or separated in machine address space, in the latter case operating independently. The standard C11 atomic operations are available in all memory spaces.

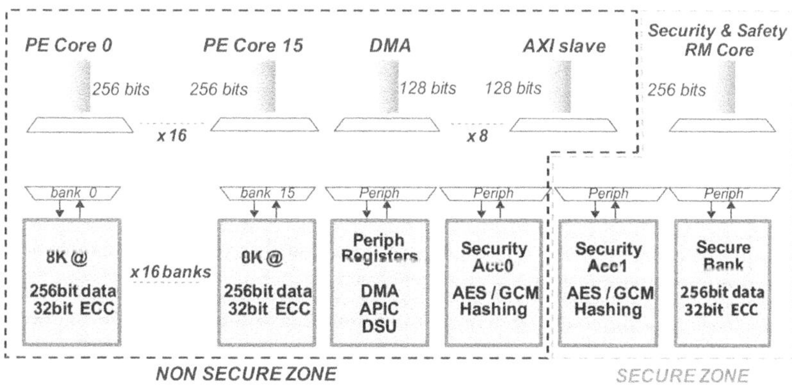

Figure 2.9. *Local interconnects of the MPPA3 processor*

2.3.3. *VLIW core*

The MPPA cores implement a 64-bit VLIW architecture, which is an effective way to design instruction-level parallel cores targeting numerical, signal and image processing (Fisher *et al.* 2005). The VLIW core has six issue lanes (Figure 2.10) that, respectively, feed a branch and control unit (BCU), two 64-bit ALUs, a 128-bit FPU, a 256-bit load–store unit (LSU) and a deep learning coprocessor. Each VLIW core has private L1 instruction and data caches, both 16 KB and four-way set associated with LRU replacement policy. All load instructions also have an L1 cache-bypass variant for direct access to the cluster SPM or L2$. These instructions improve the performance of codes with non-temporal memory access patterns, and also increase the accuracy of static analysis for computing worst-case execution time (WCET) bounds. The implementation of this VLIW core and its caches ensure that the resulting processing element is timing-compositional, a critical property with regard to computing accurate bounds on worst-case response times (WCRT) (Kästner *et al.* 2013).

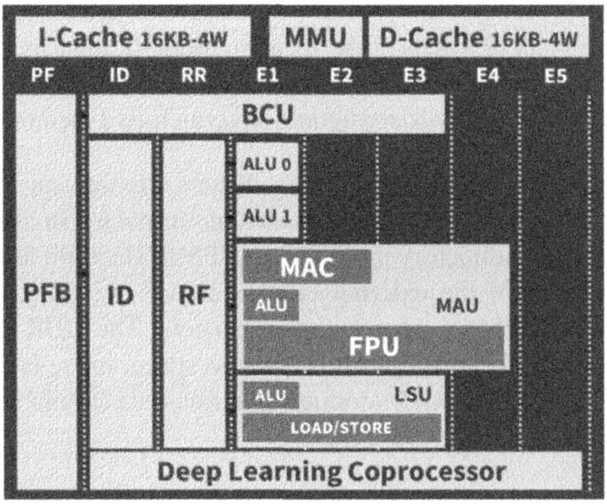

Figure 2.10. *VLIW core instruction pipeline*

Based on previous compiler design experience with different types of VLIW architectures (Dupont de Dinechin *et al.* 2000, 2004), a Fisher-style VLIW architecture has been selected, rather than an EPIC-style VLIW architecture (Table 2.3). The main features of the Kalray VLIW architecture are as follows:

– Partial predication: fully predicated architectures are expensive with regard to instruction encoding, while control speculation of arithmetic instructions performs better than if-conversion when applicable. Moreover, conditional SELECT operations are equivalent to CMOV operations with operand renaming constraints (Dupont de Dinechin 2014). Then, if-conversion only needs to be supported by conditional load/store and CMOV instructions on scalar and vector operands.

Classical VLIW architecture	EPIC VLIW architecture
SELECT operations on Boolean operand	Fully predicated ISA
Conditional load/store/floating-point operations	Advanced loads (data speculation)
Dismissible loads (control speculation)	Speculative loads (control speculation)
Clustered register files and function units	Polycyclic/multiconnect register files
Multi-way conditional branches	Rotating registers
Compiler techniques	
Trace scheduling	Modulo scheduling
Partial predication	Full predication
Main examples	
Multiflow TRACE processors	Cydrome Cydra-5
HP Labs Lx / STMicroelectronics ST200	HP-Intel IA64
Philips TriMedia	Texas Instruments VelociTI

Table 2.3. *Types of VLIW architectures*

– Dismissible loads: these instructions enable control speculation of load instructions by suppressing exceptions on address errors, and by ensuring that no side-effects occur in the I/O areas. Additional configuration in the MMU refine their behavior on protection and no-mapping exceptions.

– No rotating registers: rotating registers rename temporary variables defined inside software pipelines, whose schedule is built while ignoring register anti-dependences. However, rotating registers add significant ISA and implementation complexity, while temporary variable renaming can be done by the compiler.

– Widened memory access: widening the memory accesses on a single port is simpler to implement than multiple memory ports, especially when memory address translation is implied. This simplification enables, in turn, the support of misaligned memory accesses, which significantly improves compiler vectorization opportunities.

– Unification of the scalar and SIMD data paths around a main register file of 64×64-bit registers, for the same motivations as the POWER vector-scalar architecture (Gschwind 2016). Operands for the SIMD instructions map to register pairs (16 bytes) or to register quadruples (32 bytes).

2.3.4. *Coprocessor*

On the MPPA3 processor, each VLIW core is paired with a tightly coupled coprocessor for the mixed-precision matrix multiply-accumulate operations of deep learning operators (Figure 2.11). The coprocessor operates on a dedicated data path that includes a 48×256-bit register file. Within the six-issue VLIW core architecture, one issue lane is dedicated to the coprocessor arithmetic instructions, while the branch and control unit (BCU) may also execute data transfer operations between the coprocessor registers and the VLIW core general-purpose registers. Finally, the coprocessor leverages the 256-bit LSU of the VLIW core to transfer data blocks from/to the SMEM, at the rate of 32 bytes per clock cycle. It then uses these 32-byte data blocks as left and right operands of matrix multiply-accumulate operations.

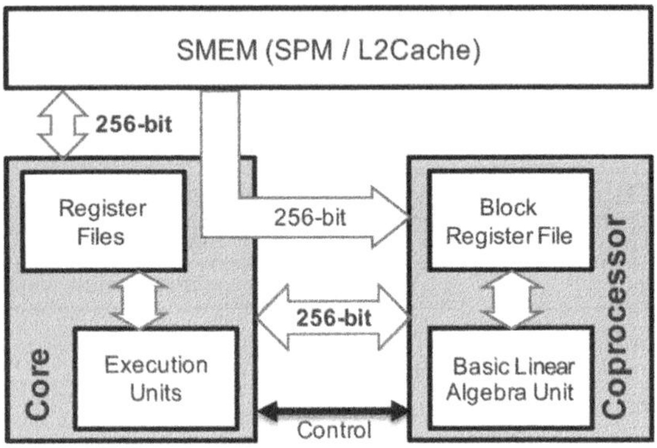

Figure 2.11. *Tensor coprocessor data path*

The coprocessor data path is designed by assuming that the activations and weights, respectively, have row-major and column-major layout in memory, in order to avoid the complexities of Morton memory indexing (Rovder *et al.* 2019). Due to the mixed-precision arithmetic, matrix operands may take one, two or four consecutive registers, with element sizes of one, two, four and eight bytes. In all cases, the coprocessor operations interpret matrix operands as having four rows and a variable number of columns, depending on the number of consecutive registers and the element size. In order to support this invariant, four 32-byte "load-scatter" instructions are provided to coprocessor registers. A load-scatter instruction loads 32 consecutive bytes from memory, interprets these as four 64-bit (8 bytes) blocks and writes each block into a specified quarter of each register that composes the destination operand (Figure 2.12). After executing the four load-scatter variants, a 4×P submatrix of a matrix with row-major order in memory is loaded into a coprocessor register quadruple.

The coprocessor implements matrix multiply-accumulate operations on INT8.32, INT16.64 and FP16.32 arithmetic[1]. The coprocessor is able to multiply-accumulate

1. Numbers in each pair denote, respectively, the bit-width of the multiplicands and the accumulator.

4×8 by 8×4 8-bit matrices into a 4×4 32-bit matrix (128 MAC operations per clock cycle), held in two consecutive registers (Figure 2.13). The 8×4 8-bit matrix operand is actually a 4×8 operand that is transposed at the input port of the multiply-accumulate operation. The coprocessor may also perform multiply-accumulate operations of two 4×4 16-bit matrices into a 4×4 64-bit matrix (64 MAC operations per clock cycle), held in four consecutive registers. Finally, the coprocessor supports multiply-accumulate of two 4×4 FP16 matrices into a 4×4 FP32 matrix, but performed by four successive operations[2] (16 FMA operations per clock cycle). The FP16.32 matrix operations actually compute exact four-deep dot products with accumulation, by applying Kulisch's principles on an $80+\epsilon$ accumulator (Brunie 2017).

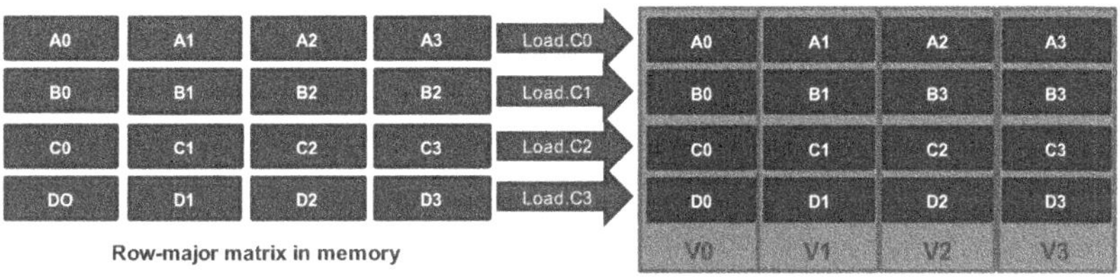

Figure 2.12. *Load-scatter to a quadruple register operand*

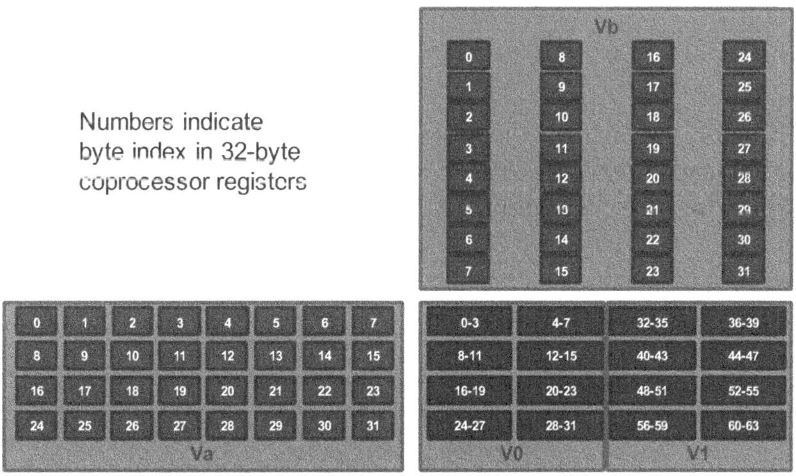

Figure 2.13. *INT8.32 matrix multiply-accumulate operation*

2. Motivated by saving the silicon area and not constrained by the architecture.

2.4. The MPPA3 software environments

2.4.1. *High-performance computing*

The programming environment used for high-performance computing on the MPPA3 processor is derived from the Portable Computing Language (PoCL) project[3], which proposes an open-source implementation of the OpenCL 1.2 standard[4] with support for some of the OpenCL 2.0 features. The OpenCL-C kernels are compiled with LLVM, which has been retargeted for this purpose to the Kalray VLIW core. In OpenCL, a host application offloads computations to an abstract machine:

– an OpenCL device is an offloading target where computations are sent using a command queue;

– an OpenCL device has a global memory allocated and managed by the host application, and shared by the multiple compute units of the OpenCL device;

– an OpenCL compute unit comprises several processing elements (PEs) that share the compute unit local memory;

– each OpenCL PE also has a private memory, and shared access to the device's global memory without cache coherence across compute units.

The OpenCL sub-devices are defined as non-intersecting sets of compute units inside a device, which have dedicated command queues while sharing the global memory.

On the MPPA3 processor, high-performance computing functions are dispatched to partitions composed of one or more compute clusters, each of which is exposed as an OpenCL sub-device. In the port of the PoCL environment, support for OpenCL sub-devices has been developed, while two offloading modes are provided:

LWI (Linearized Work Items): all the work items of a work group are executed within a loop on a single PE. This is the default execution mode of PoCL;

SPMD (Single Program Multiple Data): the work items of a work group are executed concurrently on the PEs of a compute cluster, with the `__local` OpenCL memory space shared by the PEs and located into the SMEM (Figure 2.14).

These mappings of the abstract OpenCL machine elements onto the MPPA3 architecture components are presented in Table 2.4. Although the LWI mode appears better suited to the OpenCL-C kernel code written for GPGPU processors, the SPMD mode is preferred for optimizing performance, as it allows the configuration of most of the compute cluster SMEM as OpenCL local memory shared by the work group.

3. http://portablecl.org/.

4. Passing the OpenCL 1.2 conformance with PoCL is work in progress.

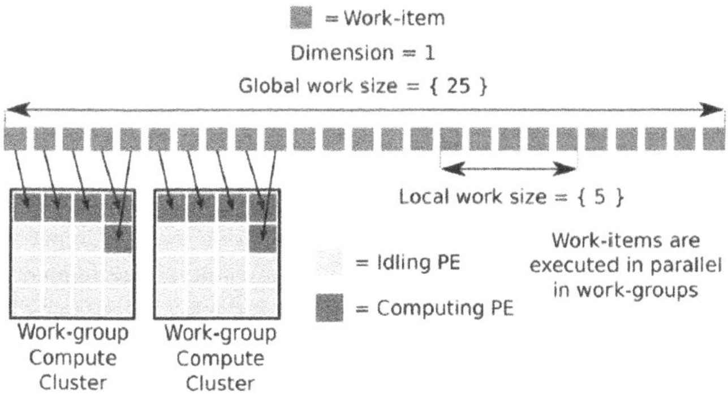

Figure 2.14. *OpenCL NDRange execution using the SPMD mode*

OpenCL	Device	Global memory	Sub-device	Compute unit
MPPA3 Component	MPPA processor or MPPA domain	External DDR memory	Group of compute cluster(s)	Compute cluster (SPMD) Processing element (LWI)

Table 2.4. *OpenCL machine elements and MPPA architecture components*

Most often, there is a need to port C/C++ code and to access the high-performance features implemented in the GCC compiler for the Kalray VLIW core. Among these, the C named address space extension defined by ISO/IEC TR 18037:2008 is used to annotate objects and addresses that are accessed using *non-temporal* (L1D cache bypass) and/or *non-trapping* loads. In order to call the code compiled by GCC and the MPPA communication libraries (Hascoët *et al.* 2017) from OpenCL-C kernels, the LLVM OpenCL-C compiler and PoCL have been extended to understand function declarations annotated with `__attribute__((mppa_native))`. Whenever such reference is seen in OpenCL-C source code, the PoCL linking stages assumes that the symbol is resolved, and the MPPA3 compute cluster run-time environment dynamically loads and links the native function, before starting the execution of the kernel.

This native function extension also enables kernels to access other services such as a lightweight lock-free POSIX multi-threading environment, fast inter-PE hardware synchronizations, dynamic local memory allocation and remoting of system calls to the host OS, including FILE I/O.

2.4.2. *KaNN code generator*

The KaNN (Kalray Neural Network) code generator is a deep learning inference compiler targeting the MPPA3 platform. It takes as input a trained neural network model, described within a standard framework such as Caffe, TensorFlow or ONNX, and produces executable code for a set of compute clusters exposed as an OpenCL sub-device (Figure 2.15). Targeting OpenCL sub-devices allows several model inferences to execute concurrently on a single MPPA3 processor. The KaNN code generator optimizes for batch-1 inference, with the primary objective of reducing latency. At the user's option, FP32 operators in the original network can be converted to FP16 operators. Integer quantization, such as the one used by TensorFlow Lite, is also supported; however, it must be expressed in the input model. Indeed, such models are assumed to be trained with fake quantization (Jacob *et al.* 2018), which must match the actual quantization applied during inference.

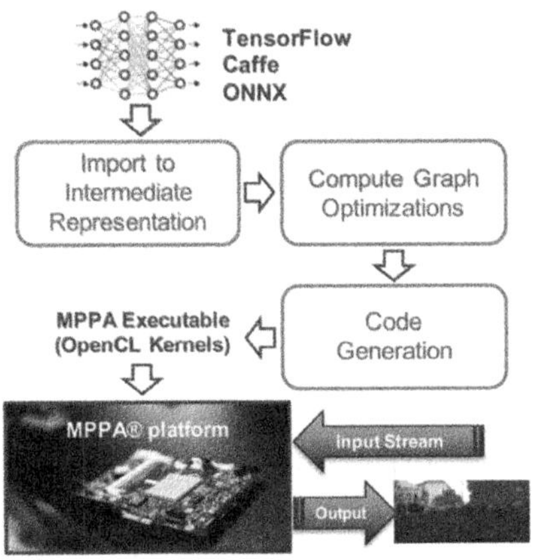

Figure 2.15. *KaNN inference code generator workflow*

Following the import of the input model into an intermediate representation, optimizations are applied to the compute graph:

– elimination of channel concatenation and slicing copies;

– padding of input activations of convolutional layers;

– folding of batch normalizations, scalings, additions, into a single pointwise fused multiply-add operator;

– fusion of convolutions with ReLU activation functions;

– adaptation of arithmetic representations.

The KaNN code generation scheme performs inference in topological sort order of the (optimized) compute graph, parallelizing the execution of each operator over all the compute clusters of the target sub-device. When executing an operator, its input and output activations are distributed across the target local memories configured as SPM, while the network parameters are read from the (external) DDR memory. Depending on the type of operator (convolutional or fully connected), the spatial dimension sizes and the channel depth, input and output activations are distributed over the compute cluster local memories by splitting either along the spatial dimensions or along the channel dimension (Figure 2.16):

– In case of spatial splitting of the output activations, each compute cluster only accesses an input activation tile and its shadow region, while all the operator parameters are required; these are read once from the DDR memory and multicasted to all the target compute clusters.

– In case of channel splitting of the output activations, the full input layer must be replicated in the local memory of each compute cluster, but only the corresponding slice of parameters is read from the DDR memory.

In all cases, activations are computed once, laid out sequentially along the channel dimension and possibly copied to other local memories.

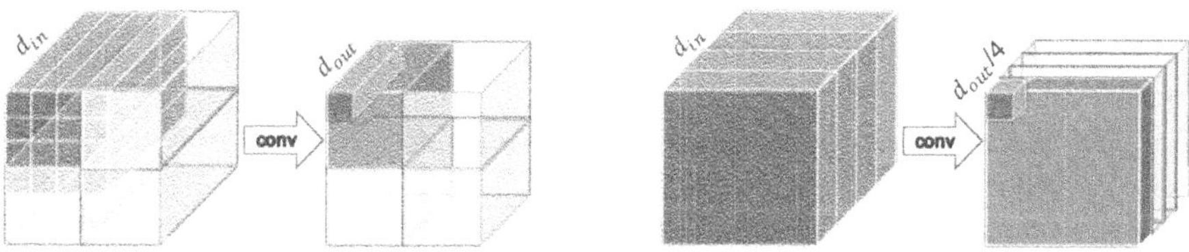

$[3][3][d_{in}][d_{out}]$ $[3][3][d_{in}][d_{out}/4]$

Figure 2.16. *Activation splitting across MPPA3 compute clusters*

For any compute cluster in the target sub-device, the code generation process defines and implements a local schedule for:

– local memory buffer allocations/deallocations;

– DDR memory read/multicast of parameters;

– execution of operator operations;

– inter-cluster activation exchanges;

– inter-cluster synchronizations.

This process is backed by the computation graph (Figure 2.17) augmented with parameter read tasks (yellow) and activation production tasks (blue).

The results of KaNN code generation is a collection of OpenCL binary kernels, where each kernel interprets the contents of a static data block composed of a sequence of records. Each record contains its length, a native compute function pointer and a structure containing arguments for the compute function. For each record, the OpenCL kernel calls the native compute function with the pointer to the structure. The kernel ends after the interpretation of the last record.

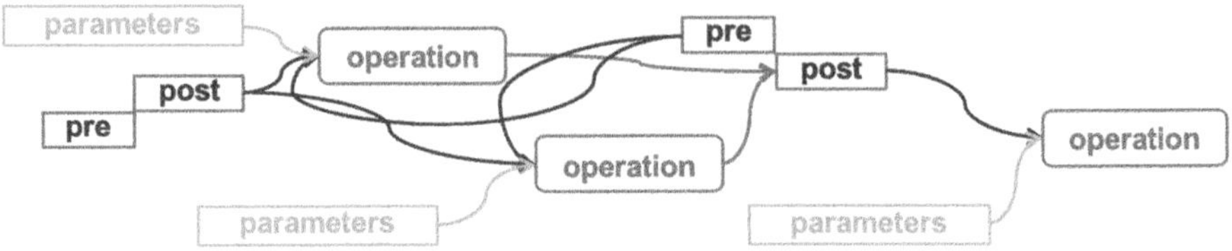

Figure 2.17. *KaNN augmented computation graph*

2.4.3. *High-integrity computing*

High-integrity computing on the MPPA3 processor refers to applications that execute in a physically isolated domain of the processor, whose functions are developed under model-based design and must meet hard real-time constraints. The Research Open-Source Avionics and Control Engineering (ROSACE) case study introduced the model-based design of avionics applications that targeted multi-core platforms (Pagetti *et al.* 2014). The model-based design for the MPPA processor focuses on mono-periodic and multi-periodic harmonic applications (Figure 2.18) that are described using the Lustre (Halbwachs *et al.* 1991) or the SCADE Suite[5] synchronous dataflow languages (Graillat *et al.* 2018, 2019). The execution environment is composed of one or more clusters configured for asymmetric multi-processing (section 2.3.2), where each core is logically associated with one SMEM bank, and where tasks run to completion.

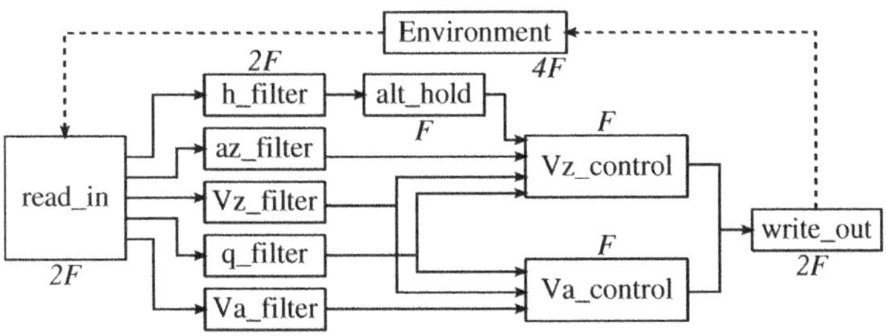

Figure 2.18. *ROSACE harmonic multi-periodic case study (Graillat* et al. *2018)*

The code generation workflow assumes that some nodes of the synchronous dataflow program are identified by the programmer as concurrent tasks, and defines the implicit top-level "root" task. A Lustre or SCADE Suite high-level compiler generates C-code for this set of tasks, communicating and synchronizing through one-to-one channels. Channels correspond to single-producer, single-consumer FIFOs of

5. https://www.ansys.com/products/embedded-software/ansys-scade-suite.

depth one, whose implementation is abstracted form the task C-code under SEND and RECV macros. The rest of the code generation workflow involves:

- providing workers, each able to execute a set of tasks sequentially;
- scheduling and mapping the set of tasks on the workers;
- implementing the communication channels and their SEND/RECV methods;
- compiling C-code with the CompCert formally verified compiler.

In the MPPA workflow, the workers are the PEs associated with a memory bank.

Timing verification follows the principles of the multi-core response time analysis (MRTA) framework (Davis *et al.* 2018). Starting from the task graph, its mapping to PEs, and given the worst-case execution time (WCET) of each task in isolation, the multi-core inference analysis (MIA) tool (Rihani *et al.* 2016; Dupont de Dinechin *et al.* 2020) refines the execution intervals of each task while updating its WCET for interference on the shared resources. The MIA tool relies on the property that the PEs, the memory hierarchy and the interconnects are timing-compositional. The refined release dates are used to activate a fast hardware release mechanism for each task. A task then executes when its input channels are data-ready (Figure 2.19).

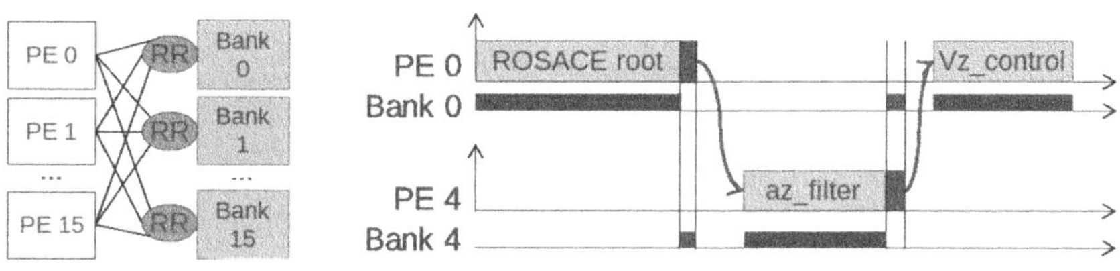

Figure 2.19. *MCG code generation of the MPPA processor*

2.5. Conclusion

We introduced the third-generation MPPA processor, which implements a many-core architecture that targets intelligent systems, defined as cyber-physical systems enhanced with high-performance machine learning capabilities and strong cyber-security support. As with the GPGPU architecture, the MPPA3 architecture is composed of a number of multi-core compute units that share the processor external memory and I/O through on-chip global interconnects. However, the MPPA architecture is able to host standard software, offers excellent time predictability and provides strong partitioning capabilities. This enables us to consolidate, on a single or dual processor platform, the high-performance machine learning and computer vision functions implied by vehicle perception, the high-integrity functions developed through model-based design, and the cyber-security functions required by secured communications.

2.6. References

Bodin, B., Munier-Kordon, A., and Dupont de Dinechin, B. (2013). Periodic schedules for cyclo-static dataflow. *The 11th IEEE Symposium on Embedded Systems for Real-time Multimedia*, Montreal, QC, Canada, 105–114.

Bodin, B., Munier-Kordon, A., and Dupont de Dinechin, B. (2016). Optimal and fast throughput evaluation of CSDF. *Proceedings of the 53rd Annual Design Automation Conference*. Austin, USA, 160:1–160:6.

Brunie, N. (2017). Modified fused multiply and add for exact low precision product accumulation. *24th IEEE Symposium on Computer Arithmetic*. London, United Kingdom, 106–113.

Carmichael, Z., Langroudi, H.F., Khazanov, C., Lillie, J., Gustafson, J.L., and Kudithipudi, D. (2019). Performance-efficiency trade-off of low-precision numerical formats in deep neural networks. *Proceedings of the Conference for Next Generation Arithmetic*. New York, USA, 3:1–3:9.

CAST (2016). Multi-core Processors, Technical Report CAST-32A, FAA [Online]. Available: https://www.faa.gov/aircraft/air_cert/design_approvals/air_software/cast/cast_papers/.

Cavicchioli, R., Capodieci, N., Solieri, M., and Bertogna, M. (2019). Novel methodologies for predictable CPU-To-GPU command offloading. *Proceedings of the 31st Euromicro Conference on Real-Time Systems*. Stuttgart, Germany, vol. 133 of *LIPIcs*, 22:1–22:22.

Chung, E., Fowers, J., Ovtcharov, K., Papamichael, M., Caulfield, A., Massengill, T., Liu, M., Ghandi, M., Lo, D., Reinhardt, S., Alkalay, S., Angepat, H., Chiou, D., Forin, A., Burger, D., Woods, L., Weisz, G., Haselman, M., and Zhang, D. (2018). Serving DNNs in real time at datacenter scale with project brainwave. *IEEE Micro*, 38, 8–20.

CNX (2019). Autoware.AI-Software-Architecture [Online]. Available: https://www.cnx-software.com/wp-content/uploads/2019/02/Autoware.AI-Software-Architecture.png.

Davis, R.I., Altmeyer, S., Indrusiak, L.S., Maiza, C., Nélis, V., and Reineke, J. (2018). An extensible framework for multicore response time analysis. *Real-Time Systems*, 54(3), 607–661.

de Dinechin, F., Forget, L., Muller, J.-M., and Uguen, Y. (2019). Posits: The good, the bad and the ugly. *Proceedings of the Conference for Next Generation Arithmetic*. Association for Computing Machinery, New York, USA.

Dupont de Dinechin, B. (2004). From machine scheduling to VLIW instruction scheduling. *ST Journal of Research*, 1(2).

Dupont de Dinechin, B. (2014). Using the SSA-Form in a code generator. *23rd International Conference on Compiler Construction*, vol. 8409 of *Lecture Notes in Computer Science*, Springer, 1–17.

Dupont de Dinechin, B., and Graillat, A. (2017). Feed-forward routing for the wormhole switching network-on-chip of the kalray MPPA2 processor. *Proceedings of the 10th International Workshop on Network on Chip Architectures*. Cambridge, USA, 10:1–10:6.

Dupont de Dinechin, B., de Ferrière, F., Guillon, C., and Stoutchinin, A. (2000). Code generator optimizations for the ST120 DSP-MCU core. *Proceedings of the 2000 International Conference on Compilers, Architectures and Synthesis for Embedded Systems, CASES*, San Jose, USA, 93–102.

Dupont de Dinechin, B., Ayrignac, R., Beaucamps, P., Couvert, P., Ganne, B., de Massas, P. G., Jacquet, F., Jones, S., Chaisemartin, N. M., Riss, F., and Strudel, T. (2013). A clustered manycore processor architecture for embedded and accelerated applications. *IEEE High Performance Extreme Computing Conference*, Waltham, USA, 1–6.

Dupont de Dinechin, B., van Amstel, D., Poulhiès, M., and Lager, G. (2014). Time-critical computing on a single-chip massively parallel processor. *Design, Automation and Test in Europe Conference and Exhibition*, Dresden, Germany, 1–6.

Dupont de Dinechin, M., Schuh, M., Moy, M., and Maïza, C. (2020). Scaling up the memory interference analysis for hard real-time many-core systems. *Design, Automation and Test in Europe Conference and Exhibition*, Grenoble, France, 1–4.

Firesmith, D. (2017). Multicore Processing [Online]. Available: https://insights.sei.cmu.edu/ sei_blog/2017/08/multicore-processing.html.

Fisher, J. A., Faraboschi, P., and Young, C. (2005). *Embedded Computing: A VLIW Approach to Architecture, Compilers and Tools*. Morgan Kaufmann Publishers Inc., San Francisco, USA.

Forsberg, B., Palossi, D., Marongiu, A., and Benini, L. (2017). GPU accelerated real-time path planning and the predictable execution model. *Procedia Computer Science – International Conference on Computational Science, Zurich, Switzerland*, 108, 2428–2432.

Graillat, A., Moy, M., Raymond, P., and Dupont de Dinechin, B. (2018). Parallel code generation of synchronous programs for a many-core architecture. *Design, Automation and Test in Europe Conference and Exhibition*, Dresden, Germany, 1139–1142.

Graillat, A., Maiza, C., Moy, M., Raymond, P., and Dupont de Dinechin, B. (2019). Response time analysis of dataflow applications on a many-core processor with shared-memory and network-on-chip. *Proceedings of the 27th International Conference on Real-Time Networks and Systems*. Toulouse, France, 61–69.

Gschwind, M. (2016). Workload acceleration with the IBM POWER vector–scalar architecture. *IBM Journal of Research and Development*, 60(2–3).

Gustafson, J.L. (2017). Beyond floating point: Next-generation computer arithmetic [Online]. Available: https://web.stanford.edu/class/ee380/Abstracts/170201-slides.pdf.

Gustafson, J.L. and Yonemoto, I.T. (2017). Beating floating point at its own game: Posit arithmetic. *Supercomputing Frontiers and Innovations*, 4(2), 71–86.

Halbwachs, N., Caspi, P., Raymond, P., and Pilaud, D. (1991). The synchronous data flow programming language LUSTRE. *Proceedings of the IEEE*, 79(9), 1305–1320.

Hascoët, J., Dupont de Dinechin, B., de Massas, P.G., and Ho, M.Q. (2017). Asynchronous one-sided communications and synchronizations for a clustered manycore processor. *Proceedings of the 15th IEEE/ACM Symposium on Embedded Systems for Real-Time Multimedia*, Seoul, Republic of Korea, 51–60.

Hascoët, J., Dupont de Dinechin, B., Desnos, K., and Nezan, J. (2018). A distributed framework for low-latency openVX over the RDMA NoC of a clustered manycore. *2018 IEEE High Performance Extreme Computing Conference* HPEC, Waltham, USA, 1–7.

Huang, M., Men, L., and Lai, C. (2013). Accelerating mean shift segmentation algorithm on hybrid CPU/GPU platforms. In *Modern Accelerator Technologies for Geographic Information Science*, Shi, X., Kindratenko, V. and Yang, C. (eds). Springer, New York.

Intel (2018). BFLOAT16 – Hardware Numerics Definition Revision 1.0. November 2018.

Jacob, B., Kligys, S., Chen, B., Zhu, M., Tang, M., Howard, A.G., Adam, H., and Kalenichenko, D. (2018). Quantization and training of neural networks for efficient integer-arithmetic-only inference. *IEEE Conference on Computer Vision and Pattern Recognition*. Salt Lake City, USA, 2704–2713.

Jia, Z., Maggioni, M., Staiger, B., and Scarpazza, D.P. (2018). Dissecting the NVIDIA volta GPU architecture via microbenchmarking. *ArXiv*, abs/1804.06826.

Johnson, J. (2018). Rethinking floating point for deep learning. *ArXiv*, abs/1811. 01721.

Kanduri, A., Rahmani, A.M., Liljeberg, P., Hemani, A., Jantsch, A., and Tenhunen, H. (2017). *A Perspective on Dark Silicon*. Springer International Publishing.

Kästner, D., Pister, M., Gebhard, G., Schlickling, M., and Ferdinand, C. (2013). Confidence in timing. *SAFECOMP 2013 - Workshop SASSUR (Next Generation of System Assurance Approaches for Safety-Critical Systems) of the 32nd International Conference on Computer Safety, Reliability and Security*, Toulouse, France.

Krishnamoorthi, R. (2018). Quantizing deep convolutional networks for efficient inference: A whitepaper. ArXiv abs/1806.08342.

Lee, E.A., Reineke, J., and Zimmer, M. (2017). Abstract PRET Machines. *IEEE Real-Time Systems Symposium, RTSS*, Paris, France, December 5–8, 1–11.

NVIDIA (2020). Programming Tensor Cores in CUDA 9 [Online]. Available: https://devblogs.nvidia.com/programming-tensor-cores-cuda-9/.

Pagetti, C., Saussié, D., Gratia, R., Noulard, E., and Siron, P. (2014). The ROSACE case study: From simulink specification to multi/many-core execution. *20th IEEE Real-Time and Embedded Technology and Applications Symposium*. Berlin, Germany, 309–318.

Perret, Q., Maurère, P., Noulard, E., Pagetti, C., Sainrat, P., and Triquet, B. (2016). Temporal isolation of hard real-time applications on many-core processors. *IEEE Real-Time and Embedded Technology and Applications Symposium*. Vienna, Austria, April 11-14, 37–47.

Resmerita, D., Farias, R.C., Dupont de Dinechin, B., and Fillatre, L. (2020). Benchmarking alternative floating-point formats for deep learning inference. *Conférence francophone d'informatique en Parallélisme, Architecture et Système*.

Rihani, H., Moy, M., Maiza, C., Davis, R.I., and Altmeyer, S. (2016). Response time analysis of synchronous data flow programs on a many-core processor. *Proceedings of the 24th International Conference on Real-Time Networks and Systems*. Brest, France, 67–76.

Rodriguez, A., Ziv, B., Fomenko, E., Meiri, E., and Shen, H. (2018). Lower numerical precision deep learning inference and training. Intel AI Developer Program, 1–19 [Online]. Available: https://software.intel.com/content/www/us/en/develop/articles/lower-numerical-precision-deep-learning-inference-and-training.html.

Rovder, S., Cano, J., and O'Boyle, M. (2019). Optimising convolutional neural networks inference on low-powered GPUs. *12th International Workshop on Programmability and Architectures for Heterogeneous Multicores*. Valencia, Spain.

Saidi, S., Ernst, R., Uhrig, S., Theiling, H., and Dupont de Dinechin, B. (2015). The shift to multicores in real-time and safety-critical systems. *International Conference on Hardware/Software Codesign and System Synthesis*. Amsterdam, The Netherlands, October 4–9, 220–229.

Wilhelm, R. and Reineke, J. (2012). Embedded systems: Many cores - Many problems. *7th IEEE International Symposium on Industrial Embedded Systems*. Karlsruhe, Germany, June 20–22, 176–180.

3

The Plural Many-core Architecture – High Performance at Low Power

Ran GINOSAR

Electrical Engineering Department, Technion – Israel Institute of Technology, Haifa, Israel

The Plural many-core architecture combines multiple CPU cores, hardware scheduling and a large on-chip shared memory. A PRAM-like task-level-parallelism programming model is used. The program is expressed as a collection of sequential task codes and a task precedence graph. The task codes are stored in shared memory, for execution by the cores, while the task precedence graph is loaded into the hardware scheduler and its execution determines the order of task executions. Microarchitectural variations include cache memories and local memories in each core and using the on-chip shared memory as a last-level cache to an off-chip memory. The hardware scheduler is interconnected with all cores by an active tree-structured network-on-chip, implemented in hardware or virtually through shared memory. The cores are interconnected with the shared memory by a bidirectional buffered packet multi-stage interconnection network. I/O processing cores also access the shared memory through the same network so that all I/O takes place to and from shared memory. The cores do not interact other than through shared memory and do not access the I/O port. The architecture is shown to be highly efficient for DSP and machine learning applications. The Plural architecture achieves a high level of core utilization and a high level of performance-to-power ratio. The architecture is highly scalable for various silicon process nodes.

For a color version of all figures in this book, see www.iste.co.uk/andrade/multi1.zip.

Multi-Processor System-on-Chip 1 – Architectures,
coordinated by Liliana ANDRADE and Frédéric ROUSSEAU.

3.1. Introduction

Multiple-core architectures are divided into multi-cores and many-cores. Multi-cores, such as ARM Cortex A9 and Intel Xeon, typically provide some form of cache coherency, and are designed to execute many unrelated processes, governed by an operating system such as Linux. In contrast, many-cores such as Tilera's TilePro, Adapteva's Epiphany, Nvidia's GPU, Intel's Xeon Phi and the Plural architecture execute parallel programs specifically designed for them and avoid operating systems, in order to achieve higher performance and higher power efficiency.

Many-core architectures come in different flavors: a two-dimensional array of cores arranged around a mesh NoC (Tilera and Adapteva), GPUs and other many-cores with clusters of cores (Kalray), and rings. This chapter discusses the Plural architecture (Bayer and Ginosar 1993, 2002; Avron and Ginosar 2012, 2015; Bayer and Peleg 2012), in which many cores are interconnected to a many-port shared memory rather than to each other (Figure 3.1).

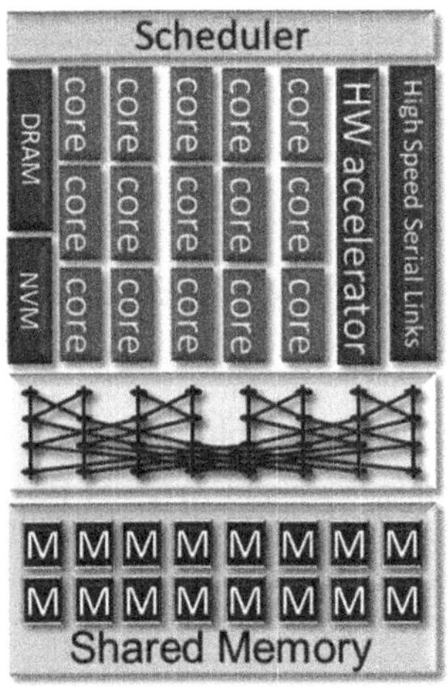

Figure 3.1. *Plural many-core architecture. Many cores, hardware accelerators and multiple DMA controllers of I/O interfaces access the multi-bank shared memory through a logarithmic network. The hardware scheduler dispatches fine grain tasks to cores, accelerators and I/O*

Many-cores also differ on their programming models, ranging from PRAM-like shared memory through CSP-like message-passing to dataflow. Memory access and message passing also relate to data dependencies and synchronization – locks, bulk-synchronous patterns and rendezvous. The Plural architecture uses a strict shared memory programming model.

The last defining issue relates to task scheduling – allocating tasks to cores and handling task dependencies. Scheduling methods include static (compile time) scheduling, dynamic software scheduling, architecture-specific scheduling (e.g. for NoC) and hardware schedulers, as in the Plural architecture, in which data dependencies are replaced by task dependencies in order to enhance performance and efficiency and to simplify programming.

3.2. Related works

Multi-core architectures include ARM Cortex A9 (Jacquet *et al.* 2014) and Intel Xeon. Many-core architectures include the mesh-tiled Tilera (Wentzlaff *et al.* 2007) and Adapteva (Varghese *et al.* 2014), NVidia GPU (Nickolls and Dally 2010), Intel ring-topology Xeon Phi (Heinecke *et al.* 2013) and dataflow clusters by Kalray (Dupont de Dinechin *et al.* 2014). The research XMT many-core (Wen and Vishkin 2008) is PRAM-inspired and employs hardware scheduling, similar to the Plural architecture. It uses declarative parallelism to direct scheduling (Tzannes *et al.* 2014). The Plural architecture is discussed in Bayer and Ginosar (1993, 2002); Avron and Ginosar (2012, 2015); Bayer and Peleg (2012). An early hardware scheduler is reported in Crummey *et al.* (1994). The baseline multi-stage interconnection network was introduced in Wu and Feng (1980).

3.3. Plural many-core architecture

This section presents the Plural architecture (Figure 3.1). The Plural architecture defines a shared memory single-chip many-core. The many-core consists of a hardware synchronization and scheduling unit, many cores and a shared on-chip memory accessible through a high-performance logarithmic interconnection network. Different types of cores may be used, including RISC cores and DSP cores. The cores may contain instruction and data caches, as well as a private "scratchpad" memory. The data cache is flushed and invalidated by the end of each task execution, guaranteeing consistency of the shared memory. The cores should be designed for low-power operation using low voltage and low clock frequency. Performance is achieved by a high level of parallelism, rather than by sheer speed, and access to the on-chip shared memory across the chip takes only a small number of cycles.

The on-chip shared memory is organized in a large number of banks to enable many ports that can be accessed in parallel by the many cores via the network. To reduce collisions, addresses are interleaved over the banks. The cores are connected to the memory banks by a multi-stage many-to-many interconnection network (section 3.6). As explained in the next section, there is no need for any cache coherency mechanism.

3.4. Plural programming model

The Plural PRAM-like programming model is based on non-preemptive execution of multiple sequential tasks. The programmer defines the tasks, as well as their dependencies and priorities, which are specified by a (directed) *task graph*. Tasks are executed by cores, and the task graph is "executed" by the scheduler.

In the Plural shared-memory programming model, concurrent tasks cannot communicate. A group of tasks that are allowed to execute in parallel may share read-only data, but they cannot share data that is written by any one of them. If one task must write into a shared data variable and another task must read that data, then they are *dependent* – the writing task must complete before the reading task may commence. That dependency is specified as a directed edge in the task graph and enforced by the hardware scheduler. Tasks that do not write-share data are defined as *independent*, and may execute concurrently. Concurrent execution does not necessarily happen at the same time – concurrent tasks may execute together or in any order, as determined by the scheduler.

Some tasks, typically amenable to independent data parallelism, may be *duplicable*, accompanied by a *quota* that determines the number of instances that should be executed (declared parallelism (Tzannes *et al*. 2014)). All instances of the same duplicable task are mutually independent (they do not write-share any data) and concurrent, and hence they may be executed in parallel or in any arbitrary order. These instances are distinguishable from each other merely by their *instance number*. Ideally, their execution time is short (fine granularity). Concurrent instances can be scheduled for execution at any (arbitrary) order, and no priority is associated with instances.

Each task progresses through, at most, four states (Figure 3.2). Tasks without predecessors (enabled at the beginning of program execution) start in the *ready* state. Tasks that depend on predecessor tasks start in the *pending* state. Once all predecessors to a task have completed, the task becomes *ready* and the scheduler may schedule its instances for execution and allocate (dispatch) the instances to cores. Once all instances of a task have been allocated, the task is *all allocated*. And once all its instances have terminated, the task moves into the *terminated* state (possibly enabling successor tasks to become *ready*).

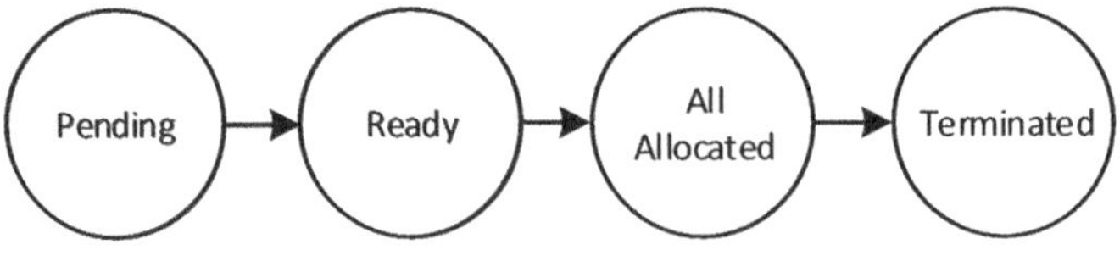

Figure 3.2. *Task state graph*

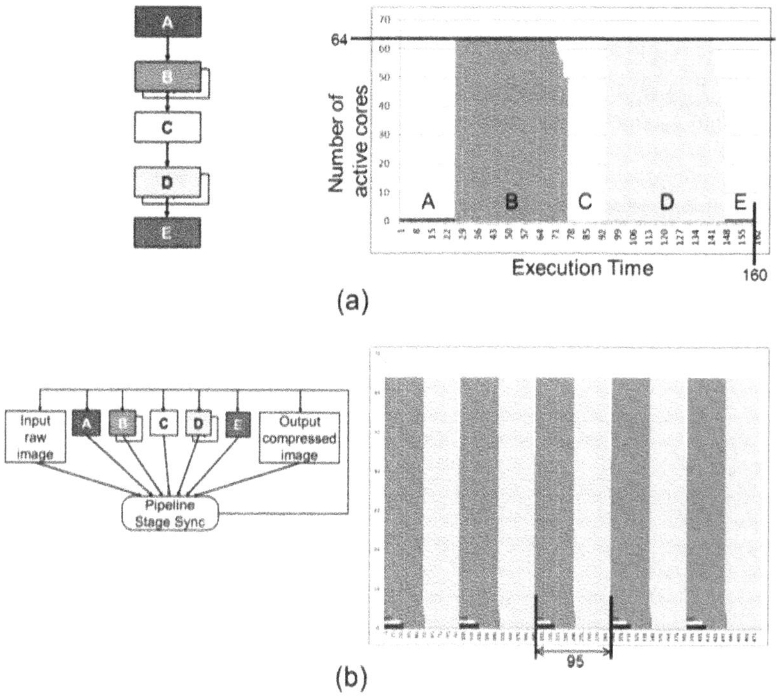

Figure 3.3. *Many-flow pipelining: (a) Task graph and single execution of an image compression program; (b) Many-flow task graph and its pipelined execution*

Many-flow pipelining facilitates enhanced core utilization in streamed signal processing. Consider the task graph examples for executing JPEG2000 image compression and the processor utilization charts of Figure 3.3. In (a), five tasks A–E are scheduled in sequence. Tasks B and D are duplicable with a large number of instances, enabling efficient utilization of 64 cores. Tasks A, C and E, on the other hand, are sequential. The execution time of compressing one image is 160 time units, and overall utilization, reflected by the ratio of colored area to the 64×160 rectangle, is 65%. The core utilization chart (on the right) indicates the number of busy cores over time, and different colors represent different tasks. In the many-flow task graph (Figure 3.3(b)), a pipeline of seven images is processed. During one iteration, say iteration k, the output stage sends compressed image k, task E processes image $k+1$, task D computes the data of image $k+2$ and so on. Note that the sequential tasks A, C and E are allocated first in each iteration, and duplicable instances occupy the remaining cores. A single iteration takes 95 time units, and the latency of compressing a single image is extended to five iterations, but the throughput is enhanced and the core utilization chart now demonstrates 99% core utilization.

Data dependencies are expressed (by the programmer) as task dependencies. For instance, if a variable is written by task t_w and must later be read, then reading must occur in a group of tasks $\{t_r\}$ and $t_w \rightarrow \{t_r\}$. The synchronization action of completion of t_w prior to any execution of tasks $\{t_r\}$ provides the needed barrier.

3.5. Plural hardware scheduler/synchronizer

The hardware scheduler assigns tasks to cores for execution. The scheduler maintains two data structures, one for managing cores (Figure 3.4) and the other for managing tasks (Figure 3.5). Core and task state graphs are shown in Figure 3.6 and Figure 3.2, respectively.

Core #	State	Task #	Instance #	...	...
0					
1					
2					
...					

Figure 3.4. *Core management table*

Task #	Duplication quota	Dependencies	State	# Allocated instances	# Terminated instances
0					
1					
2					
...					

Data from task graph

Figure 3.5. *Task management table*

The hardware scheduler operates as follows. Initially, all cores are listed as *idle* and the task graph is loaded into the first three columns of the task management table. The scheduler loops forever over its computation cycle. On each cycle, the scheduler performs two activities: allocating tasks for execution and handling task completions.

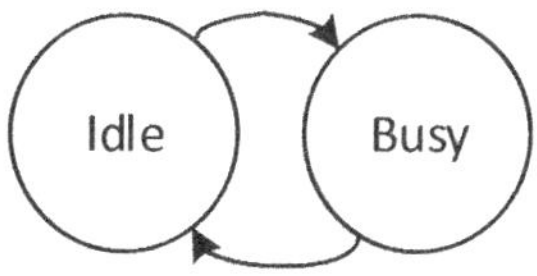

Figure 3.6. *Core state graph*

To allocate tasks, the scheduler first selects ready tasks from the task management table. It allocates each such task to idle cores by changing the task state to *All Allocated* (if the task is regular, or if all duplicable instances have been dispatched), by increasing the count of allocated instances in the task management table and by noting the task number (and instance number, for duplicable tasks) in the core management table. Finally, task/instance activation messages are dispatched to the relevant cores. The activation message for a specific core includes the code entry address and (in case of a duplicable instance) the instance ID number.

To handle task completions, the scheduler collects termination messages from cores that have completed task executions. It changes the state of those cores to *Idle*. For regular tasks, the task state is changed to *Terminated*. For duplicable tasks, the counter of terminated tasks in the task management table is incremented, and if it has reached the quota value, then the state of that task is changed to *Terminated*. Next, the scheduler updates the dependencies entry of each task in the table, which depends on the terminated task: the arrival of that token is noted, the dependency condition is recomputed, and if all predecessors of any task have been fulfilled, then the state of that task is changed to *Ready*, enabling allocation and dispatch in subsequent scheduler computation cycles.

The *scheduler capacity*, namely the number of simultaneous tasks which the scheduler is able to allocate or terminate during each computation cycle, is limited. Any additional task allocations and task termination messages beyond scheduler capacity wait for subsequent cycles in order to be processed. A core remains idle from the time it issues a termination message until the next task allocation arrives. That idle time comprises not only the delay at the scheduler (wait and processing times) but also any transmission latency of the termination and allocation messages over the scheduler-to-cores network.

The allocation and termination algorithms are shown in Figure 3.7.

```
ALLOCATION
1. Choose a Ready task (according to priority, if specified)
2. While there is still enough scheduler capacity and there
   are still Idle cores
   a. Identify an Idle core
   b. Allocate an instance to that core
   c. Increase counter of allocated task instances
   d. If # allocated instances == quota, change task state
      to All Allocated and continue to next task (step 1)
   e. Else, continue to next instance of same task (step 2)

TERMINATION
1. Choose a core which has sent a termination message
2. While there is still enough scheduler capacity
   a. Change core state to Idle
   b. Increment # terminated instances
   c. If # terminated instances == quota, change task state
      to Terminated
   d. Recompute dependencies for all other tasks that depend
      on the terminated task, and where relevant change
      their state to Ready
```

Figure 3.7. *Allocation (top) and termination (bottom) algorithms*

Scheduling efficiency depends on the ratio of scheduling latency (reflected in idle time of cores) to task execution time. Extremely fine grain tasks (e.g. those executing for 1~100 cycles) call for very short scheduling latencies (down to zero cycles) to be efficient. Alternatively, speculative advanced scheduling may fill queues attached to each core so that the core can start executing a new instance once it has completed the previous instance (see Avron and Gnosar (2015) for such an analysis). However, typical tasks tend to incur compiler overhead (prologue and epilogue code sequences generated by even the most efficient optimizing compilers), and typical programming practices of parallel tasks tend to avoid the shortest tasks, resulting in average task duration exceeding 100 cycles. With average scheduling latency of only 10–20 cycles, enabled by hardware implementation, we obtain execution efficiency close to 99%.

The hardware scheduler may be implemented as custom logic. Two other possibilities may be considered, one based on two content-addressable memory (CAM) arrays implementing the two management tables and another implementation as software executing on a dedicated fast core with its dedicated high-throughput memory.

A special section of the scheduler schedules high-priority tasks (HPTs), which are designed as "interrupt handling routines" to handle hardware interrupts. As explained in section 3.7, all I/O interfaces (including interfaces to accelerators) are based on DMA controllers that issue interrupts once they have completed their action. The most urgent portion of handling the interrupt is packaged as an HPT, and less urgent parts are formulated as a normal task. HPT is dispatched immediately and pre-emptively by the scheduler. Each core may execute one HPT, and one HPT does not pre-empt another HPT. Thus, a maximum of 64 HPTs may execute simultaneously on a 64-core Plural architecture. The use of HPT is further explained in section 3.8.

3.6. Plural networks-on-chip

The Plural architecture specifies two specialized networks-on-chip (NoCs), one connecting the scheduler to all cores and other schedulable entities (DMA controllers and hardware accelerators) and a second NoC connecting all cores and other data sources (DMA controllers) to the shared memory.

3.6.1. *Scheduler NoC*

The scheduler-to-cores NoC uses a tree topology. This NoC off-loads two distributed functions from the scheduler, task allocation and task termination.

The distributed task allocation function receives clustered task allocation messages from the scheduler. In particular, a task allocation message related to a duplicable task specifies the task entry address and a range of instance numbers that should be dispatched. The NoC partitions such a clustered message into new messages specifying the same task entry address and sub-range of instance numbers, so that the sub-ranges of any two new messages are mutually exclusive and the union of all new messages covers the same range of instance numbers as the original message. The NoC nodes maintain core and task management tables which are subsets of those tables in the scheduler (Figure 3.4 and Figure 3.5, respectively), to enable making these distributed decisions.

The distributed task termination process complements task allocations. Upon receiving instance terminations from cores or subordinate nodes, a NoC node combines the messages and forwards a more succinct message specifying ranges of completed tasks.

3.6.2. *Shared memory NoC*

The larger NoC of the Plural architecture connects the many cores, multiple DMA controllers and multiple hardware accelerators to many banks of the shared memory. For instance, a 64-core Plural architecture is best served with 256 or more memory banks. To simplify layout, floor-planning and routing, we use a baseline logarithmic-depth multi-stage interconnection network (Wu and Feng 1980), symbolically drawn in Figure 3.1. Some of the NoC switch stages are combinational, while others use registers and operate in a pipeline. Two separate networks are used, one for reading and another for writing. The read network accesses and transfers 16 bytes (128 bits) in parallel, matching cache line size and serving cache fetch in a single operation. The write network is limited to 32 bits, compatible with the write-through mechanism employed in the DSP cores. Writing smaller formats (16 and 8 bits) is also allowed.

Two alternative types of NoC are useful in the Plural architecture, a circuit switching network and a packet switching network. The circuit switching network establishes a connection once a memory access request is presented by a core. The connection assures a deterministic latency through the network, and consequently a deterministic memory access time, if there is no access conflict. The network detects access conflicts contending on the same memory bank, proceeds to serve one of the requests and notifies the other cores to retry their access. The affected cores immediately retry a failed access. Two or more concurrent read requests from the same address are served by a single read operation and a multicast of the same value to all requesting cores.

Circuit switching networks are difficult to scale for growth in the number of cores and the respective growth in the number of memory banks. Packet switching is more scalable, and the network always completes all access requests without asking the cores to retry their requests. Such networks may use asynchronous interconnects, enabling the splitting of the network, as well as the entire processor, into many clock domains, further contributing to scalability. However, packet switching requires buffer storage for stalled packets at each stage of the network. The penalties in terms of power, latency and congestion are strongly influenced by the programming model and by the application.

3.7. Hardware and software accelerators for the Plural architecture

Certain operations cannot be performed efficiently on programmable cores. Typical examples require bit-level manipulations that are not provided for by the instruction set, such as those used for error correcting codes (LDPC, Turbo code, BCH, etc.) and for encryption. The Plural architecture offers two solutions: on-chip and off-chip hardware accelerators. On-chip hardware accelerators are not directly accessible from the cores. Rather, they are accessible only through shared memory, as follows. First, the data to be processed by the accelerator is deposited in shared memory. Next, the accelerator is invoked. Data is fetched to the accelerator by a dedicated DMA controller, and the outcome is sent back to shared memory by a complementing second DMA controller. This mode of operation decouples the accelerator from the cores and eliminates busy waiting of cores.

The second possibility is to use an external acceleration on either an FPGA or an ASIC. High-speed serial links on any implementation of the Plural architecture would enable efficient utilization of such external acceleration. This mode offers scalability and extendibility to the Plural architecture.

All input/output interfaces operate asynchronously to the cores. Each interface is managed by one DMA controller for input and a second DMA controller for output. Many different types of I/O interfaces should be made available in any implementation of the Plural architecture, including high-rate DRAM, high-rate NVM (error detection and correction is carried out at the I/O interfaces, offloading that compute load from the cores) and high-speed serial links.

All DMA controllers are scheduled by the scheduler, submit interrupt signals to the scheduler (as explained in section 3.5) and read and write data directly to the shared memory through the NoC (see section 3.6). The system software required for managing I/O is described in section 3.8.

3.8. Plural system software

The system run-time software stack is shown schematically in Figure 3.8. The boot sequence library is based on the boot code of the core used. It is modified to enable execution by many cores in parallel. Only one of the cores performs the shared memory content initialization. The boot code includes core self-test, cache clearing, memory protection configuration and execution status notification to an external controlling host.

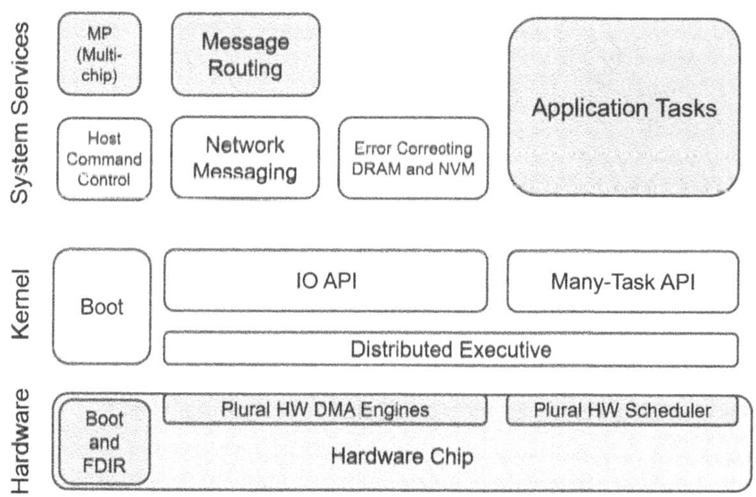

Figure 3.8. *Plural run-time software. The kernel enables boot, initialization, task processing and I/O. Other services include execution of host commands, networking and routing, error correction and management of applications distributed over multiple Plural chips*

The runtime kernel (RTK) performs the scheduling function for the core. It interacts with the hardware scheduler, receives task allocation details, launches the task code and responds with task termination when the task is finished. The RTK also initiates the power-down sequence when no task is received for execution. The first task allocated by the scheduler is responsible for loading the application task graph into the scheduler. This code is automatically generated during a pre-compile stage according to the task graph definition. Application tasks are allocated after the initialization task is finished.

Certain library routines manage EDAC for memories, encapsulate messaging and routing services for off-chip networking (especially over high-speed serial links), respond to commands received from an external host (or one of the on-chip cores, playing the role of a host), perform fault detection, isolation and recovery (FDIR) functions and offer some level of virtualization when multiple Plural architecture chips are used in concert to execute coordinated missions.

Other components of the RTK manage I/O and accelerators. Configuring the interfaces requires special sequences, such as link detection and activation, clock enabling, DMA configuration, etc. Each interface has its own set of parameters according to the required connectivity, storage type, data rate and so on.

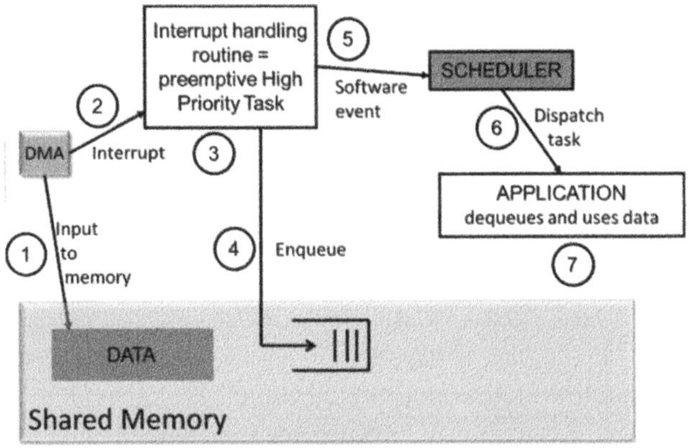

Figure 3.9. *Event sequence performing stream input*

Figure 3.9 demonstrates the hardware–kernel–application sequence of events in the case of an input of a predefined data unit over a stream input link. The DMA controller, previously scheduled, stores input data in a pre-allocated buffer in memory (step 2). Upon completion, it issues an interrupt (step 2). An HPT is invoked (step 3, see section 3.5) and stores pointers and status in shared memory, effectively enqueuing the new arrival (step 4). It ends up by issuing a *software event* to the scheduler (step 5). Eventually, the scheduler dispatches a task that has been waiting for that event (step 6).

That task can consume the data and then dequeue it, releasing the storage where the data was stored (step 7). Other I/O operations are conducted similarly.

3.9. Plural software development tools

The Plural SDK enables software development, debug and tuning, as shown in Figure 3.10. The IDE tool chain includes a C/C++ compiler for the core, an assembler, a linker and a library of functions customized for the core, taking full advantage of any available VLIW capability (computing and moving data at the same time) and SIMD (performing several multiply and accumulate operations in parallel).

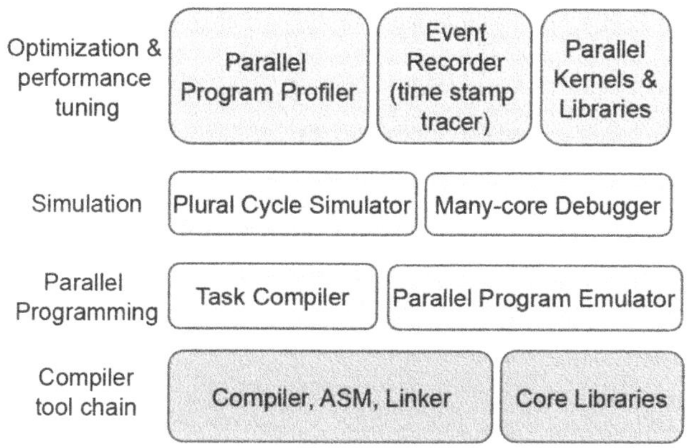

Figure 3.10. *Plural software development kit*

Plural parallel programming is supported by the task compiler, which translates the task graph for the scheduler, a many-task emulator (MTE) that enables efficient development of parallel codes on personal computers and a many-core debugger, which synchronizes debug operations of all cores. The Plural parallel simulator is cycle accurate, fully simulating the cores as well as all other hardware components on the chip.

The profiler provides a complete record of parallel execution on all cores. The event recorder generates traces with time stamps of desired events. The kernel and libraries are described in section 3.8.

3.10. Matrix multiplication algorithm on the Plural architecture

Signal processing computations, as well as machine learning tasks, are often based on linear algebra operators such as dot product, matrix–vector multiplication and matrix–matrix multiplication. While some applications deal with sparse vectors and matrices, many practical and useful cases use dense matrices. This section presents a Plural algorithm for matrix–matrix multiplication.

In general, $C = A \times B$, where A, B and C are $N \times N$ matrices. In more detail, $C_{i,j} = \sum_m A_{i,m} \times B_{m,j}$. Each result element is independent of all other result elements and can be computed by a separate instance if it is a duplicable task. The instances each perform a dot product by reading one row and one column from shared memory, and the result is written back into shared memory. The different instances share reading but write exclusively. Sample code is shown in Figure 3.11. The first and last regular tasks are meaningless place-holders. The duplicable task computes its row and column numbers out of the instance ID parameter, performs dot product and writes the result back to share memory.

```
#define M 100
float A[M][M], B[M][M], C[M][M];

/* REGULAR */
int mm_start()
{
    int i, j;
    for (i=0; i < M; i++)
        for(j=0; j < M; j++)
            { A[i][j] = 13; B[i][j] = 9;}
}

/* DUPLICABLE */
void mm(unsigned int id)
{
    int i, j, m;
    float sum = 0;
    i = id % M;
    j = id / M;
    for (m=0; m < M; m++)
        sum += A[i][m] * B[m][j];
    C[i][j] = sum;
}

/* REGULAR */
int mm_end()
{ printf(“finished mm\n”); }
```

Figure 3.11. *Matrix multiplication code on the Plural architecture*

Note that shared data structures (the matrices) are declared globally outside the scope of the tasks. Note especially that tasks are coded as C functions; this choice is made in order to enable the use of standard compilers and avoid the need for special parsers or parallelizing compilers. Task codes are never invoked by other functions. They are initiated only by the hardware scheduler. Regular tasks do not take any parameters but may produce a return code (enabling choice in the task graph). Duplicable tasks take a single argument, the instance ID, and must not generate any return code. The instance ID parameter is automatically generated by the hardware scheduler, as explained in section 3.5.

The corresponding task graph is shown in Figure 3.12 in both text and graph forms. Note that the quota for the duplicable task is specified in the task graph text. Alternatively, it can be set by the code, in any task that is precedent to it.

```
#define M 100
#define MM 10000

regular     mm_start()
duplicable  mm(mm_start) MM
regular     mm_end(mm)
```

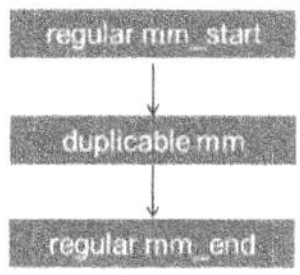

Figure 3.12. *Task graph for matrix multiplication*

3.11. Conclusion

The Plural many-core architecture combines multiple CPU cores, hardware scheduling and a large on-chip shared memory. A PRAM-like task-level-parallelism programming model is used. The program is expressed as a collection of sequential task codes and a task precedence graph. The task codes are stored in shared memory, for execution by the cores, while the task precedence graph is loaded into the hardware scheduler and its execution determines the order of task executions. Data parallelism is expressed as task parallelism, which is more efficient in terms of performance and power. The cores are interconnected with the shared memory by a logarithmic multi-stage interconnection network. The cores do not interact other than through shared memory and do not access the I/O port. The architecture is shown to be highly efficient for signal processing and machine learning applications, both based on simple linear algebra operators that exhibit high levels of data parallelism. The Plural architecture achieves a high level of core utilization and a high performance-to-power ratio. The architecture is highly scalable for various silicon process nodes.

3.12. References

Avron, I. and Ginosar, R. (2012). Performance of a hardware scheduler for many-core architecture. In *IEEE 14th International Conference on High Performance Computing and Communication & 2012 IEEE 9th International Conference on Embedded Software and Systems (HPCC-ICESS)*, 151–160.

Avron, I. and Ginosar, R. (2015). Hardware scheduler performance on the plural many-core architecture. *Proceedings of the 3rd International Workshop on Many-Core Embedded Systems, MES'15*. Association for Computing Machinery, New York, 48–51.

Bayer, N. and Ginosar, R. (1993). High flow-rate synchronizer/scheduler apparatus and method for multiprocessors. U.S. Patent 5,202,987.

Bayer, N. and Ginosar, R. (2002). *Tightly Coupled Multiprocessing: The Super Processor Architecture*. Springer, Berlin, 329–339.

Bayer, N. and Peleg, A. (2012). Shared memory system for a tightly-coupled multiprocessor. U.S. Patent 8,099,561.

Crummey, T., Jones, D., Fleming, P., and Marnane, W. (1994). A hardware scheduler for parallel processing in control applications. In *International Conference on Control, 1994. Control'94.*, 2, 1098–1103.

Dupont de Dinechin, B., Amstel, D., Poulhies, M., and Lager, G. (2014). Time-critical computing on a single-chip massively parallel processor. *Proceedings of Design, Automation and Test in Europe, DATE*, 1–6.

Heinecke, A., Vaidyanathan, K., Smelyanskiy, M., Kobotov, A., Dubtsov, R., Henry, G., Shet, A., Chrysos, G., and Dubey, P. (2013). Design and implementation of the linpack benchmark for single and multi-node systems based on Intel® Xeon Phi™ coprocessor. In *IEEE 27th International Symposium on Parallel and Distributed Processing, IPDPS*, 126–137.

Jacquet, D., Hasbani, F., Flatresse, P., Wilson, R., Arnaud, F., Cesana, G., Gilio, T.D., Lecocq, C., Roy, T., Chhabra, A., Grover, C., Minez, O., Uginet, J., Durieu, G., Adobati, C., Casalotto, D., Nyer, F., Menut, P., Cathelin, A., Vongsavady, I., and Magarshack, P. (2014). A 3 GHz dual core processor ARM Cortex TM -A9 in 28 nm UTBB FD-SOI CMOS with ultra-wide voltage range and energy efficiency optimization. *IEEE Journal of Solid-State Circuits*, 49(4), 812–826.

Nickolls, J. and Dally, W. (2010). The GPU computing era. *IEEE Micro*, 30(2), 56–69.

Tzannes, A., Caragea, G., Vishkin, U., and Barua, R. (2014). Lazy scheduling: A runtime adaptive scheduler for declarative parallelism. *ACM Transactions on Programming Languages and Systems*, 36, 1–51.

Varghese, A., Edwards, B., Mitra, G., and Rendell, A. (2014). Programming the adapteva epiphany 64-core network-on-chip coprocessor. *Proceedings of the International Parallel and Distributed Processing Symposium, IPDPS*, 31.

Wen, X. and Vishkin, U. (2008). FPGA-based prototype of a PRAM-On-Chip processor. In *Conference on Computing Frontiers – Proceedings of the 2008 Conference on Computing Frontiers, CF'08*, 55–66.

Wentzlaff, D., Griffin, P., Hoffmann, H., Bao, L., Edwards, B., Ramey, C., Mattina, M., Miao, C.-C., Brown, J., and Agarwal, A. (2007). On-chip interconnection architecture of the tile processor. *IEEE Micro*, 27(5), 15–31.

Wu, C.-L. and Feng, T.-Y. (1980). On a class of multistage interconnection networks. *IEEE Transactions on Computers*, 29(8), 694–702. Available at: https://doi.org/10.1109/TC.1980.1675651.

4

ASIP-Based Multi-Processor Systems for an Efficient Implementation of CNNs

Andreas BYTYN, René AHLSDORF and Gerd ASCHEID

Institute for Communication Technologies and Embedded Systems, RWTH Aachen University, Germany

Convolutional neural networks (CNNs) that are used for the analysis of video signals are very compute-intensive. Tasks such as scene segmentation for autonomous driving need CNNs with a greater number of large-sized layers. Since the task needs to be performed within the time constraint given by the frame rate, the hardware must support hundreds of GOP/s. Such embedded applications require very cost- and thermal/power-efficient implementations, which cannot be achieved with fully programmable solutions such as graphic processors. On the contrary, a dedicated hardware implementation does not offer the flexibility to implement algorithmic improvements that are often found in research. Therefore, we propose a solution based on an application-specific instruction set processor (ASIP). This approach represents a compromise between efficiency and programmability optimized for the given task. This chapter begins with the study of suitable ASIP architectures for single-core solutions, subsequently extending the approach to an MPSoC solution. The ASIP is based on a VLIW–SIMD architecture. It will be discussed how to maximally parallelize the execution by also considering memory bandwidth constraints. The goal is to maximize the throughput with minimum power consumption. The performance

For a color version of all figures in this book, see www.iste.co.uk/andrade/multi1.zip.

Multi-Processor System-on-Chip 1 – Architectures,
coordinated by Liliana ANDRADE and Frédéric ROUSSEAU.

of the proposed architecture is analyzed based on power simulations of synthesized designs and compared to the state-of-the-art. Based on the insights from the ASIP architecture exploration, an extension to an MPSoC-solution is studied. This study includes exploration of suitable NoC and memory architectures.

4.1. Introduction

In recent years, a certain dominance of algorithms coming from the field of deep learning has established itself for tasks such as object classification and detection or semantic scene segmentation. This increased rate of adoption can be explained by the large performance increases achieved using neural networks compared to classical methods, as evident by the results observed in the annually held ILSVRC challenge (Russakovsky *et al.* 2015). More specifically, convolutional neural networks (CNNs) have established themselves at the forefront of aforementioned image processing tasks. While providing increased algorithmic performance, these methods also require an increased amount of computational power that cannot be delivered by general-purpose processors. Consequently, both research and industry are continuously proposing new solutions to tackle this challenge, for example, by designing application-specific integrated circuits (ASICs) (Cavigelli and Benini 2017; Chen *et al.* 2017; Aimar *et al.* 2019) or by modifying graphics processing units (GPUs) in order to accelerate these new workloads (NVIDIA 2019). While the former, on the one hand, usually provide the best-in-class performance and energy efficiency, they lack flexibility and therefore reusability, which is important, especially in commercial products. GPUs, on the other hand, provide a large degree of programmability that comes with an increased power consumption, compared to the ASICs. Another solution is to use FPGAs (DiCecco *et al.* 2016; Gokhale *et al.* 2017; Fowers *et al.* 2018), which, although offering compelling performance, show an inherent efficiency deficit compared to ASICs because hardware is emulated using pre-defined structures such as lookup tables and DSP units. It therefore seems desirable to combine the traits of these different concepts in a way that yields a more favorable trade-off between efficiency and programmability. Due to these observations, we propose to use an application-specific instruction set processor (ASIP) that provides optimized arithmetic units and memory interfaces for the specific workload while still providing a significant degree of flexibility due to its instruction-set architecture. While the use of a single core provides sufficient performance for executing object classification or detection tasks (Bytyn *et al.* 2019), tasks such as scene segmentation require at least one order of magnitude more performance due to their high resolution inputs. The scaling of a single processor is, however, limited by VLSI design considerations such as clock tree generation, congestion in the register file ports and general routing issues of critical paths, for example, signals for stalling the pipeline or routing data to or from the on-chip SRAM memories. Based on these implementation challenges, it is necessary for workloads with higher demands to consider scaling beyond the single-core level. Consequently, we extend our approach towards a multi-processor system-on-chip

(MPSoC) with network-on-chip (NoC) interconnects, which allows us to scale beyond the aforementioned limitations. To this end, a NoC with mesh structure is used to form a multi-core platform, which can be scaled to the application demands by increasing or decreasing the mesh size. While this approach allows us to use optimally sized ASIP instances, it becomes necessary to explore overall system parameters, such as the required flit-width of the NoC and the number of flits per packet that is transmitted. By using a system-level simulation of the proposed architecture, key parameters are evaluated and their impact on overall performance is discussed.

This chapter is structured as follows. Section 4.2 introduces related works from the area of hardware accelerators for deep learning and NoC-based multi-core systems. In section 4.3, the architecture of the proposed ASIP is summarized, while section 4.4 investigates the scalability of the core and identifies limiting factors based on fully synthesized circuits in a 28 nm TSMC technology. Afterwards, section 4.5 introduces the relevant NoC components used for the MPSoC and gives an overview of the multi-core system. In section 4.6, several parameter studies of the MPSoC are executed and section 4.7 concludes this chapter with some final remarks.

4.2. Related works

Over the years, many different CNN accelerators have been proposed in research. *Eyeriss* (Chen *et al.* 2017) is a well-known architecture that uses a fixed-size 12x14 array of MAC processing elements (PEs), that can be reconfigured to execute the different layers of a CNN. The resulting throughput and energy efficiency are 60 GOP/s at 1.17 V core voltage and 246 GOP/s/W at 0.82 V, respectively, when evaluated for the well-known CNN AlexNet. The *Origami* architecture (Cavigelli and Benini 2017) uses four sum of product (SoP) units with a pre-defined number of 49 multipliers within each unit to calculate several output channels of a CNN in parallel. These units are fed with data from on-chip buffers and cannot be programmed. Overall, the proposed architecture achieves an effective average throughput of 145 GOP/s and an energy efficiency of 437 GOP/s/W at 1.2 V core voltage executing a custom CNN. Using voltage scaling, the energy efficiency can be increased to 803 GOP/s/W at 55 GOP/s throughput. Another architecture is *Envision* (Moons and Verhelst 2017), which is an ASIP that leverages a 16x16 array of MAC units for its underlying computations. Its maximum effective throughput is 63 GOP/s at a layer-dependent core voltage that is between 0.85 V and 0.92 V for AlexNet, resulting in an energy efficiency of 815 GOP/s/W. A common property of these different architectures is that they use fixed-point arithmetic with wordwidths between 8 and 16 bit to save dynamic energy consumed by the arithmetic units. The deep-learning specific instruction-set processor (DSIP) (Jo *et al.* 2018) is a programmable architecture with a master–slave concept that can be scaled across two dimensions. First, several pairs of master and slave cores can be chained in a 1D fashion to form a continuous pipeline for the execution of sequential layers in a CNN. Second, each such pair can be extended to a vertical stack in which each master core manages multiple slave cores to increase its computational abilities. This concept provides a certain degree of scalability; however, the authors mention that the processing time for larger systems is determined by the final store operations that must be performed sequentially. Overall, the DSIP accelerator provides an average throughput of 23 GOP/s for AlexNet when configured with a total of 64 MAC units. The corresponding energy efficiency is 251 GOP/s/W. By using different configurations, this throughput can be scaled. A brief summary of these different architectures is given in Table 4.1, which also lists the peak throughputs as opposed to the CNN-specific ones mentioned earlier.

	Envision	Eyeriss	Origami	DSIP
Implementation	Silicon	Silicon	Silicon	Silicon
CMOS technology	40 nm	65 nm	65 nm	65 nm
MAC units	256	168	196	64
Clock frequency [MHz]	204	200	500	250
Die area [kGates]	1600	1176	697	282
SRAM [kB]	148	181.5	43	139.6
Core voltage [V]	0.7–1.1	0.82–1.17	0.8–1.2	1.2
Peak power [mW]	287	332	449	153
ALU wordwidth [Bit]	4–16	16	12	16
Peak throughput [GOP/s]	104.4	67.2	196	32

Table 4.1. *Comparison of state-of-the-art CNN accelerators*

Regarding the topic of NoC-based multi-processor systems, a large amount of studies are available in the literature. The interested reader is referred to Bjerregaard and Mahadevan (2006) for an extensive in-depth survey of NoC topologies, modeling and analysis methodologies and a review of state-of-the art frameworks such as HERMES (Moraes *et al.* 2004), which is used in this work. The authors of Karkar *et al.* (2016) presented information on NoCs using emerging technologies such as wireless and optical interconnects. In Sahu and Chattopadhyay (2013), a survey on the mapping of applications onto NoC-enabled multi-core systems was conducted. The authors of Shin *et al.* (2018) proposed using a heterogeneous multi-core system with specialized cores for both CNNs and recurrent neural networks (RNNs), together with other cores, for example, for data pre-processing and control flow organization. However, they did not provide any information about the design of their NoC or its parameters. In Choi *et al.* (2018), a multi-core platform comprising of general-purpose CPUs and GPUs was proposed for both the inference and training of CNNs. The authors optimized each link in the NoC by formulating a constrained optimization problem that took into account task-specific traffic patterns. Unfortunately, the information on the processing cores was very limited and only two very small CNNs were benchmarked.

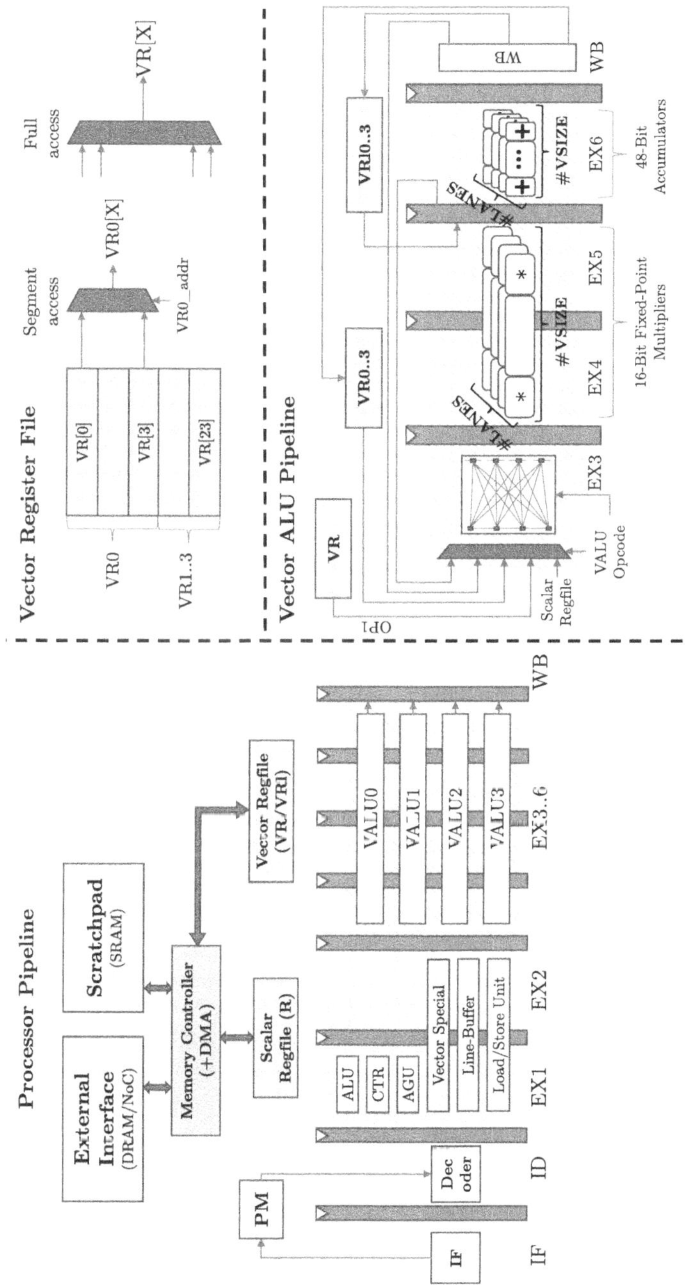

Figure 4.1. *Overview of the ASIP pipeline with its vector ALUs and register file structure*

4.3. ASIP architecture

The pipeline architecture of the ASIP used in this work is shown in Figure 4.1, together with the vector register file structure and the pipeline view of each single vector ALU (VALU0..3). We base the ASIP used in this work on the *ConvAix* core, which was previously published in (Bytyn *et al.* 2019). A summary of the micro-architecture is given below.

Considering the massive need for parallel execution of, especially, MAC operations for any CNN, the custom processor uses an eight-slot VLIW instruction set that is capable of executing up to four vector instructions, two load/store instructions and two general-purpose instructions, such as control flow and address generation, in parallel. The program microcode is read from a 32 kB uncached program memory (PM), which is sufficiently large to store routines for, for example, convolutions, pooling operations and general dataflow control. Using a retargetable C-compiler, it is possible to easily adapt these computational kernels to the demands of different CNN topologies. Since executing a convolution requires loading of at least one filter and several input pixels, the core's pipeline is designed to provide load capabilities in the first two pipeline stages (EX1/EX2) with the option of directly forwarding this result into the following four-stage vector arithmetic units to avoid any interlocking of the pipeline stages during this type of operation. A dedicated line buffer is used as an application-specific cache, which can provide rows of input pixels in a strided fashion without any memory overhead, as this is often required by strided convolutions. The *vector special* unit is used for calculating, for example, activation functions such as the well-known ReLU. Only one of these units is instantiated because this operation is much less frequent than others. A software-managed scratchpad memory with design-time configurable size is used to store intermediate data, such as partial sums or filter weights and input/output pixels. The direct memory access (DMA) controller embedded within the memory controller unit is capable of moving data from the external interface into the scratchpad and vice versa. These transfers are executed in a fully parallel fashion without interrupting the core, unless a barrier is specified in the C program code. This allows a fully interleaved execution of both computation and memory transactions. For the multi-core system, instead of connecting this external interface directly to a DRAM memory, it is connected to the NoC's network interface, which then handles the request. Each of the four slots reserved for vector operations contains one vector ALU, thereby offering instruction-level parallelism. Of these units, each single one has two dimensions of parallelism in itself. First, a number of design-time configurable lanes (*#LANES*) work in parallel, executing the same instruction but using different data, as configured by the opcode in the EX3 stage of the VALU. It is therefore possible to, for example, broadcast a row of input pixels across multiple lanes or to broadcast a scalar value. Second, each lane implements a SIMD-like vector operation for a vector with *#VSIZE* elements. In the case of CNNs, we usually broadcast different filter values to different lanes and assign to all lanes the same input row with length equal to *#VSIZE*. The corresponding output pixels are then accumulated in an output row stationary fashion. One important design aspect is the two vector register files VR and VRl, which are used to store input operands and accumulated results, respectively. Because our architecture provides a maximum of 16 lanes (four lanes in four slots), the VR register file has space for up to 24 vectors, where each element of every vector is 16 bit in size. A maximum of 16 of these 24 entries are required for storing input operands, while the additional eight entries are used for data movement, for example. If the core is scaled down, i.e. fewer lanes per VALU, the size of this register file is reduced accordingly. The larger register file VRl is used to store accumulated results, which are 48 bit wide per element. Since it is only needed for partial sums, it is always sized equal to the total number of lanes. While having this very flexible setup allows for many degrees of freedom in programming the core, it also creates a significant burden in terms of the required data multiplexing in the register files. To alleviate this problem, the register file provides a segment access and a full access scheme. The former permits only access to certain sub-blocks of the entire file, while the latter allows full access. Because we do not want to impede the compilers' options for register allocation during load/store and move operations, these instructions always have access to the entire register file. However, the VALUs have a more restricted access: while the first operand (OP1) of any VALU can always access the entire register file, the second can only access one of the aforementioned sub-blocks (VR0..3). Furthermore, the very wide accumulator units always have access to only a sub-block of the register file (VRl0..3). These measures help reduce congestion in the multiplexing of the register file's ports while still assuring sufficient freedom for the programmer and the compiler.

4.4. Single-core scaling

As mentioned in section 4.3, the base processor architecture was designed with multiple design-time configurable scaling parameters, namely the number of lanes *#LANES* per VALU and the number of elements per vector *#VSIZE* that can be processed in parallel. Before introducing the results of scaling the proposed architecture, major design factors that impact the scalability are first discussed.

The dual-ported on-chip memory of the ASIP is implemented as banked vector memory with its number of banks equal to the vector parallelism *#VSIZE,* in order to make it possible to read or write two complete vectors within one cycle. A schematic of the memory layout is presented in Figure 4.2. Based on the application demands, the on-chip memory interface must provide unaligned vector accesses, i.e. it must be possible to read a full-sized vector with its first element residing in any of the banks. Due to this requirement, the memory interface requires complex multiplexing of addresses and data. Unfortunately, the complexity of these multiplexers grows with the vector parallelism, and although – at least in theory – this growth is logarithmic, it is one of the limiting factors for hardware scalability. A similar scaling problem is encountered in the case of vector register files. As mentioned earlier, the number of entries per register file must be greater than or equal to the total number of lanes. This means that the depth of the multiplexers required for accessing it grows, albeit only logarithmically, as for the memory. However, if the vector parallelism is increased, more multiplexers at the register file ports are required, because the vector element count per entry of the register file is increased. This can cause serious congestion and ultimately prohibit successful timing closure. Another factor inhibiting the arbitrary scalability of the single core is the simple physics of any VLSI design. It does not come as a surprise that chip area grows by adding more vector units and increasing the register and memory sizes. Because our design uses a single clock domain, routing the clock and other critical signals becomes more difficult and, consequently, increases wire delays. This, of course, could be alleviated by partitioning the design into multiple clock domains, which, however, requires synchronization at the boundaries. In our proposed design, it is, particularly, the stall signal that is responsible for halting the pipeline in case any data hazards are encountered, which becomes a problem for larger designs. While some of the aforementioned scaling limitations are architecture, specific and could probably be further alleviated, for example, by introducing more pipeline stages into the memory interface, by manually optimizing the multiplexing of the register file and by using delayed stall signals, these limitations will, at some point, still kick in and effectively prohibit any further scaling.

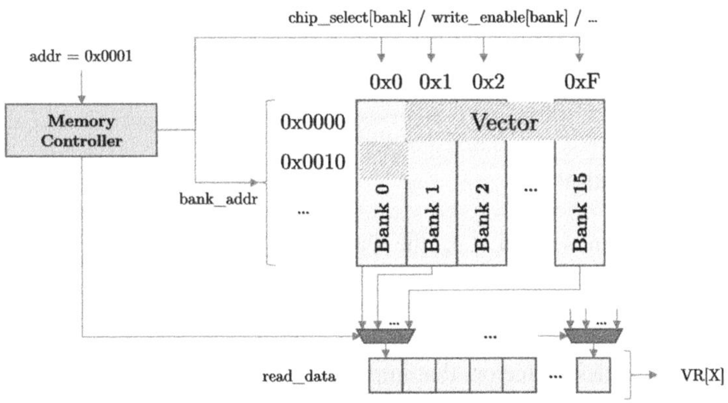

Figure 4.2. *On-chip memory subsystem with banked vector memories and an example of unaligned access at address 0x1*

To investigate the practical limitations for the proposed ASIP, the core is synthesized for several different configurations of varying *#VLANES* and *#VSIZE*. For logic synthesis, Synopsys Design Compiler P-2019.03 is used, along with a 28 nm technology library from TSMC that operates at a nominal core voltage of 1.0 V. All syntheses are performed using a typical process corner at 25°C. The clock tree and reset signal are assumed to be ideal during synthesis and are later added during the place and route (PnR) step. A plot showing the resulting core area (excluding SRAM memories) over the clock period for the different configurations is shown in Figure 4.3. These results, together with some additional figures, are summarized in Table 4.2.

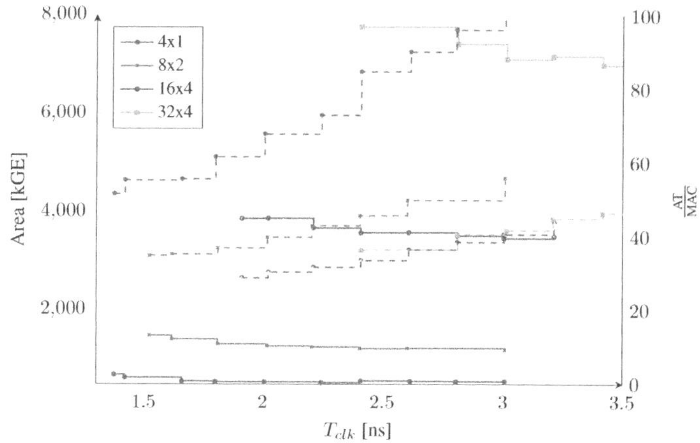

Figure 4.3. *Cell area of the synthesized cores' logic for different clock periods T_{clk} (solid lines) and area-timing product divided by the number of MAC units (dashed lines)*

Configuration[a]	4×1	8×2	16×4	32×4
MAC units	16	64	256	512
Clock frequency [MHz]	728	657	523	415
Die area [kGE][b]	687	1486	3870	7771
SRAM [kB]	48	80	144	272
Peak average power [mW][c]	193	444	1132	1872
Peak throughput [GOP/s]	23.3	84.1	267.8	425

[a] Each configuration is denoted by *#VSIZE*x*#VLANES*.
[b] Logic only where one gate equivalent (GE) is assumed to be a NAND-2 gate.
[c] This is determined based on a stress test that fully uses all VALUs with completely non-sparse data.

Table 4.2. *Synthesis results for different configurations of the ASIP*

The curves in Figure 4.3 clearly show that an increased number of processing entities per core degrades the achievable timing significantly. Furthermore, based on the relatively small slope of the area curves, it can be concluded that this specific architecture is not bounded by the arithmetic units that tend to explode in size if they lie in the critical path. To evaluate which of the different configurations provides the best trade-off between additional processing capabilities and degraded timing, the area-timing product normalized to the number of MAC units is also plotted. Consequently, the 16x4 configuration with a total of 256 MAC units is identified as the best configuration for this design. The smaller configurations do not contain enough PEs to amortize the overhead of the program control unit, the memory interface and the pipeline registers, for example. In the case of larger configurations, the degraded timing starts to overtake the effect of amortizing the aforementioned area overhead, resulting in a higher normalized AT product.

It can be concluded that scaling at the single-core level quickly reaches its limits. While the presented ASIP provides more than sufficient processing capabilities for small- and medium-sized CNNs, such as AlexNet (Krizhevsky *et al.* 2012), it cannot cope with very large networks, such as SegNet (Badrinarayanan *et al.* 2017), which is used for semantic scene segmentation. One possible approach to tackle this performance limitation is to put more effort into the architectural design of the ASIP. However, the achievable return of investment is rather slim compared to extending the design towards a multi-processor system-on-chip (MPSoC), which is presented in the next section.

4.5. MPSoC overview

Several different concepts can be used for the interconnection of multiple cores, for example, crossbar switches, ring buses and networks-on-chip (NoC). For the proposed ASIP, a NoC interconnect is used since it can be arbitrarily scaled regardless of the number of cores attached to it (at least from a VLSI design perspective). As described in (Bjerregaard and Mahadevan 2006), many different topologies, such as torus, binary tree and mesh, can be used to form the NoC. Due to the homogeneity of our system, we decide to use a mesh topology. Another factor that motivates this choice is that an X/Y routing protocol can be used for a mesh that results in a predictable routing scheme. Having predictability can help in various application-specific dataflow optimizations, which are, however, not discussed here. A very well-known framework for creating such mesh-style NoCs is called *HERMES* (Moraes *et al.* 2004), and its application in the proposed context is briefly depicted below.

The main components of the NoC are its router and the network interface connecting each core to the router. Furthermore, a master core is needed to configure the different ASIPs by assigning CNN-specific execution parameters to each instance. Finally, a block that acts as an interface to the off-chip memory, i.e. the DRAM controller, must be included in the NoC so that the ASIP instances can issue memory transactions. These different components are modeled using system-level simulation techniques in which each processing core is abstracted by a task-oriented model that emulates approximately the same external behavior in terms of memory transactions as the original ASIP. Critical blocks such as the NoC router are modeled in a cycle-accurate fashion to allow the capture of fine-grain effects such as link congestion or port buffer overflows. Figure 4.4 shows a 3x3 mesh configuration with a total of seven ASIP instances, one master core and a central DRAM interface. As can be seen in Figure 4.4, each ASIP is embedded into a processing tile together with a direct memory access network interface (DMANI) that handles all off-chip transactions by directly accessing the core's scratchpad SRAM memories. Consequently, the aforementioned DMANI replaces the original DMA of the single core. The routers then forward transactions based on the X/Y routing protocol and can also eventually stall a core in case an outstanding transaction cannot be finished in time. As in the case of the single core, several design parameters must be optimized to achieve the highest possible overall performance. Most importantly, the flit-width, i.e. the number of bits transmitted between two adjacent routers within one cycle, and the packet length, i.e. the number of flits per packet, must be optimized. It seems reasonable that selecting a larger flit-width is always desirable in the case of a bandwidth-limited application; however, it is still important to evaluate the actual achievable performance gain through such a modification. However, the optimum packet length is a more tricky problem, as choosing a short packet length enables fairer arbitration between different requests, and can also create a large amount of routing overhead. Choosing a long packet length amortizes this routing overhead better, but can result in long stalling times of single cores. In addition, the size of the port buffers in the individual routers plays a crucial role, since these buffers can help compensate short-time peaks in traffic. For these reasons, we explore different settings of the aforementioned parameters in the following section. For all investigations, the NoC is configured as 3x3 mesh and a hypothetic DRAM interface with a line width of 128 bit is assumed, meaning that the interface can load or store 128 bit per cycle. The NoC and the DRAM interface are run at a clock frequency of $f_{noc} = 1$ GHz, as opposed to the ASIPs that are run at $f_{asip} = 500$ MHz, based on the previous single-core results. Each ASIP is configured according to the previously introduced 16x4 setup, as this configuration has proven to be the most efficient one.

4.6. NoC parameter exploration

For an effective exploration, it is important to use a realistic workload, i.e. executing a CNN. For this reason, exemplary layers from the relatively large VGG-16 network (Simonyan and Zisserman 2015) are used below. More specifically, a very memory bandwidth-limited layer and a compute-limited layer are selected: layers 9 (*conv_4_2*) and 2 (*conv_1_2*) of VGG-16, respectively. During the following exploration, they are referred to as *I/O heavy* and *comp. heavy* for the sake of clarity. Since the mapping of these layers onto the MPSoC is a topic on its own, it is not possible to discuss it in detail here. The interested reader is referred to Bytyn *et al.* (2020) for more information on the mapping process, both at a single-core level and for the complete MPSoC.

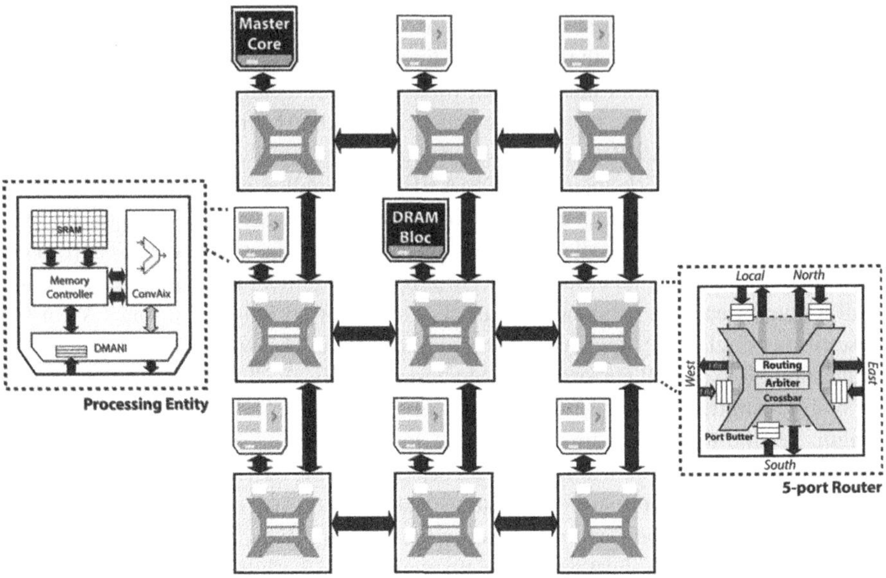

Figure 4.4. *3x3 NoC based on the HERMES framework*

We begin by investigating the impact of the flit-width on the resulting runtimes for both layers. Figure 4.5 shows that, as expected, the runtime can be significantly reduced by enlarging the flit-width. However, it should also be noted that no linear relationship between the two exists: ranging from 16 bit to 256 bit equals a 16x increase in NoC bandwidth, which, however, only generates a 5.7x (*comp. heavy*) and a 3.8x (*I/O heavy*) improvement in terms of the runtime. This is surprising since, especially, the *I/O heavy* layer should benefit the most from an increased NoC bandwidth. Comparing the same 16 bit baseline to 128 bit (as opposed to 256 bit) shows a 3.5x speed-up for the *comp. heavy* and a 3.0x speed-up for the *I/O heavy* layer, which lie much closer together. This rather curious observation can be explained by the fact that the *I/O heavy* layer has a very large parameter count of 2.36M compared to 36.9K for the compute heavy layer. Consequently, even at an increased overall bandwidth, the ASIPs have to wait for a fixed, rather long, time at the beginning for filters to be loaded before processing can begin. Thus, the overall smaller speed-up for the *I/O heavy* layer is explained. The port buffer size shows a similar trend of increased speed-up for larger buffers. The slope of this curve is, however, smaller, which makes sense because, while an increased flit-width results in more average bandwidth, a larger port buffer size primarily serves as a latency hiding technique. This allows requests to continue, even though no actual arbitration has happened yet, but does not help in terms of the average bandwidth observed over a longer period of time.

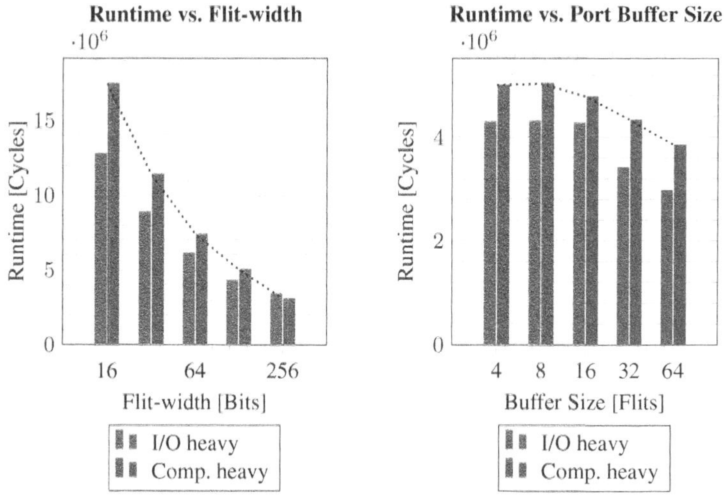

Figure 4.5. *Runtime over flit-width and port buffer size for two exemplary layers of VGG-16 that are bandwidth (I/O heavy) and compute (comp. heavy) limited*

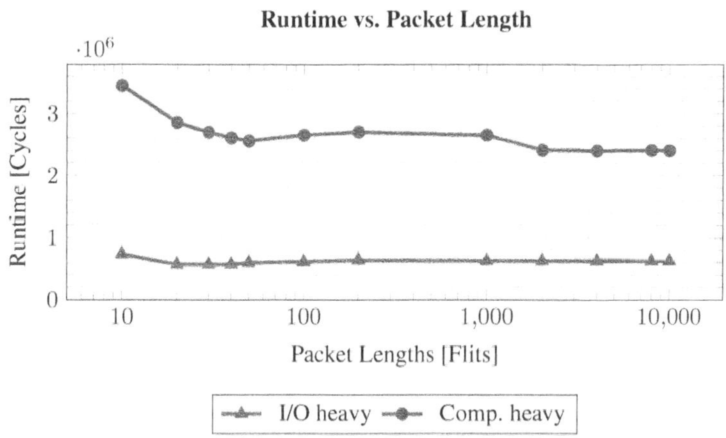

Figure 4.6. *Runtime for different packet lengths*

Figure 4.6 shows the runtime for different packet lengths. Choosing the smallest packet length, i.e. 10 flits per packet, results in an increased runtime of 1.4x (*comp. heavy*) and 1.3x (*I/O heavy*) compared to the fastest configurations of 4000 flits per packet for the *comp. heavy* and 20 for the *I/O heavy*. Interestingly, it is especially the *I/O heavy* layer that performs better when using a shorter packet length. The reason behind this observation is that said layer's memory transactions are spread more widely in terms of their sizes. Together with the increased bandwidth requirement, especially large packet lengths can block smaller transactions, thereby inducing increased stalling times in ASIPs that have outstanding transactions of short lengths. Conversely, the *comp. heavy* layers are primarily compute bound with much smaller overall transaction sizes and therefore are less impeded by similar congestion phenomena. Considering both layers jointly, it seems to be a reasonable choice to select approximately 50 flits per packet as a compromise, achieving near-optimal performance in both cases.

4.7. Summary and conclusion

In this chapter, we have studied the scaling of a (programmable) MPSoC for processing CNNs. Unlike a GPU that targets a variety of applications, the proposed application-specific instruction set processor (ASIP)-based MPSoC is designed and optimized specifically for CNNs. This enables architecture optimizations that yield higher throughput at lower power consumption (more precisely, energy consumption per CNN operation), which is essential for use cases such as scene segmentation in autonomous driving.

The studies of a single ASIP have shown that there are limits to parallelization with a single core. In particular, data access and transfer, register bank sizes and multiplexing are key factors that reduce the gains for increasing degrees of processing parallelism. For the proposed ASIP architecture, a vectorsize-x-vectorlanes configuration of 16x4 delivered a performance optimum. The structure of a CNN is very suitable for multi-processor architectures, as it is organized in layers. Therefore, in the second step, scaling was studied at an MPSoC level. Here, the main factor determining the performance is the data transfer and thus the interconnection network of the MPSoC. The key parameters that can be optimized in the considered NoC architecture are the flit-width, packet length and port buffer size, since they directly influence throughput and latency. The studies showed that a simple theoretical scaling does not reflect the reality. For instance, with increasing flit-width, the runtime improvements decrease. The optimum packet length depends on the configuration of the processed CNN layer, and thus a compromise is required for the processing of the complete CNN. Overall, the studies provided some important insights into the various effects that influenced the performance gain by parallelization, both at a single-core and a multiple-core level.

Of course, the resulting overall performance and, in particular, a comparison to dedicated ASICs, as well as to other programmable architectures, are of high relevance. This, however, requires a discussion of the software mapping strategies (since we have an interconnected multi-processor architecture) and the application-relevant neural networks. This would have gone far beyond the available space, but can be found in (Bytyn *et al.* 2020).

4.8. References

Aimar, A., Mostafa, H., Calabrese, E., Rios-Navarro, A., Tapiador-Morales, R., Lungu, I., Milde, M.B., Corradi, F., Linares-Barranco, A., Liu, S., and Delbruck, T. (2019). NullHop: A flexible convolutional neural network accelerator based on sparse representations of feature maps. *IEEE Transactions on Neural Networks and Learning Systems*, 30(3), 644–656.

Badrinarayanan, V., Kendall, A., and Cipolla, R. (2017). SegNet: A deep convolutional encoder–decoder architecture for image segmentation. *IEEE Transactions on Pattern Analysis and Machine Intelligence*, 39(12), 2481–2495.

Bjerregaard, T. and Mahadevan, S. (2006). A survey of research and practices of network-on-chip. *ACM Computing Surveys*, 38(1).

Bytyn, A., Leupers, R., and Ascheid, G. (2019). An application-specific VLIW processor with vector instruction set for CNN acceleration. *IEEE International Symposium on Circuits and Systems (ISCAS)*, 1–5.

Bytyn, A., Ahlsdorf, R., Leupers, R., and Ascheid, G. (2020). Dataflow aware mapping of convolutional neural networks onto many-core platforms with network-on-chip interconnect. Available at: https://arxiv.org/pdf/2006.12274.pdf.

Cavigelli, L. and Benini, L. (2017). Origami: A 803-GOp/s/W convolutional network accelerator. *IEEE Transactions on Circuits and Systems for Video Technology*, 27(11), 2461–2475.

Chen, Y., Krishna, T., Emer, J.S., and Sze, V. (2017). Eyeriss: An energy-efficient reconfigurable accelerator for deep convolutional neural networks. *IEEE Journal of Solid-State Circuits*, 52(1), 127–138.

Choi, W., Duraisamy, K., Kim, R.G., Doppa, J.R., Pande, P.P., Marculescu, D., and Marculescu, R. (2018). On chip communication network for efficient training of deep convolutional networks on heterogeneous manycore systems. *IEEE Transactions on Computers*, 67(5), 672–686.

DiCecco, R., Lacey, G., Vasiljevic, J., Chow, P., Taylor, G., and Areibi, S. (2016). Caffeinated FPGAs: FPGA framework for convolutional neural networks. *International Conference on Field-Programmable Technology (FPT)*, 265–268.

Fowers, J., Ovtcharov, K., Papamichael, M., Massengill, T., Liu, M., Lo, D., Alkalay, S., Haselman, M., Adams, L., Ghandi, M., Heil, S., Patel, P., Sapek, A., Weisz, G., Woods, L., Lanka, S., Reinhardt, S.K., Caulfield, A.M., Chung, E.S., and Burger, D. (2018). A configurable cloud-scale DNN processor for real-time AI. *ACM/IEEE 45th Annual International Symposium on Computer Architecture (ISCA)*, 1–14.

Gokhale, V., Zaidy, A., Chang, A.X.M., and Culurciello, E. (2017). Snowflake: An efficient hardware accelerator for convolutional neural networks. *IEEE International Symposium on Circuits and Systems (ISCAS)*, 1–4.

Jo, J., Cha, S., Rho, D., and Park, I. (2018). DSIP: A scalable inference accelerator for convolutional neural networks. *IEEE Journal of Solid-State Circuits*, 53(2), 605–618.

Karkar, A., Mak, T., Tong, K., and Yakovlev, A. (2016). A survey of emerging interconnects for on-chip efficient multicast and broadcast in many-cores. *IEEE Circuits and Systems Magazine*, 16(1), 58–72.

Krizhevsky, A., Sutskever, I., and Hinton, G.E. (2012). ImageNet classification with deep convolutional neural networks. *Advances in Neural Information Processing Systems*, 1–9.

Moons, B. and Verhelst, M. (2017). An energy-efficient precision-scalable convNet processor in 40-nm CMOS. *IEEE Journal of Solid-State Circuits*, 52(4), 903–914.

Moraes, F., Calazans, N., Mello, A., Möller, L., and Ost, L. (2004). HERMES: An infrastructure for low area overhead packet-switching networks on chip. *Integration, the VLSI Journal*, 38(1), 69–93.

NVIDIA (2019). NVIDIA Turing GPU Architecture [Online]. Available at: https://www.nvidia.com/en-us/design-visualization/technologies/turing-architecture/.

Russakovsky, O., Deng, J., Su, H., Krause, J., Satheesh, S., Ma, S., Huang, Z., Karpathy, A., Khosla, A., Bernstein, M., Berg, A.C., and Fei-Fei, L. (2015). ImageNet large scale visual recognition challenge. *International Journal of Computer Vision (IJCV)*, 115(3), 211–252.

Sahu, P.K. and Chattopadhyay, S. (2013). A survey on application mapping strategies for network-on-chip design. *Journal of Systems Architecture*, 59(1), 60–76.

Shin, D., Lee, J., Lee, J., Lee, J., and Yoo, H.-J. (2018). DNPU: An energy-efficient deep-learning processor with heterogeneous multi-core architecture. *IEEE Micro*, 38(5), 85–93.

Simonyan, K. and Zisserman, A. (2015). Very deep convolutional networks for large-scale image recognition. In *3rd International Conference on Learning Representations, ICLR 2015*, San Diego, CA, USA, May 7-9, 2015. Bengio, Y. and LeCun, Y. (eds) [Online]. Available: http://arxiv.org/abs/1409.1556, https://dblp.org/rec/journals/corr/SimonyanZ14a.bib.

PART 2

Memory

5

Tackling the MPSoC Data Locality Challenge

Sven RHEINDT, Akshay SRIVATSA, Oliver LENKE, Lars NOLTE, Thomas WILD and Andreas HERKERSDORF

Chair of Integrated Systems, Department of Electrical and Computer Engineering, Technical University of Munich, Germany

Data access latencies and bandwidth bottlenecks frequently represent major limiting factors for the computational effectiveness of multi- and many-core processor architectures. This chapter deals with two conceptually complementary approaches to ensure that the data to be processed and the processing entities remain spatially confined, even under fluctuating workload and application compositions: region-based cache coherence and near-memory acceleration.

A 2D array of compute tiles with multiple RISC cores, two levels of caches and a tile-local SRAM memory serves as the target processing platform. Compute tiles, I/O tiles and globally shared DDR SDRAM memory tiles are interconnected by a meshed network-on-chip (NoC) with support for multiple quality of service levels. Overall, this processing architecture can follow both a distributed shared memory and a distributed memory model with message passing. The limited degree of parallelism in many embedded computing applications also bounds the number of compute tiles potentially sharing associated data structures. Therefore, we envision region-based cache coherence (RBCC) among a subset of compute tiles over global coherence approaches. Coherence regions can be dynamically reconfigured at runtime and comprise a number of arbitrary (adjacent or non-adjacent) compute tiles. These

For a color version of all figures in this book, see www.iste.co.uk/andrade/multi1.zip.

Multi-Processor System-on-Chip 1 – Architectures,
coordinated by Liliana ANDRADE and Frédéric ROUSSEAU.

tiles are interconnected through regular NoC channels for load/store requests to remote memory and the exchange of coherency protocol messages. We will show that, in comparison to global coherence, region-based coherence allows substantially smaller coherence directories (i.e. reduced by approximately 40% in size for a 16-tile system with up to four tiles per region) to be maintained, and shorter latencies for checking possible sharers. In addition, we propose near-memory acceleration (NMA) as another concept to increase data-to-task locality. NMA positions processing resources for specific forms of data manipulations as close as possible to the data memory. The evident benefits are reduced global interconnect usage, shorter access latencies and, thus, increased compute efficiency. In distributed shared memory architectures, where accelerator units can be affiliated with tile local SRAMs as well as with globally shared DDR SDRAMs, near-memory acceleration requires thorough consideration of task mapping as well as task and data migration into and among compute tiles.

5.1. Motivation

We have just begun to experience the opportunities and challenges of the Fourth Industrial Revolution. Indeed, the disruption that will follow from the introduction of smart, i.e. machine learning driven data analytics, ubiquitously connected sensors, IoT (Internet of Things) devices, cyber-physical systems and high-performance cloud computing centers, is nothing less than an ongoing industrial revolution. Already today, it affects and transforms many aspects of private, business and societal life in national economies around the globe. The term – which presumably is as central for characterizing the current transformation process as "steam engine" was in the 18th Century for the First Industrial Revolution – is "data", and the capability to effectively handle huge amounts of data: "big data". "The world's most valuable resource is no longer oil, but data", titled the May 2017 issue of The Economist (Parkins 2017). IDC market research forecasts that we currently are still in the infancy of data volumes and that the majority of data in 2025 is expected not to originate from large supercomputing and data centers, but rather from the above-mentioned massively distributed sensors, IoT devices and machine learning-enabled edge and mobile systems (Turek 2017).

Heterogeneous multi- and many-core processors, as well as dedicated ASIC- or FPGA-based accelerators, are the main workhorses of today's and future embedded computing systems. Such systems are regularly deployed in industrial manufacturing, financial services, energy generation and management, public health care as well as automotive and public transport domains. As with IT systems at large, many-core processors can be considered to be equally challenged by the necessity to handle large data volumes and high data rates, up to several hundred Gbit/s. In particular, access interference at shared memory interfaces and on network-on-chip interconnect links became serious bottlenecks for the effective exploitation of the nominal compute performance under a given TDP (Thermal Design Power) constraint.

This phenomenon, also referred to as the data locality wall, has been confirmed by several independent sources from industry and academia. As early as 2006, Bill Dally, at that time with Stanford University, emphasized and quantitatively proved that it is data movement, not arithmetic computation, that matters in computer architecture in the many-core era (Dally 2006). Energy cost and access latencies vary by a factor of 1,000 depending on whether local registers or global memory is accessed. One decade later, at the MEMSYS 2017 conference, Peter Kogge, University of Notre Dame, identified memory-intensive apps with hardly predictable data access patterns, and little data reuse during computation, as the major reason for the still persisting locality challenge (Kogge 2017). In the Memory-Centric Architecture session of the DAC 2018 conference, AMD declared managing of data locality for high-performance computing applications a yet unsolved problem. Lack in understanding the affinity between data and computation has been identified as a major reason for this assessment (Falsafi *et al.* 2016; Jayasena 2018).

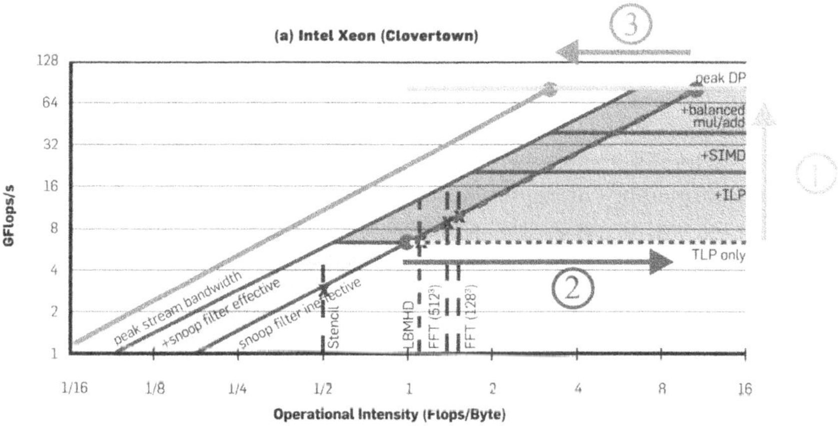

Figure 5.1. *Roofline: a visual performance model for multi-core architectures. Adapted from Williams* et al. *(2009)*

The inherent problem of worsening the data locality with increasing compute performance can be well analyzed and visualized with the roofline model from Berkeley (Williams *et al.* 2009). The roofline model allows the visualization of the estimated ceilings for both the MPSoC compute performance (enumerated in Mega or Giga-FLOPs/s – floating-point operations per second) and its characteristic memory bandwidth. Applications do have specific arithmetic intensities, i.e. ratios of instructions executed over a number of main memory data movements (FLOPs/byte). The roofline model allows for given operational intensities to determine whether an MPSoC platform is compute or memory bound, depending on which ceiling is first traversed vertically. By (exponentially) increasing the nominal compute performance with various architectural innovations (arrow (1) in Figure 5.1), the intersection point with the operational intensity line of global memory moves to the right (2) toward (exponentially) larger Flops/Byte. In consequence, with increasing compute performance, more applications risk becoming memory constraints. In order for lower FLOPs/byte applications to utilize the increased compute performance, an increase in peak data bandwidth or memory access efficiency would be necessary (3).

Both storing application data closer to a limited number of relevant processing cores and reducing/preventing latency-costly NoC transfers to a minimum contribute to the green shift in operation intensity in Figure 5.1. Region-based cache coherence reduces the administration overhead for coherency management in directory-based coherency protocols and is justified by the limited degree of parallelism, which are characteristic for embedded compute applications. Near-memory computing increases data-to-task locality by positioning processing resources for specific forms of data manipulations as close as possible to the data memory. The reduction in global interconnect usage and shorter access latencies thus leads to an increase in compute efficiency. Both RBCC and NCA/NMA contribute to tackling the MPSoC data locality challenge for native as well as hybrid forms of distributed shared memory and shared memory models and architectures. From our perspective, data locality is – having already been for almost two decades and will remain in the foreseeable future – a serious and pressing issue in computer architecture because no "one-size-fits-all" solution exists for different application domains and different compute architectures. More likely, a plurality of approaches targeting the same objective (i.e. data-to-task locality) is urgently necessary in the scope of today's and future heterogeneous application characteristics and evolving compute architectures.

5.2. MPSoC target platform

Concept development for the RBCC as well as NMA approaches were performed on the basis of a reference architecture, as depicted in Figure 5.2. Each compute tile consists of four SPARC-V8 RISC cores with private level 1 caches, a shared level 2 cache, a tile local SRAM memory (TLM) of 8 MByte capacity and a network adapter for interfacing to the processor-/chip-level NoC interconnect. In addition are the proposed macro extensions in the form of a coherency region manager (CRM), synchronization support (Sync), message queue management (Q Mgmt), near-cache accelerator (NCA) and near-memory accelerator (NMA). The latter being located within one or two dedicated memory tiles. The NoC interconnect has specific provisions for low-latency exchange of coherency messages among compute tiles (Masing *et al.* 2018). The experimental validations described in section 5.4 and section 5.5 have been conducted with a SystemC TLM-2.0 model, or on a re-configurable hardware platform, implementing a 4×4 tile structure with 64 RISC cores in total, on four XILINX Virtex 7 FPGAs. This architecture supports distributed memory (DM) via message passing and distributed shared memory (DSM) programming paradigms. These paradigms can be applied to both the physically distributed SRAM-based TLMs and the global DDR SDRAM(s).

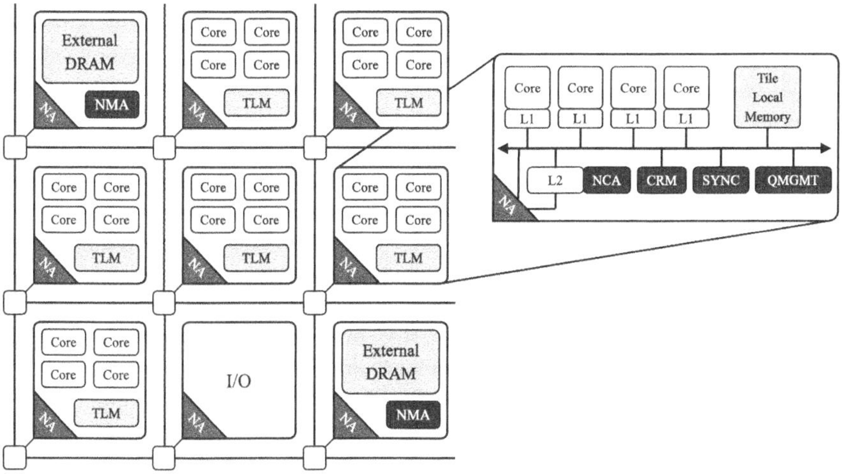

Figure 5.2. *Proposed tile-based many-core architecture*

5.3. Related work

The efficiency and scalability of cache coherence plays an important role for tile-based MPSoCs. This is addressed by our work on region-based cache coherence (RBCC), described in section 5.4. Hardware coherence mechanisms were introduced by the Stanford Dash group (Lenoski *et al.* 1992) distributed shared memory architectures, followed by the recent Tilera (Wentzlaff *et al.* 2007) and Cavium Octeon (Kessler 2011) architectures. They support global coherence, which often has scalability issues owing to increasing coherence directory sizes. Therefore, the community is unclear whether global coherence is here to stay (Boyd-Wickizer *et al.* 2010; Martin *et al.* 2012). Research on directory optimization focuses on reducing the number of directory entries (height) using sparse directories (Gupta *et al.* 1990). The Select (Yao *et al.* 2015) and Cuckoo (Ferdman *et al.* 2011) directories optimize sparse directories for size and replacement policies, respectively. We believe that global coherence for large tile-based many-cores is unnecessary as applications tend to saturate on their degrees of parallelism (maximum of 32–48 threads), as exhibited by the PARSEC benchmarks (Southern and Renau 2016). The Intel®Xeon Phi (Sodani *et al.* 2016) is an example system that moves away from global coherence by enabling coherence for certain NUMA domains. However, this is a boot time decision and limited to fixed quadrants of the chip, which may not be favorable for all applications. Coherence domain restriction (CDR) (Fu *et al.* 2015) is another technique that restricts coherence in large MPSoCs with unified shared memory.

In recent years, many in- and near-memory computing approaches tackled the data-to-task locality challenges of MPSoCs. Application-specific processing-in-memory (PIM) accelerators, especially, are widely used. They have been proposed for simple memory-intensive tasks (Yitbarek *et al.* 2016), as well as for more complex operations, such as matrix multiplications (Neggaz *et al.* 2018), or the training of neural networks (Schuiki *et al.* 2019). Moreover, there are accelerators that process irregular data structures such as graphs (Ahn *et al.* 2015; Ozdal *et al.* 2016; Li *et al.* 2018). Many of these application-specific accelerators are implemented on the hybrid memory cube (HMC) (Hybrid Memory Cube Consortium 2019). We, however, focus on accelerating crucial runtime support functionalities of operating systems. Our near-memory accelerators tackle inter-tile synchronization and queue management as well as graph copy operations. Much research has been conducted in the area of improving software queue implementations for shared memory architectures (Michael and Scott 1996; Sanghoon *et al.* 2011; Wang *et al.* 2016). However, efficient inter-tile queues and atomic synchronization primitives within NoC-based systems are very rare (Lenoski *et al.* 1992; Brewer *et al.* 1995; Michael and Scott 1995). More complex atomic operations and the transfer of the queue element payload have mostly been neglected, which will lead to data-to-task locality issues in tile-based architectures. There is also previous work that improves object graph copies in software. Nguyen *et al.* (2018) presented an approach to transfer objects between Java Virtual Machine heaps without full serialization. The Pegasus (Mohr and Tradowsky 2017) approach was proposed for systems with a partitioned address space, however, without addressing the data-to-task locality issue arising from the far-from-memory graph copy.

5.4. Coherence-on-demand: region-based cache coherence

Modern MPSoCs can comfortably host multiple applications simultaneously. In order for such systems to execute applications that adhere to the classical shared memory programming paradigm, cache coherence is an essential aspect. Global cache coherence covering all tiles of such a large many-core system does not scale well. Furthermore, embedded computing applications rarely benefit from degrees of parallelism exceeding ≈(32–48) parallel threads (Southern and Renau 2016). This deems global coherence overly costly and even unnecessary, motivating investigations towards alternative solutions. We propose region-based cache coherence (RBCC), a hardware solution that enables scalable and flexible cache coherence for large many-core systems (Srivatsa *et al.* 2017, 2019). RBCC confines cache coherence support to a selectable subset of tiles known as a coherency region. The coherency regions can be dynamically configured or even re-configured at runtime. Coherency region attributes such as the number, location, shape and size and memory range are solely controlled on behalf of the applications based on their requirements. For example, there can be several applications simultaneously running on a large many-core system, working within their respective custom-sized coherency regions. RBCC provides a framework through which OS or runtime services dynamically request a cluster of coherent processing and/or memory resources for an application, that are provisioned if feasible.

5.4.1. *RBCC versus global coherence*

The scalability aspect of RBCC comes with the fact that coherence support is limited to a selectable cluster of tiles. This inherently reduces the overall size of the directory structures that are needed to store the sharing information, i.e. a directory using the RBCC concept would only need to track sharers within a given region. For example, Figure 5.3a illustrates a typical sparse directory structure where the sharing information is stored as full bit-vectors. Assuming the many-core system contains N tiles within a 32 bit address space, the size of a global coherence directory would be $\{(N-1)+T+F\}\cdot 2^I$, where T, F and I are the tag, flag and index bits, respectively. In contrast, the size of a directory with the RBCC concept would be $\{(M_{max}-1)+T+F\}\cdot 2^I$, where M_{max} is the maximum number of tiles within a region. M_{max} is decided at design time based on the application requirements and the size of the many-core system. Therefore, RBCC reduces the directory size by a factor R, as shown in equation [5.1].

$$R = 1 - \frac{(M_{max}-1)+T+F}{(N-1)+T+F}, \qquad [5.1]$$

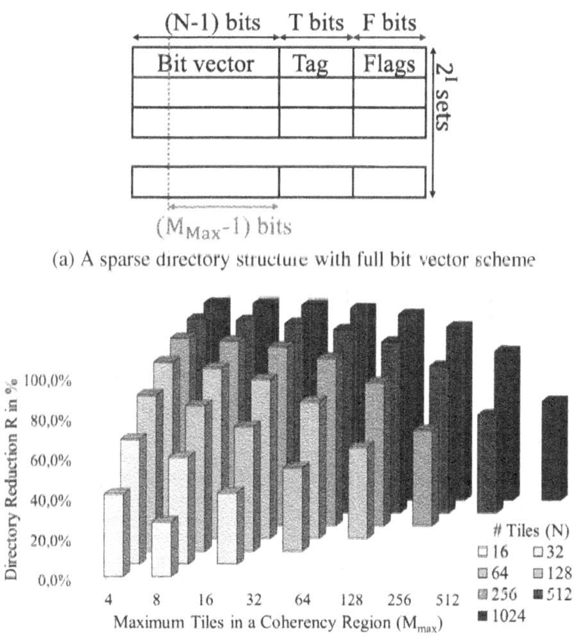

(a) A sparse directory structure with full bit vector scheme

(b) Plot of R for varying N and M_{max}

Figure 5.3. *Directory savings using the RBCC concept compared to global coherence*

Figure 5.3b illustrates this reduction in percent for varying N and M_{max}. The directory size decreases further if $M_{max} \ll N$, allowing RBCC to scale well compared to global coherence for large many-core systems.

5.4.2. *OS extensions for coherence-on-demand*

Coherence support can be further confined to actually shared application working sets, even within a coherency region. Typically, a tile local memory (TLM) houses both shareable and non-shareable data. Tracking tile-private data such as processor stacks, processor instructions and OS data results in unnecessary coherence actions. This can be alleviated by only tracking actually shared working sets in the TLM. This sub-section introduces *RBCC-malloc()*, an operating system (OS)-assisted extension to RBCC to confine coherence support within a coherency region to actually shared application working sets. When the OS initially configures a coherency region, it is aware of the memory footprint of compile-time allocated data. Therefore, the memory range for coherency region configuration can be provided accordingly. However for run-time application data allocations, which occur frequently, the OS has no knowledge regarding the memory range, and this is where *RBCC-malloc()* becomes useful (Figure 5.4). *RBCC-malloc()* is a wrapper around the commonly used *malloc()*, through which an application can indicate 1) if the data to be allocated needs to be cache coherent. Depending on this, the OS dynamically adapts 2) the memory range of the existing coherency region. Then, the OS allocates data on the heap and the control is returned 3) to the application. Through the *RBCC-malloc()* extension, coherence coverage can be optimally tailored to actually shared application working sets in a transparent manner.

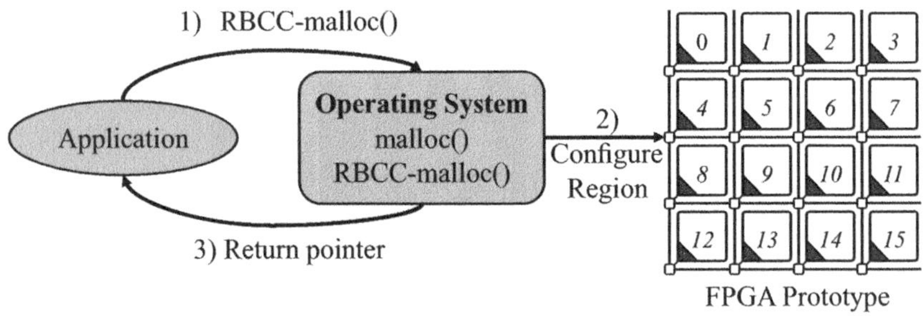

Figure 5.4. *RBCC-malloc() example*

5.4.3. *Coherency region manager*

The RBCC and *RBCC-malloc()* concepts are realized using a hardware coherency region manager (CRM) module present in every tile of our many-core system (Figure 5.2). While intra-tile coherence for the level 1 cache is supported using a common bus snooping protocol, inter-tile coherence for both level 1 and level 2 caches is enabled by the CRM. The CRM is a configurable module accessible via memory-mapped registers whose primary task is to create/dissolve coherency regions and to enforce a coherent view over all memories within each coherency region. Figure 5.5 depicts the internal block diagram of the CRM. The following sections briefly describe the functionality of each CRM sub-module and their interactions to guarantee cache coherence.

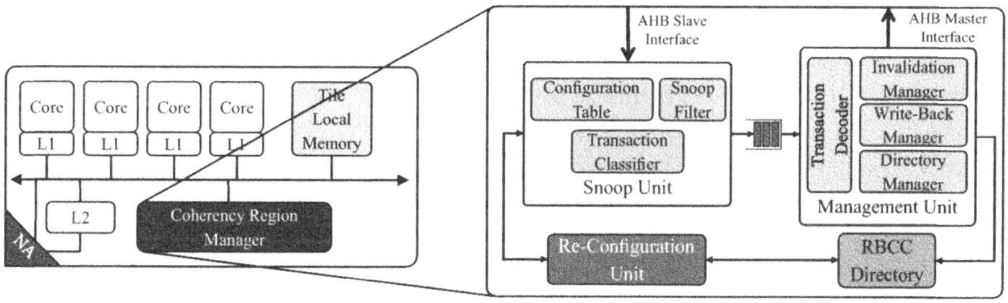

Figure 5.5. *Internal block diagram of the coherency region manager*

Snoop unit: this is the front end of the CRM through which coherency region configurations or re-configurations are initiated. The snoop unit uses a configuration table to track the coherency region information of its local tile. The table has five fields that specify the attributes of a coherency region. The *ID* field acts as a unique identifier to distinguish between multiple coherency regions. The *start* and *end address* fields define the memory range of the TLM to be kept coherent within the coherency region. The *sharers* field indicates the number of remote tiles that are allowed a coherent view over the TLM range defined by the *start* and *end address* fields. The *direction* field expresses whether the sharing is unidirectional or bidirectional. The snoop unit filters incoming bus transactions on the basis of information found in the configuration table, and classifies them into four major message types. These messages are sent to the management unit through a FIFO for further processing.

Management unit: this is the back end of the CRM which carries out inter-tile coherence actions. Four major actions are taken by the management unit based on the messages received from the snoop unit to guarantee coherence. A *directory update* is performed when a remote tile reads data from the tile's local memory. Through this action, the CRM keeps track of "which" and "how many" remote tiles share a local memory block. The size of a memory block is equal to the size of a cache line. When a tile writes to a remote memory in the coherency region, the memory block is held in the local L2 cache as a dirty block (assuming a writeback cache). The management unit *forcefully writes back* this memory block from the local L2 cache to the remote TLM. This action ensures that memory at the source is always up-to-date. When the snoop unit detects a write to the local memory, the management unit triggers the *invalidation generate* command. This sends out an invalidation message to all remote tiles sharing the modified memory block. Incoming invalidation messages from remote tiles trigger an *invalidation execute* command. This invalidates the corresponding memory block from both the local L2 and L1 caches.

Re-configuration unit: this sub-module is responsible for run-time coherency region re-configurations. In case the OS wants to add, remove or even relocate tiles of an existing coherency region for a particular application, this unit detects the changes made to each field in the configuration table and performs the required actions. For example, if a new tile is added to the coherency region, then the *sharers* field is updated. If a tile is deleted from the coherency region, then the sharing information for the deleted tile is selectively cleared in the directory. The time penalties for these actions depend on how many tiles were removed/re-located and the size of the updated memory range.

Resource utilization and timing: the CRM is synthesized and implemented as part of our target MPSoC platform. Figure 5.6 reports the resource demand of the CRM for a 4×4 architecture for different coherency region sizes. The snoop unit performs complex filter and classification functions and, hence, requires more resources than the management unit with its simple decode and coherence execution actions.

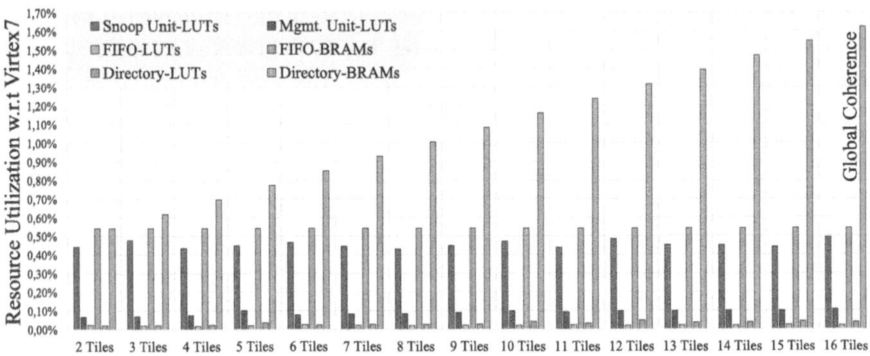

Figure 5.6. *Breakdown of the CRM's resource utilization for increasing region size*

The FIFO and directory are the only sub-modules that use block-ram (BRAM) resources. The FIFO consumes a constant amount of BRAMs, whereas the BRAM demand for the directory increases linearly with the region size. These hardware directory utilization results also follow equation [5.1], re-emphasizing RBCC's scalability compared to global coherence. Table 5.1 shows the number of clock cycles required for the CRM configuration, as well as the overheads for the four major coherence actions of a single tile. Synthesis reports reveal a logic delay of 3 ns and a routing delay of 5 ns on a Virtex-7 FPGA, which correspond to an operation frequency of 125 MHz.

Operations	Snoop	Mgmt.
CRM configuration	26 clks	4 clks
Directory update	7 clks	4 clks
Inv. generation	4 clks	10 clks
Inv. execution	2 clks	24 clks
Force writeback	2 clks	9 clks

Table 5.1. *CRM operations' latency*

5.4.4. *Experimental evaluations*

The feature extraction sub-function of a video streaming application has been investigated in order to evaluate the benefits of shared memory programming enabled by RBCC, in comparison to an alternative message passing-based implementation. The extracted features are object edges/boundaries present in a video frame. A continuous video stream is sent from a host PC via an Ethernet interface to the FPGA prototype (IO tile), where they are transferred to the local memory of a compute tile. Image processing is partitioned onto several tiles such that the processors of each tile work on a certain part of the image only. With RBCC enabled, every remote tile accesses its part of the image directly from the TLM of the master tile. This requires cache coherence which is guaranteed by the CRM. If RBCC is disabled, the image is transferred from the master tile to TLMs of the involved remote tiles via a message passing mechanism. After processing, the features from different tiles are collected at the master tile and returned to the host PC to be merged into the source image for visualization. An MPSoC with 16 tiles in total and a maximum region size of 4 is used to execute the application in both shared memory (*rbcc*) and message passing (*mp*) modes. In addition to this, we also investigate the impact of inter-tile communication on both modes by introducing artificial background traffic between the coherency regions during application execution.

Figure 5.7 shows the per-frame execution time for both the *rbcc* and *mp* modes of operation. The second half of the graph plots the execution time with background traffic active. We can see that the execution time for the *rbcc* mode is about 30% to 40% better than that for the *mp* mode throughout both video clips. In the *rbcc* mode, each tile directly accesses the image via remote read operations that are cached. The cached data are kept coherent by the CRM. In the *mp* mode, the image needs to be explicitly distributed to the remote TLMs via messages, which implies an extra software overhead. After image transfer, the processing is performed locally.

In Figure 5.8, the breakdown of the overall execution time into image distribution time and image processing time clearly shows that the image distribution overheads in the *mp* mode are a major reason for the lower performance, in comparison to the *rbcc* mode. The *rbcc* mode processing fundamentally consists of "on-demand" image read operations followed by image processing. Figure 5.8 also depicts different video clips (dk: Donkey Kong, si: Space Invaders, pm: Pac-Man) with varying degrees of background traffic (s, m, l, xl) interfering with the inter-tile communication of the application. For the *mp* mode, the image distribution time further increases with increases in background traffic, emphasizing the importance of efficient inter-tile communication. The raw *mp* mode image processing time is shorter than that in the *rbcc* mode as all cores access tile-local TLMs only.

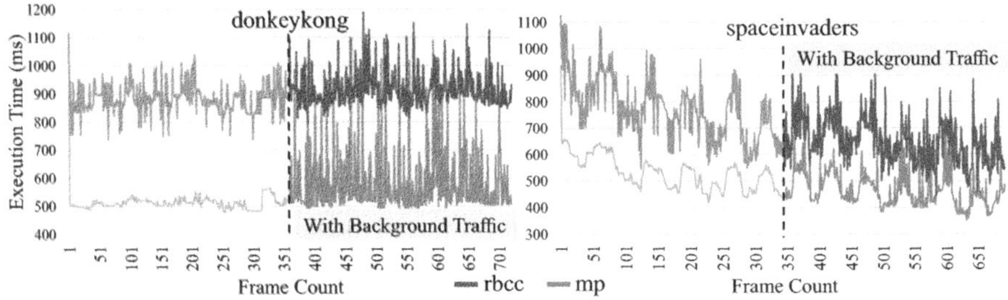

Figure 5.7. *Execution time per-frame: rbcc mode and mp mode*

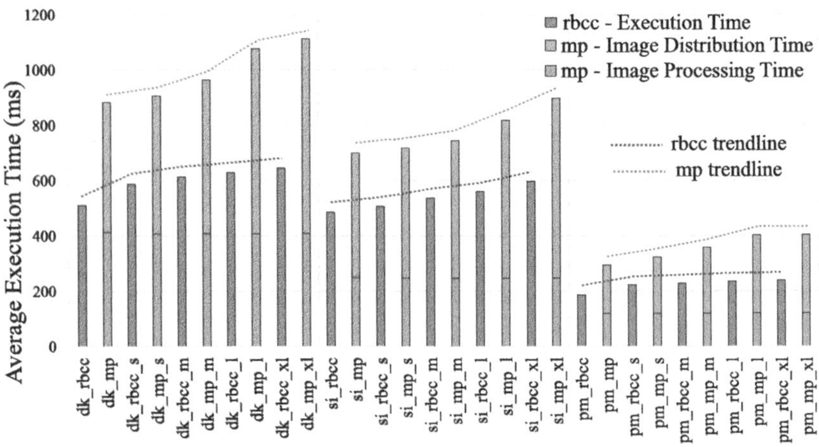

Figure 5.8. *Breakdown of execution time for different clips with increasing background traffic*

5.4.5. *RBCC and data placement*

RBCC allows the OS to specify the number, shape and locality of the tiles comprising a coherency region. From a data locality perspective, applications benefit if the tiles of a coherency region are spatially close to each other. Even within a coherency region, data placement plays a major role as it determines the amount of the inter-tile communication traffic between processing and memory resources. Our target architecture consists of both tile local SRAM and global DRAM memories exhibiting non-uniform memory access (NUMA) latencies. Ideally, an application would prefer all its data to reside in the tile local memory. Owing to size limitations, a hierarchy, together with remote TLM as the second preference, and lastly global DRAM memory, is established.

Using SystemC simulations, we investigated how different data placement techniques within a coherency region affect application performance. We explore two compile-time data placement techniques, *first touch* and *most accessed*, by statically analyzing the memory access pattern of applications. Both data placement techniques work on the granularity of memory blocks, where each memory block is equal to a cache line. Initially, all application data is assumed to be in the global DRAM memory. With *first touch*, the memory blocks are placed in the TLM of the tile whose processor first touches it. If the preferred TLM is full, then the memory block remains in the DRAM. With *most accessed*, the memory blocks are placed in the TLM of the tile whose processors access it the most. If the preferred TLM is full, the memory block is placed into the TLM of the next *most accessed* tile. Figure 5.9 shows the execution time of four applications from the PARSEC benchmark suite (Bienia 2011) with three coherency region sizes (1 tile, 2 tiles and 4 tiles in a region) for both data placement techniques. The execution time of all benchmarks is normalized to their respective single-tile *first touch* versions. For blackscholes, we see that data placement has no impact. This is because of the small working set of blackscholes, which comfortably fits into the preferred TLM. Similarly, the swaptions benchmark also remains unaffected by the data placement technique due to its working set size. Increasing the coherency region size increases the processing and memory resources available for the benchmark. For both blackscholes and swaptions, the execution time reduces linearly as the benchmarks are embarrassingly parallel. The canneal benchmark uses large working sets and exhibits a lot of inter-tile communication. We see that the execution time for *first access* data placement suffers heavily due to sub-optimally filled TLMs. The *most accessed* placement strategy greatly optimizes inter-tile traffic as a result of better placed data, thereby reducing execution time. An increase in coherency region size affects both placement techniques differently. The execution time for the *first touch* strategy deteriorates further, as it sub-optimally places data into the different TLMs. In contrast, the *most accessed* version decreases inter-tile communication by placing data into the respective TLMs, linearly reducing execution time because of the inherent parallelism of the benchmark. The fluidanimate benchmark exhibits a similar trend to canneal as it also works on large working sets. These simulations emphasize the importance of appropriate data placement, also within coherency regions.

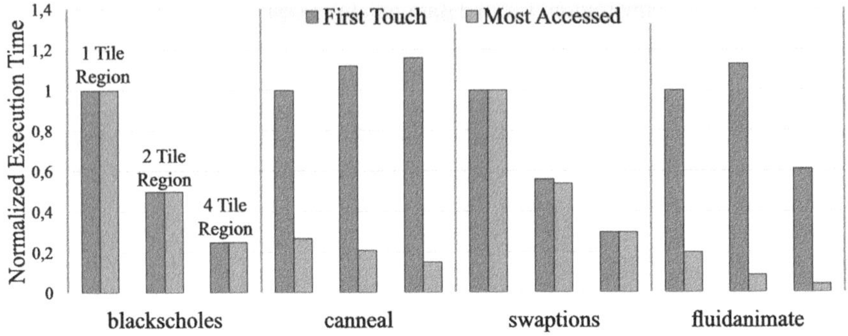

Figure 5.9. *Normalized benchmark execution time for different coherency region sizes for both data placement techniques*

5.5. Near-memory acceleration

Over the last decades, many different approaches to tackle the data-to-task locality challenge have been presented. Compute-centric solutions try to increase locality by bringing data closer to the processing elements, which is conventionally achieved by leveraging sophisticated multi-level cache hierarchies. Region-based coherence (section 5.4) already tackles many performance and scalability obstacles in this context. However, datasets tend to become much larger than the cache capacity, as well as irregular in the sense of missing temporal and spatial locality during execution. Hence, they get increasingly cache-unfriendly (Kogge 2017). Therefore, many recent approaches leverage in- or near-memory computing as an orthogonal idea to caching, in order to address the data-to-task locality problem. Instead of transferring the data to the processors via a cache hierarchy, this memory-centric approach moves the computation close to the memory where the data is stored. Thus, a high memory bandwidth in combination with a low access latency enables more efficient memory-intensive processing, while also reducing the energy consumption via shorter data transfers with lower capacitive loads.

Figure 5.10 shows a taxonomy of different in- and near-memory computing approaches. As a first dimension, we distinguish between in-memory computing, commonly referred to as "processing-in-memory" (PIM), and near-memory computing, where the processing elements are not physically part of the memory module but reside very close to the memory controller. As this is not bound to a specific memory technology, it can be applied to a wide range of compute systems. In a second classification step, near-memory processing approaches can be further divided into software-programmable near-memory cores or dedicated hardware accelerators, often referred to as *near-memory accelerators* (NMA). The dashed lines indicate that the same classification is possible for PIM. However, it is easier to integrate a near-memory than an in-memory core due to the different processing technologies. A third classification step differentiates between the acceleration of application specific tasks (AS-NMA) compared to accelerators to support the runtime and operating system (RTS-NMA). A wide range of applications could thus benefit from the latter, which are the focus of this section. Note that some RTS-NMAs can serve as AS-NMAs as well, which is indicated by the dotted lines. We, however, investigate them in the runtime support scenario. Due to limited space, AS-NMA and PIM are only covered briefly in section 5.3.

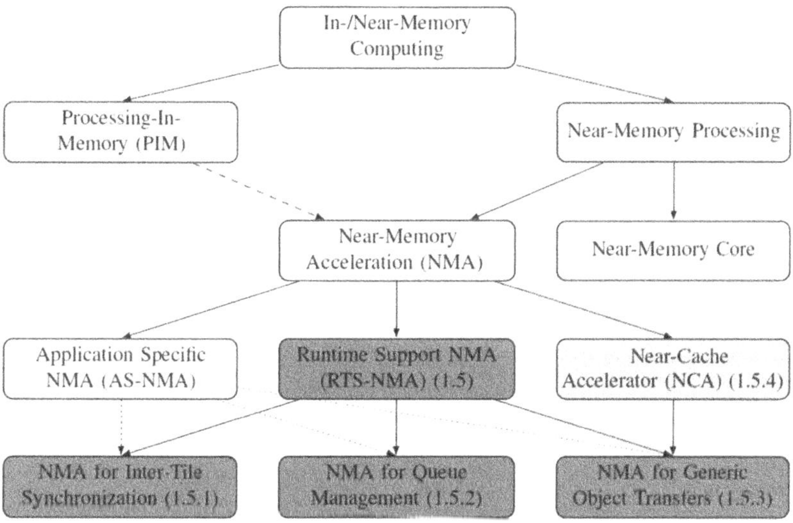

Figure 5.10. *Taxonomy of in-/near-memory computing (colored elements are covered in this section)*

In this section, we present three contributions in the context of runtime support for near-memory acceleration and show how they help to mitigate data-to-task locality challenges. The first contribution, presented in section 5.5.1, targets inter-tile synchronization via remote near-memory atomic operations such as compare-and-swap (CAS), test-and-set (TAS) or even more complex operations such as stack push/pop. In section 5.5.2, a near-memory accelerator for efficient queue management in tile-based systems is presented as our second contribution, illustrating its need and potential, particularly in architectures with physically distributed memory. Third, near-memory accelerated graph copy operations are addressed in section 5.5.3, with special focus on a comparison between a near-memory and far-from-memory implementation. All three contributions and corresponding use cases show the improvement in data-to-task locality, as well as the performance advantages of NMA in comparison to pure software implementations, particularly when used remotely and in bigger systems. In addition, so-called near-cache accelerators (NCA) are proposed and investigated in section 5.5.4. As the cache hierarchy is often bypassed when integrating in- or near-memory solutions, they need to be properly synchronized with the rest of the system, i.e., maintaining coherence and consistency between normal cores and NMAs. The NCAs are attached near the level 2 caches to perform efficient inter-tile cache operations (e.g. invalidations or writebacks), even on more complex data structures such as object graphs, and thereby relieve the cores of this duty.

5.5.1. *Near-memory synchronization accelerator*

Thread synchronization is a challenging topic in multi-processor programming because the atomic access to concurrent data structures needs to be guaranteed and realized in a performant manner. There are many well-known synchronization primitives that are present as hardware ISA extensions in today's MPSoC platforms, such as *test-and-set*, *compare-and-swap* (CAS), *linked-load/store-conditional* (LL/SC), *fetch-and-increment*, *mutex* and *semaphores*. However, on tile-based MPSoCs, those atomic primitives are often provided only locally inside of a tile. Their inter-tile support is mostly omitted since a packet-based NoC interconnect is not inherently supporting atomic memory accesses from the view of all cores. In addition, its non-uniform memory access (NUMA) properties give a new weight to different aspects of synchronization. For example, the acquisition of a remote spinlock or successful execution of a remote CAS suffers from much higher remote access latencies and retry penalties than its local pendants.

Concept: we envision efficient inter-tile synchronization primitives that are able to tackle the challenges of tile-based many-core architectures (Rheindt *et al.* 2018). We propose a near-memory synchronization accelerator which is located close to each memory, TLM within compute tiles, as depicted in Figure 5.11, as well as DRAM in memory tiles. It is tightly integrated into the network adapter and connected to the local bus; thus, it can be easily called by any local or remote core in the system and performs the atomic operations on behalf of those cores. Our study of useful candidates, which also extends the state of the art, led to the implementation of the following operations: 1) test-and-set, 2) compare-and-swap, 3) the class of fetch-and-op (increment, decrement, and, or, add, sub), 4) pointer enqueue and dequeue, and 5) stack push and pop.

These primitives are basic functional blocks to handle concurrent data structures (e.g. queues, stacks, semaphores, etc.) in many important software tasks. They are heavily used in parallel applications and libraries, as well as operating systems, and thus require a performant implementation.

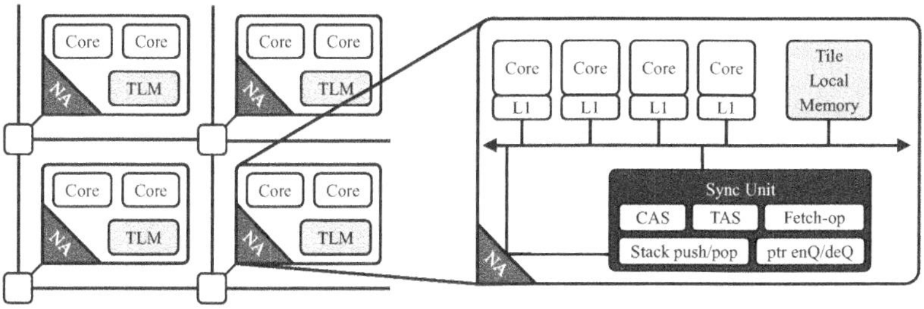

Figure 5.11. *Architecture of the remote near-memory synchronization accelerator*

As depicted in Figure 5.11, the proposed accelerator is atomically accessing the memory via the bus and by 1) locking the bus, 2) executing the desired operation and 3) unlocking the bus, it is consistent with any other atomic operation directly issued by the cores themselves. The near-memory integration is necessary for the atomic access via bus locking, otherwise the locking of the NoC interconnect or of the initiating remote core would inevitably result in a deadlock of the system. Furthermore, this near-memory integration is, at the same time, very desirable as it minimizes access latencies, guarantees minimal software involvement and performs the atomic operation with immediate success by design. We thus classify this module as a simple yet crucial runtime support near-memory accelerator (RTS-NMA).

The remote stack operations can, for example, be used to manage an efficient inter-tile memory allocator. Ordinarily, a remote allocation would require a task to be scheduled on the remote tile, which performs the allocation on behalf of the initiating core and sends back the pointer to the allocated memory. This introduces an extra round trip with its associated operating system overhead. We eliminate this round trip for buffer allocation using the remote stack push and pop atomic operations. Each tile allocates a pool of buffers and stores the pointers to them on a stack in its tile local memory. Another tile can obtain a pointer to a buffer by atomically popping from the stack via the near-memory accelerator, without any software involvement on the remote tile. When the buffer is no longer needed, it can be returned to the pool with a remote atomic push.

Evaluation: this lightweight inter-tile allocator has been integrated into the operating system (Oechslein *et al.* 2011) running on the LEON3 cores of our FPGA prototype. The evaluation of all proposed near-memory synchronization accelerators yielded the following results: 1) The near-memory synchronization accelerator outperforms pure software managed variants of the same data structure by 9.5× up to 23.9× and is even intensive to rising concurrency on the data structure. This even holds when the software variants are using ISA hardware extensions for local locks and CAS. 2) The relative advantage of the NMA for remote operations in comparison to local operations ranges from 3.3× to 6.5×. Hence, tile-based systems, in particular, benefit from the proposed extensions. 3) The modular nature of the module enables the integration of new features without much effort. 4) The additional resource cost of the dedicated hardware module is roughly 5% of a LEON3 core. The stand-alone module is able to run at 419 MHz on our FPGA prototype.

Efficient inter-tile synchronization primitives are crucial for the performance of parallel applications. Remote near-memory accelerators for these central runtime support tasks help to tackle data-to-task locality challenges. So far, these remote atomic operations only dealt with pointer operations, without taking care of the actual payload transfer. In the next section, the latter will be incorporated.

5.5.2. *Near-memory queue management accelerator*

Inter-tile task scheduling, message passing or data distribution in parallel applications, as well as libraries and operating systems, commonly require data transfers and subsequent processing on remote tiles. In these scenarios, queues are beneficial and thus play a crucial role in central places of applications, as well as the runtime support system. Figure 5.12 depicts two of many example use-cases for queues. For example, queues are used as message passing buffers for the communication and data transfers between several threads. Or considering the thread pool design pattern, where several threads get their jobs from a common queue. Queues are further utilized for scheduling and work stealing, in streaming applications, network stacks or asynchronous IOs such as the Linux *io_uring*. As queues play such an important role, it is evident that efficient queue implementation is essential to avoid performance bottlenecks.

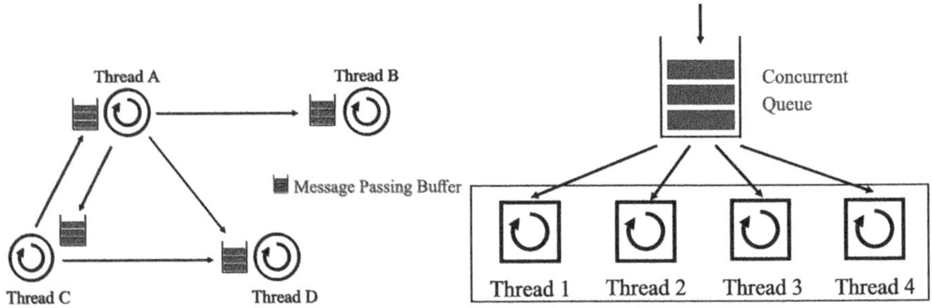

Figure 5.12. *Queues are widely used as message passing buffers*

For shared memory architectures, a good amount of research has been conducted. Many efficient queue implementations exist, such as the scalable and well-known Michael-Scott queue. On such architectures, it is often sufficient to handle pointers to the actual payload as the queue resides in the shared memory. On tile-based architectures, however, the implementation of a well-performing queue is more involved, yet has only received limited research effort. If, in the scenario depicted in Figure 5.13a, only the pointer to a queue element is enqueued or dequeued, then the element payload remains where it was. This would not solve the data-to-task locality issue for queue operations if the producer and consumer reside in different tiles. The physically distributed memory makes the transfer of the actual payload necessary to maintain data-to-task locality. Therefore, the question of how to provide scalable, efficient and yet flexible inter-tile queue management has to be answered.

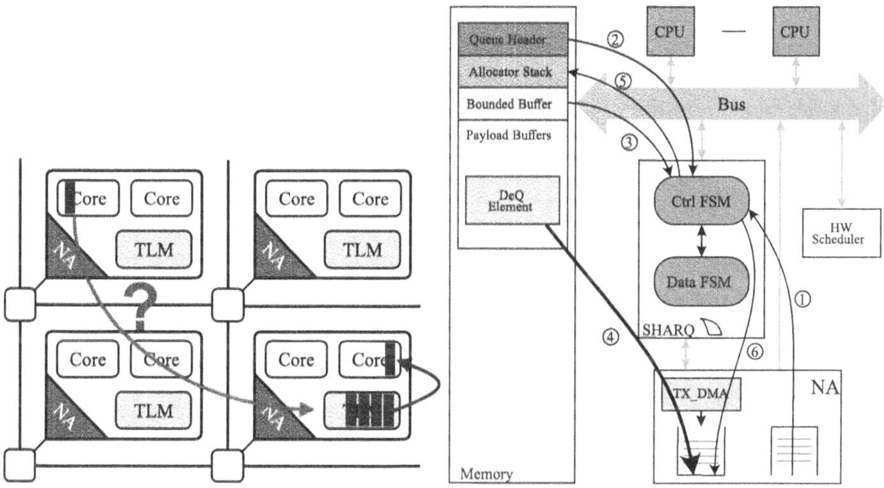

(a) Queues in distributed memory system (b) Remote dequeue operation

Figure 5.13. *Mechanism for a remote dequeue operation (right) for queues in distributed memory systems (left)*

Concept: we propose software-defined hardware-managed queues to retain the flexibility known of software queues, in combination with the performance of near-memory hardware acceleration for the queue management (Rheindt *et al.* 2019b). The performance requirements, frequency and complexity of a specific operation determine whether it is performed in hardware or software. The dynamic creation of user-defined queues with arbitrary lengths and element sizes is performed in software at runtime. This rather rare and complex operation allocates the queues in any of the available memories without consuming hardware resources to store queue elements or metadata. While the flexibility is guaranteed through software definition, performance is achieved by a near-memory queue management accelerator. The performance-critical atomic queue and memory management for the multi-producer multi-consumer (MPMC) queues, as well as the (remote) element payload transfer, is handled entirely in hardware. The so-called queue descriptor serves as the interface between the user and the hardware accelerator, as it consists of the queue metadata and points to the bounded buffer, allocator stack and payload buffers of the queue (as depicted in Figure 5.13 (right)). It is defined in software and contains all necessary information for the hardware to independently manage the queues. As, for example, free element buffer slots are initially provided in the allocator stack at queue creation, no costly system calls for dynamic memory allocation are required during queue operations. One near-memory queue management accelerator is present in each tile, which is accessible from any core in the system. It handles all local and remote enqueue and dequeue operations for all the queues residing in this tile's memory. The NMA is tightly coupled with the DMA engine of the network adapter to enable efficient inter-tile payload transfers. As depicted in Figure 5.13, a remote dequeue operation follows the steps: 1) receive the remote dequeue command, 2) get the meta-information from the queue header, 3) get the pointer to the element which will be dequeued next from the bounded buffer, 4) perform the remote data transfer of the queue element payload, 5) push the now free buffer slot onto the allocator stack and 6) signal the completion to the initiating core.

The local dequeue, as well as the enqueue operations, work accordingly. The hardware module is further able to conditionally schedule a so-called handler task, which can be user-defined and is meant for processing the queued elements. The proposed near-memory queue management accelerator therefore increases data-to-task locality with minimal software involvement and thus enables efficient inter-tile communication.

Evaluation: we integrated the queue management accelerator, a distributed operating system as well as the communication primitives (send/recv) of the MPI library into our 4 × 4 tile FPGA-based MPSoC prototype. To showcase their benefits, we choose the evaluation with the MPI-based NAS parallel benchmarks. We compare the NMA variant against a software-based queue handling approach for all five NAS benchmarks in the 16-tile variant, with a total of 64 application cores. To guarantee a fair comparison, the software-based variant leverages state-of-the-art hardware support, such as a DMA engine with efficient task spawning capabilities. Figure 5.14 (left) shows that the communication intense IS kernel (Subhlok *et al.* 2002), in particular, profits from the acceleration. The execution time was reduced by 46%. In contrast, the CG, FT and MG kernels, which are communicating less consequently, only experience a performance gain between 9% and 18%. Finally, no speed-up is observed for the EP (embarrassingly parallel) kernel as it is essentially not communicating. In Figure 5.13 (right), the execution time of the communication intense IS kernel is compared for different system sizes. The tile count is varied from 1 over 2, 4 to 16, and thus 4, 8, 16 or 64 application cores are used. First of all, the scalability of the near-memory acceleration approach is highlighted since bigger tile-based systems, particularly, benefit a lot. As stated above, the software-based approach is only competitive in single-tile systems. The significant performance improvement is achieved by offloading data transfers and queue operations to hardware, while freeing up the cores to actual computation workload.

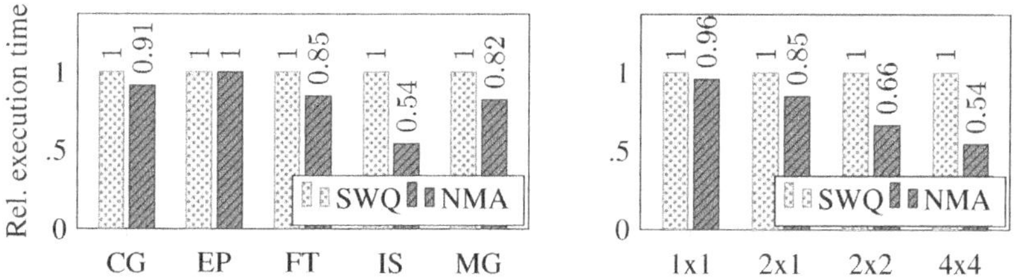

Figure 5.14. *NAS benchmark 4×4 results (left) and IS scalability (right) for different system sizes*

When synthesized, the hardware resource usage of the near-memory queue management accelerator is roughly 13% of the four LEON3 cores of each tile. The resources are, by design, independent of the software-defined queues, their element sizes and the number of allocated queues. With growing system sizes, each NMA only requires 80 additional bits per tile in order, to store the necessary meta-information to manage concurrent queue access from several source tiles. The accelerator is able to run at roughly 180 MHz.

The study of this NMA revealed that in systems with physically distributed memory, it is thus crucial for data-to-task locality to couple the accelerated queue management with efficient payload transfers.

5.5.3. *Near-memory graph copy accelerator*

In the message queue scenarios presented above, the inter tile communication was performed via message-passing between the distributed tile local memories. There are, however, many architectures that solely contain one or a few global DDR SDRAM memories located in dedicated memory tiles. Besides shared memory or message-passing programming, those MPSoCs can also follow the partitioned global address space (PGAS) programming paradigm, using languages such as X10 (Saraswat *et al.* 2019) or Chapel (Chapel Inc. 2019). The whole memory range is divided into memory partitions which are associated with the individual processing tiles. As inter-tile cache coherence is often omitted on these architectures, the PGAS paradigm requires explicit data transfers between the memory partitions. To support efficient transfers, these architectures are usually equipped with direct memory access (DMA) engines. Meanwhile, programming languages have evolved from simple data structures, such as arrays or strings, to more complex object-oriented data structures, such as object graphs. Today's high-level object-oriented programming (OOP) languages, such as Java, C++ or Python, make use of those and organize their data as objects referencing each other via pointers. When copying such object graphs, all pointers in the destination graph have to be properly adjusted. A classical DMA engine cannot be applied directly anymore because it can only copy non-pointered data correctly.

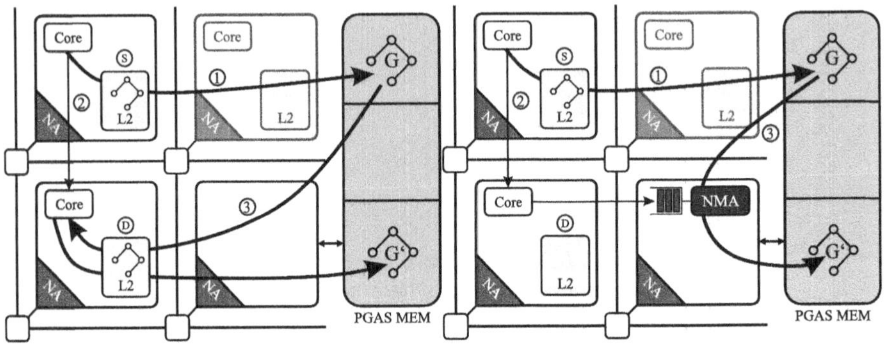

Figure 5.15. *Far-from memory (left) versus near-memory (right) graph copy*

When combining these aspects, it becomes evident that an efficient solution for object graph transfers is required for tile-based MPSoCs. Many message-passing approaches suffer from costly (de-)serialization of the graph, while others reduce the need for a serialization buffer or even use direct cloning (Mohr and Tradowsky 2017). Since the PGAS programming paradigm implies a division of the global memory into partitions, the transfer of the source graph G (see Figure 5.15) from the sender S's memory partition S to G' in the receiver D's memory partition D becomes necessary. This includes several steps, depicted in Figure 5.15 (left) for a state-of-the-art mechanism (Mohr and Tradowsky 2017). To maintain coherence and consistency, the source graph G first has to be written back from the sender's caches to its memory partition so that the NMA works on up-to-date data. This step requires a complete graph traversal to be able to write back every single object. Second, a core on the receiving tile is signaled including some metadata for the transfer. In a third step, the receiver D performs a copy of G to G' while adjusting all pointers to their new values in the destination graph G'. Although this approach has many benefits over serialization-based mechanisms, it does not tackle the data-to-task locality challenge. The graph copy operation (3) is performed "far-from-memory" via remote load-store over the NoC and the cache of the receiving tile. This is first of all slow and in addition, pollutes the cache on tile D with the source graph G unnecessarily. Furthermore, no accelerator (such as a DMA unit for non-pointered data) is used that would relieve the cores of this duty.

Concept: we propose a near-memory graph copy accelerator which tackles the above-mentioned deficiencies and thus increases data-to-task locality (Rheindt *et al.* 2019a). The same three steps still have to be performed. However, the cache-unfriendly and memory-intensive graph copy is offloaded to a near-memory accelerator located in the respective memory tile. The graph is thus copied near-memory without remote data transfers over the NoC or the cache hierarchy.

The NMA is triggered by a core on the receiving tile, copies the object graph asynchronously with the necessary "on-the-fly" pointer adjustments and signals the receiving core upon completion. The NMA is equipped with a FIFO to buffer incoming requests from any CPU in the system. It is further capable of spawning user-defined tasks to the receiving core after each graph copy to enable individualized processing of the destination graph G'. The near-memory accelerator has the ability to copy arbitrary graphs using the hardware implementable concept of pointer reversal, combined with a minor extension of the software-defined object model. It is thus a generalization and potential enhancement of a DMA engine to support the efficient transfer of pointered data structures, such as object graphs.

Tile-based MPSoCs often contain more complex memory architectures with multiple memory tiles to avoid access hotspots. If a graph needs to be copied between two memory partitions located in physically distributed memory tiles, a slight modification of the proposed mechanism is required. A near-memory graph copy is not directly possible in this case since the access to either G or G' would lead to inefficient remote accesses of the NMA. This would essentially be undesired "far-from-memory acceleration". Therefore, we propose an efficient inter-memory graph copy approach using an intermediate buffer $G*$ in the source memory partition. The NMA in the source memory tile then first copies G to the intermediate graph $G*$ while already adjusting all pointers to point to the final destination graph G'. Hence, in a second step, it is possible to simple transfer $G*$ to G' in the destination memory via a normal DMA. Upon arrival at the destination, no post-processing or de-serialization is necessary. The benefits of near-memory graph copy are thus retained even for inter-memory graph transfers.

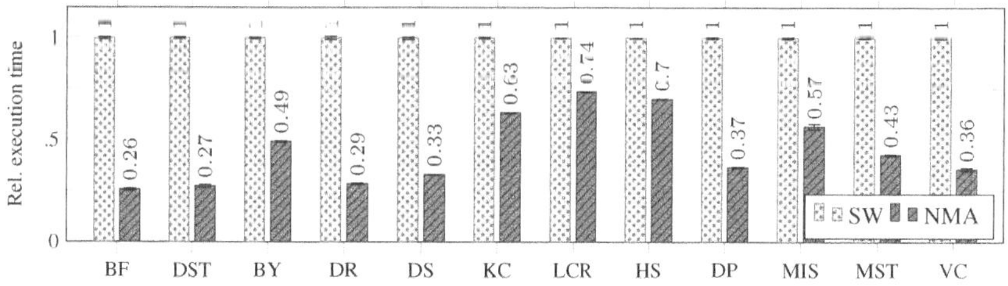

Figure 5.16. *IMSuite benchmark results on a* 4×4 *tile design with 1 memory tile*

Evaluation: we evaluated the influence of this NMA by executing the IMsuite (Gupta and Nandivada 2015), consisting of 12 different benchmark applications, and analyzing their runtime behavior. Since the IMsuite is a mixture of different, typical parallel applications, an average over all 12 benchmarks is a good indication for the real-world behavior of the system. The total speed-up of each benchmark is between $1.35\times$ and $3.8\times$ for the total application, which implies a much higher speed-up of graph copy duty itself. A detailed performance analysis is presented in Figure 5.16 for a system with one memory tile and Figure 5.17 for a system with two memory tiles, thus partially requiring inter-memory graph copies as well. Regarding the inter-memory variant, the performance gain of the NMA is slightly decreased, since it incorporates the additional DMA transfer between the two distributed memories. They are not present in the software-based variant as there is no difference in the implementation when using several memory tiles. We furthermore observe that the number of remote memory accesses in the standard variant is reduced by 30% on average, because the memory-intensive graph copy task is performed locally near-memory instead of via the NoC, as in the software variant. In the inter-memory variants, this number is only reduced by 12%. In both NMAs variant, the number of memory accesses per time also increases, since the NMA can use a much higher memory bandwidth than the far-from-memory cores. Apart from this, we varied the number of cores per tile and observed that the total runtime is only insignificantly affected by this, leading to the conclusion that the graph copy task is more memory- than compute-bound. We further synthesized our prototype with different L2 cache sizes and analyzed the influence of this parameter. As the NMA bypasses the L2 cache, an architecture with the NMA is more independent of the L2 cache size and the performance variation due to different cache sizes is rather small compared to the software implementation. When synthesized onto the FPGA prototype, the pure NMA requires about 25% less FPGA resources than a LEON3 core and 61% less than the external DRAM memory controller. Our implementation can run at a maximum frequency of 164 MHz.

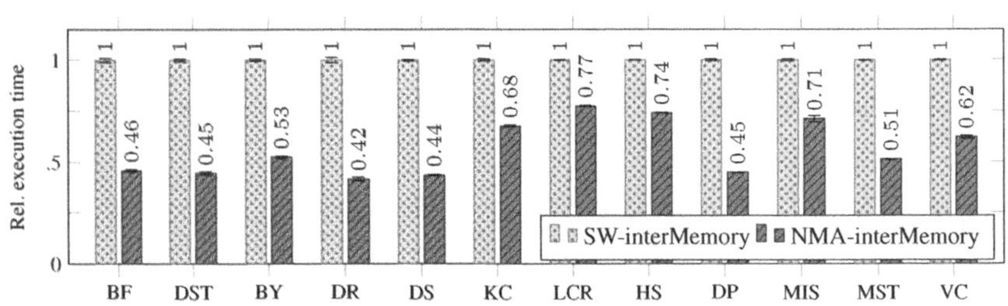

Figure 5.17. *IMSuite benchmark results for inter-memory graph copy*

5.5.4. *Near-cache accelerator*

As the cache hierarchy is often bypassed when integrating in- or near-memory solutions, they need to be properly synchronized with the rest of the system, i.e., maintaining coherence and consistency between normal cores and NMAs. The latter can only start working on data when it is ensured that no cache in the system still holds any updates of this data. Therefore, a complete cache writeback of the relevant data has to be performed before outsourcing the memory-intensive tasks to near-memory accelerators. This can be quite complicated when data structures spread over many cachelines or even contain pointers (e.g. graphs). In this case, a complete graph traversal is necessary to be able to write back every object in the graph. It also has to be considered to invalidate the caches after the NMA has finished processing so that the cores read the updated data that was modified by any near-memory accelerator.

Concept: we therefore propose a near-cache accelerator unit (NCA), similar to an NMA, located in or near the cache controller, as depicted in Figure 5.2, capable of performing two different operations on behalf of the cores: 1) traversing an arbitrary object graph and issuing cache writeback commands for each of the objects. Upon completion, it can directly dispatch a user-defined task to the receiving tile without additional system calls on sender side. 2) Performing range-based cache operations (writeback, invalidation or both, similar to Mohr and Tradowsky (2017)) with an optional subsequent triggering of the graph copy accelerator with user-defined parameters.

Evaluation: Figure 5.18 shows the normalized total runtime of the IMSuite benchmarks with and without the NCA unit. We also observe that the already high-performing NMA implementation can be accelerated further with a speed-up factor of up to 1.3×. The cores are relieved of the graph writeback and copy duty and are thus able to execute more actual application workload.

Summary: hence, tackling data-to-task locality challenges in tile-based MPSoCs is a holistic endeavor. Near-memory and near-cache accelerators, tightly integrated into the runtime system show a promising outcome as the functions are executed in the right place, while relieving the cores at the same time. Unnecessary data movement throughout the system is reduced and thus the compute efficiency increases.

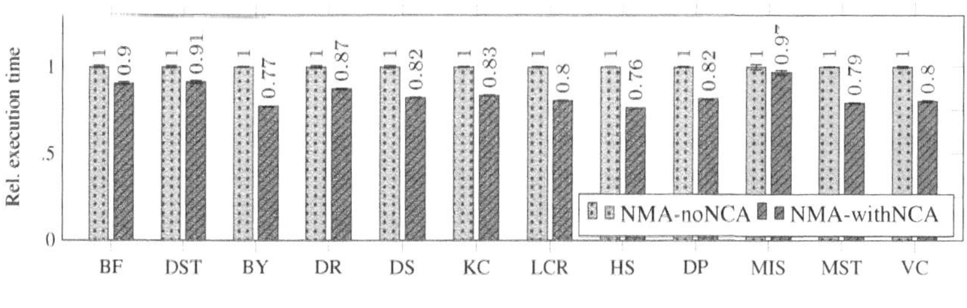

Figure 5.18. *Effect of NCA*

5.6. The big picture

The previous sections showed how RBCC dynamically extends shared memory programming beyond tile boundaries and how near-memory acceleration is used to mitigate data-to-task locality challenges in a distributed memory programming paradigm. Our target architecture houses physically distributed TLMs and global DRAM memories and supports the use of both programming paradigms. This section addresses how the RBCC and NMA concepts can be applied together, using a mix of both programming models. Figure 5.19 shows the "big picture", indicating that both RBCC and NMA co-exist outside each others' current operation domains. The next paragraphs describe the necessary adjustments required to leverage them together.

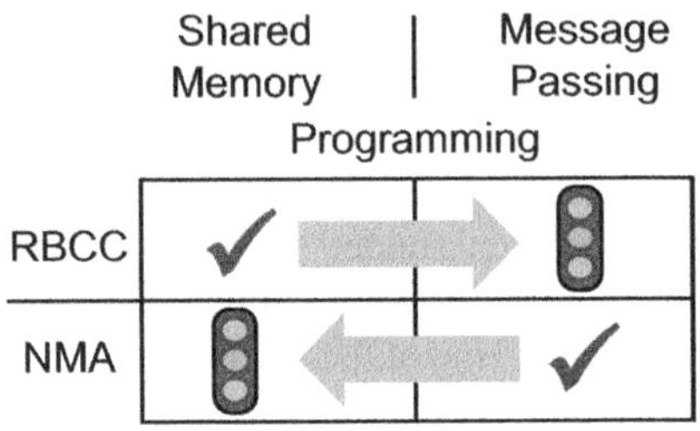

Figure 5.19. *Interplay of RBCC and NMA for shared and distributed memory programming*

Beyond a coherency region: the maximum size of a coherency region is decided at design time based on the size of the many-core system and the type of applications expected to run on it. If an application requests a coherency region greater than the maximum size, we propose a hybrid solution consisting of both shared and distributed memory programming. Application tasks running within a coherency region can communicate via the shared memory and can rely on RBCC to guarantee a coherent view of all the memories within the coherency region. Application tasks running on tiles beyond region borders need to communicate via message passing. For an efficient realization of these "inter-coherency region" interactions, our proposed NMA extensions can be used. Through this, RBCC (within coherency region) and NMA (between coherency regions) can be combined in a synergistic way.

Application-specific and coherent NMAs: near-memory acceleration described in section 5.5 was introduced in the context of distributed programming using message passing and an acquire-release consistency model. Although we use the proposed NMAs as runtime support accelerators, they can also be directly called by user applications. In Figure 5.10, these are referred to as application-specific NMAs. Besides the proposed NMAs, further AS-NMAs can be implemented. However, when leveraging NMAs, data coherency and consistency, in conjunction with memory accesses by NMAs, have to be considered as well. In libraries and runtime support systems, this is often easier as synchronization points are more obvious. Furthermore, when performing these accelerations in a shared memory programming environment, all NMA-related memory accesses need to be exposed to the CRM, in order to guarantee a coherent and consistent system. For example, the NMA for inter-tile synchronization of section 5.5.1 is a hardware module connected to a tile's main bus. Therefore, all memory operations of the NMA are visible to the CRM, which also snoops on the main bus. The NMA instruction typically conveys meta-information describing the accelerated function and a memory range on which to operate.

Data and task migration: in addition to the approaches described, data-to-task locality can also be optimized by migrating data closer to where processing takes place or tasks closer to where data reside. If this involves data received from I/O, the received data buffer may be moved to a nearer memory, or the respective task may be relocated to a nearer core, so that newly arriving data are processed in closer proximity. In general for shared datasets, explicit migration of data or tasks at runtime can also be considered if the performance improvement by shorter access latencies outweighs the migration overheads. In conjunction with NMAs, only data migration to a memory that is equipped with a required accelerator is an option, as the position of the NMA is of course fixed.

Design time decisions: this issue is also related to the design of the MPSoC architecture and its memory subsystem. Depending on the target application domain(s) and its/their requirements, an appropriate number and size of internal SRAM memories as well as external DRAM memories and their placement in the mesh have to be determined. Highly relevant to this question is also the decision on the allocation/instantiation of NMAs and their distribution over the tiled architecture. Not every type of accelerator will usually be required at every memory block, and it will even not be affordable to equip all memories with instances of all different NMAs. Depending on the mix of the target applications to be run on the MPSoC, the designer has find a sweet spot. This design time decision will then make up a boundary condition for the OS runtime management system in its target to exploit the architecture as much as possible. That is, data and task placement, or even dynamic migration, have to be done in a way that NMAs can really profit from the proximity of the relevant data and that regions are set up such that they contain the required NMAs for a given application.

5.7. Conclusion

The lack of data locality in many-core processor platforms is a persistent and multidimensional issue affecting both compute performance as well as energy efficiency. Different hardware and software architectures, programming models and application domains require different approaches in order to either keep the data local to the most effective processing engine(s) or bring the processing as close as possible to the target data. While doing so, proper coherency as well as consistency among shared data items and synchronization during data accesses and during critical section execution, must be preserved. The multi-faceted nature of the data locality issue is why it has been addressed over decades yet has still not yielded a comprehensive and conclusive solution. From the perspective of continuously evolving architecture progress, the question is whether it ever will.

This chapter presented two conceptual approaches, region-based cache coherence and near-memory accelerators for OS services, as well as a generic hardware technique to support the atomic execution of critical sections. Region-based cache coherence and near-memory acceleration can either be deployed individually or in combination. Both approaches are specifically tailored to help tackle the data locality challenge in tile-based MPSoC processing architectures, following either a distributed shared memory or distributed-memory programming model. They were implemented as AMBA High-Performance Bus-compatible SoC macros in order to make these forms of locality enablement also applicable to a wider spectrum of MPSoC architectures. Hardware costs for the SoC extensions amount to $\approx 5.5\%$ for RBCC and 6.9% for NCA and queue management of a regular compute tile with four cores. Measured/simulated execution time reductions for IMSuite, NAS and PARSEC benchmarks range between 9% and 74%.

5.8. Acknowledgments

This work was supported by the Deutsche Forschungsgemeinschaft (DFG, German Research Foundation) – project number 146371743 – TRR 89: Invasive Computing. We are particularly thankful for the fruitful collaboration with our project partners at FAU Erlangen-Nürnberg: T. Langer, S. Maier, F. Schmaus, W. Schröder-Preikschat, KIT Karlsruhe, L. Masing, A. Nidhi, J. Becker, A. Fried, G. Snelting; and at TUM: D. Gabriel, A. Schenk.

5.9. References

Bienia, C. (2011). Benchmarking modern multiprocessors. PhD Thesis, Princeton University.

Brewer, E.A. (1995). Remote queues: Exposing message queues for optimization and atomicity. *Proceedings of the ACM Symposium on Parallel Algorithms and Architectures*, 42–53.

Chapel Inc. (2019). Chapel Language Specification [Online]. Available: https://chapel-lang.org/docs/language/spec/index.html

Dally, W.J. (2006). Computer architecture in the many-core era. *Proceedings from the 24th International Conference on Computer Design*, San Jose, USA, 1, 1–4 October.

Falsafi, B., Stan, M., Skadron, K., Jayasena, N., Chen, Y., Tao, J., Nair, R., Moreno, J.H., Muralimanohar, N., Sankaralingam, K., and Estan, C. (2016). Near-memory data services. *IEEE Micro*, 36(1), 6–13.

Ferdman, M., Lotfi-Kamran, P., Balet, K., and Falsafi, B. (2011). Cuckoo directory: A scalable directory for many-core systems. *Proceedings of the 2011 IEEE 17th International Symposium on High Performance Computer Architecture*, 169–180.

Fu, Y., Nguyen, T.M., and Wentzlaff, D. (2015). Coherence domain restriction on large scale systems. *Proceedings of the 48th International Symposium on Microarchitecture*, New York, USA, 686–698.

Gupta, A., Dietrich Weber, W., and Mowry, T. (1990). Reducing memory and traffic requirements for scalable directory-based cache coherence schemes. *Proceedings of the International Conference on Parallel Processing*, 312–321.

Hybrid Memory Cube Consortium (2019). Hybrid Memory Cube Specification 2.1 [Online]. Available: https://www.nuvation.com/sites/default/files/Nuvation-Engineering-Images/Articles/FPGAs-and-HMC/HMC-30G-VSR_HMCC_Specification.pdf.

Jayasena, N. (2018). Memory-centric accelerators in high-performance systems. *Proceedings of the 55th Design Automation Conference*, San Francisco, USA, 24–28 June.

Kessler, R.E. (2011). The cavium 32 core OCTEON II 68xx. *Proceedings of the 2011 IEEE Hot Chips 23 Symposium*, 1–33.

Kogge, P. (2017). Memory intensive computing, the 3rd wall, and the need for innovation in architecture [Online]. Available: https://memsys.io/wp-content/uploads/2017/12/The_Wall.pdf.

Lenoski, D., Laudon, J., Gharachorloo, K., Weber, W.-D., Gupta, A., Hennessy, J., Horowitz, M., and Lam, M.S. (1992). The stanford dash multiprocessor. *Computer*, 25(3), 63–79.

Masing, L., Srivatsa, A., Kreß, F., Anantharajaiah, N., Herkersdorf, A., and Becker, J. (2018). In-noc circuits for low-latency cache coherence in distributed shared-memory architectures. *Proceedings of the 2018 IEEE 12th International Symposium on Embedded Multicore/Many-core Systems-on-Chip*, 138–145.

Michael, M.M. and Scott, M.L. (1995). Implementation of atomic primitives on distributed shared memory multiprocessors. *Proceedings of the First IEEE Symposium on High-Performance Computer Architecture*, 222–231.

Mohr, M. and Tradowsky, C. (2017). Pegasus: Efficient data transfers for PGAS languages on non-cache-coherent many-cores. *Proceedings of the Conference on Design, Automation and Test in Europe*, 1785–1790.

Nguyen, K., Fang, L., Navasca, C., Xu, G., Demsky, B., and Lu, S. (2018). Skyway: Connecting managed heaps in distributed big data systems. *ACM SIGPLAN Notices*, 53(2), 56–69.

Oechslein, B., Schedel, J., Kleinöder, J., Bauer, L., Henkel, J., Lohmann, D., and Schröder-Preikschat, W. (2011). OctoPOS: A parallel operating system for invasive computing. *Proceedings of the International Workshop on Systems for Future Multi-Core Architectures*, 9–14.

Parkins, D. (2017). Regulating the internet giants: The world's most valuable resource is no longer oil, but data. *The Economist*, May 6, The Economist Group Limited, London.

Rheindt, S. (2018). CaCAO: Complex and compositional atomic operations for NoC-based manycore platforms. *Proceedings of the 31st International Conference on Architecture of Computing Systems*, Braunschweig, Germany, April 9–12, 139–152.

Rheindt, S. (2019a). NEMESYS: Near-memory graph copy enhanced system-software. *Proceedings of the International Symposium on Memory Systems*, Washington, DC, USA, 3–18.

Rheindt, S. (2019b). SHARQ: Software-defined hardware-managed queues for tile-based manycore architectures. *Proceedings of the 19th International Conference of Embedded Computer Systems: Architectures, Modeling, and Simulation*, Samos, Greece, 212–225.

Sanghoon, L., Devesh, T., Yan, S., and James, T. (2011). HAQu: Hardware-accelerated queueing for fine-grained threading on a chip multiprocessor. *Proceedings of the Conference on High-Performance Computer Architecture*, 99–110.

Saraswat, V., Bloom, B., Peshansky, I., Tardieu, O., and Grove, D. (2019). X10 Language Specification [Online]. Available: http://x10.sourceforge.net/documentation/languagespec/x10-latest.pdf.

Sodani, A., Gramunt, R., Corbal, J., Kim, H., Vinod, K., Chinthamani, S., Hutsell, S., Agarwal, R., and Liu, Y. (2016). Knights landing: Second-generation Intel Xeon Phi product. *IEEE Micro*, 36(2), 34–46.

Southern, G. and Renau, J. (2016). Analysis of PARSEC workload scalability. *Proceedings of the IEEE International Symposium on Performance Analysis of Systems and Software*, 133–142.

Srivatsa, A., Rheindt, S., Wild, T., and Herkersdorf, A. (2017). Region based cache coherence for tiled MPSoCs. *Proceedings of the 2017 30th IEEE International System-on-Chip Conference*, 286–291.

Srivatsa, A., Rheindt, S., Gabriel, D., Wild, T., and Herkersdorf, A. (2019). CoD: Coherence-on-demand – Runtime adaptable working set coherence for DSM-based manycore architectures. In *Embedded Computer Systems: Architectures, Modeling, and Simulation*, Pnevmatikatos, D.N., Pelcat, M., and Jung, M. (eds). Springer International Publishing, Cham.

Subhlok, J., Venkataramaiah, S., and Singh, A. (2002). Characterizing NAS benchmark performance on shared heterogeneous networks. *Proceedings of the 16th International Parallel and Distributed Processing Symposium*, USA, 91.

Turek, D. (2017). The transformation of HPC: Simulation and cognitive methods in the era of Big Data [Online]. Available: https://www. slideshare.net/insideHPC/the-transformation-of-hpc-simulation-and-cognitive-methods- in-the-era-of-big-data.

Wang, Y., Wang, R., Herdrich, A., Tsai, J., and Solihin, Y. (2016). CAF: Core to core Communication Acceleration Framework. *Proceedings on the 2016 International Conference on Parallel Architectures and Compilation*, 351–362.

Wentzlaff, D., Griffin, P., Hoffmann, H., Bao, L., Edwards, B., Ramey, C., Mattina, M., Miao, C., Brown III, J.F. and Agarwal, A. (2007). On-chip interconnection architecture of the tile processor. *IEEE Micro*, 27(5), 15–31.

Williams, S., Waterman, A. and Patterson, D.A. (2009). Roofline: An insightful visual performance model for multicore architectures. *Communications of the ACM*, 52(4), 65–76.

Yao, Y., Wang, G., Ge, Z., Mitra, T., Chen, W. and Zhang, N. (2015). SelectDirectory: A selective directory for cache coherence in many-core architectures. *Design, Automation and Test in Europe Conference and Exhibition*, 175–180.

6

mMPU: Building a Memristor-based General-purpose In-memory Computation Architecture

Adi ELIAHU[1], Rotem BEN HUR[1], Ameer HAJ ALI[2] and Shahar KVATINSKY[1]

[1] *Electrical Engineering Department, Technion – Israel Institute of Technology, Haifa, Israel*
[2] *Electrical Engineering and Computer Science Department, University of California, Berkeley, USA*

The majority of computer systems use the von Neumann architecture, in which the memory and the computing units are separate. As a result, when performing calculations, data must be transferred from the memory to the processing unit, which demands massive energy expenditure. Newly emerging non-volatile memory technologies, widely referred to as memristors, have been developed to enable in-memory data processing, avoiding the energy waste associated with data movement. This chapter describes the general-purpose architecture of the memristive memory processing unit (mMPU). The unit is designed to use memristors to execute different logic operations within the memory, and it is especially efficient for single instruction multiple data (SIMD), where massive parallelism can be achieved using the memristive crossbar array. We first describe memristor-based logic gates, then explain manual and automatic methods in order to apply them to execute complex logic functions. We then describe a memory controller for our memristor-based architecture.

For a color version of all figures in this book, see www.iste.co.uk/andrade/multi1.zip.

Multi-Processor System-on-Chip 1 – Architectures,
coordinated by Liliana ANDRADE and Frédéric ROUSSEAU. © ISTE Ltd 2020.

6.1. Introduction

Most computer architectures today are based on the von Neumann architecture, where the memory and the processing unit are completely separated from each other. This structure often demands data movement from the memory to the processing unit when a calculation is performed. A data transfer like this is very slow and consumes a large amount of power, especially in data-intensive applications.

Recently, new non-volatile resistive memory technologies, widely referred to as memristors, have been developed. The concept was first introduced by Leon Chua (1971), and was later associated with the resistive switching phenomenon by Hewlett-Packard labs (Strukov *et al.* 2008). Since then, numerous non-volatile resistive memory technologies have been developed, for example resistive RAM (RRAM) (Angel Lastras-Montaño and Cheng 2018) and phase change memory (PCM) (Wong *et al.* 2010). The memristor device can function as both a memory element and a processing unit, thus enabling performance of in-memory computation and avoiding massive data transfer.

The memristor can change its resistance between two values: R_{ON} and R_{OFF}, which represent "1" and "0" logic values, respectively. The memristor state changes in accordance with the voltage drop on the memristor or the current that flows through it, and saves this value even in the absence of a power source, rendering it a non-volatile memory element.

Memristors have been integrated into many architectures. To accelerate neural networks, Shafiee *et al.* (2016) developed an architecture that multiplies a vector and a matrix using analog multiplication within a memristive crossbar . Accelerators for other applications, for example graph processing (Song *et al.* 2017) and DNA sequencing (Kaplan *et al.* 2017), have also been described and implemented.

Most memristor-based architectures discussed in the literature focus on accelerating a specific task. This chapter presents a memristor-based, general-purpose architecture, called the memristive memory processing unit (mMPU), and describes its structure in a bottom-up approach, from the manner in which the calculation is performed in-memory, to the memory controller architecture. We first present MAGIC (Memristor-Aided loGIC) (Kvatinsky *et al.* 2014), a logic gate family that enables the execution of logic within a memory crossbar array. We focus on the MAGIC NOR gate and explain how any logic can be executed using this gate. Then, we elaborate how different algorithms from different disciplines can be implemented using MAGIC gates to efficiently perform logic in-memory. We then explain how an algorithm can be mapped onto the memristive crossbar and automatically executed using MAGIC NOR gates, in SIMPLE (Ben Hur *et al.* 2017) and SIMPLER (Ben Hur *et al.* 2019) mapping techniques. Based on these techniques, macro-instructions of the architecture are mapped and executed in the memory, and a controller is designed to enable efficient execution of these commands, exploiting the unique properties of memristive technologies (Ben Hur and Kvatinsky 2016; Talati *et al.* 2019).

There are three modes of mMPU architecture: memory-only (where the memristive memory acts as a standard memory), accelerator (where the memory is used to process data and accelerate calculations), and hybrid (which combines both the memory-only and accelerator modes). In this chapter, we will focus on the accelerator mode.

6.2. MAGIC NOR gate

There are several logic families, each of which uniquely uses memristors to design logic gates. In the stateful logic family (Reuben *et al.* 2017), the logic gate inputs and outputs are represented by memristor resistances. One representative of the stateful logic family is the MAGIC NOR logic gate (Kvatinsky *et al.* 2014), whose structure is illustrated in Figure 6.1(a). The initial resistance of the two memristors, denoted as IN_1 and IN_2, are the logic gate inputs, and the final resistance of the memristor, denoted as OUT, is the output. To execute the MAGIC NOR gate, first, the output memristor resistance is initialized to R_{ON}. Then, the input memristors are connected to V_G, the operation voltage, and the output memristor is connected to the ground. Thereafter, the output memristor adjusts its resistance according to the voltage drop it senses. To enable reliable functionality of the logic gate, it is assumed that $R_{OFF} \gg R_{ON}$. For example, when the inputs are both "1", i.e. the R_{ON} resistance value, the voltage drop on the input memristor branch is relatively low, and therefore the voltage drop on the output memristor is greater than $\frac{V_G}{2}$, causing it to change its state from R_{ON} to R_{OFF} and matching the expected output of a NOR gate for these inputs.

The MAGIC NOR gate is memristor crossbar-compatible, i.e. its structure can be easily adapted to the memory array, and can thus be integrated into the memory without requiring many modifications, as depicted in Figure 6.1(b). Furthermore, MAGIC gates can achieve massive parallelism, as demonstrated in Figure 6.1(c). The same operation can be executed in parallel for different data in different rows by simply applying the operating voltage on the rows included in the calculation, as long as the data are aligned. Hence, computations within the mMPU are efficient, especially for single instruction multiple data (SIMD) instructions.

We developed a tool that evaluates the power consumed by MAGIC operations within the memristive crossbar. The tool calculates the voltages and currents in each memristor by performing nodal analysis and solving a matrix-vector multiplication equation. The tool considers the sneak path current phenomenon (David and Feldman 1968), occurring in crossbar arrays, in which current flows in paths other than the reading path, adding resistance in parallel to the resistance of the read cell. Hence, the tool produces a reliable result. The main advantage of the tool is that it enables the evaluation of many MAGIC operations, without building a model for them, by simply configuring several parameters. Many system parameters are configurable with this tool, for example the structure of the crossbar array (1T1R, 1S1R or pure memristive), transistor resistance, initial memristor resistances, the voltages on the wordlines and bitlines, and others. Figure 6.2 shows the outcomes of the initial voltages applied on the memristors in a 16×16 1 Transistor–1 Memristor (1T1R) crossbar array. The figure demonstrates the static analysis, i.e. performance of a single iteration. In this example, a V_G voltage was applied on columns 0 and 4, and a zero voltage was applied on column 9; other bitlines and wordlines were floating. All the memristors were initialized with a R_{ON} resistance, except for the memristors in column 0 of the even rows. The tool can also perform dynamic analysis and calculate the voltages continuously.

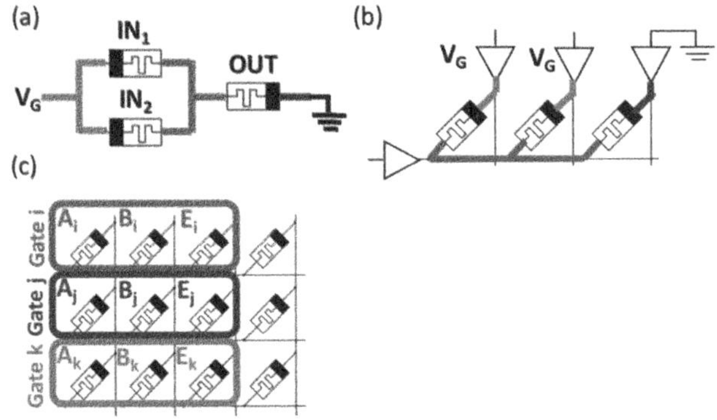

Figure 6.1. *The MAGIC NOR gate. (a) MAGIC NOR gate schematic; (b) MAGIC NOR gate within a memristive crossbar array configuration; (c) MAGIC gates operated in parallel in a memristive crossbar*

6.3. In-memory algorithms for latency reduction

Since NOR is a functionally complete system, any logic and arithmetic operation can be implemented using MAGIC NOR gates. MAGIC NOR gate-based algorithms have been developed to implement common logic functions. In Talati *et al.* (2016), a MAGIC-based N-bit full adder algorithm was proposed. The full adder is a basic logic function that lays a foundation for the implementation of more complicated useful functions, for example multiplication.

In Haj-Ali *et al.* (2018), four manually crafted algorithms for the efficient execution of fixed-point multiplication using MAGIC gates were proposed, including a full-precision, fixed-point multiplication, a limited-precision, fixed-point multiplication and area-optimized versions of the two aforementioned algorithms. These algorithms are partially based on the full adder implementation in Talati *et al.* (2016). The area-optimized versions reuse the memristors differently to reduce the area, at the cost of increasing the latency. These algorithms are used as building blocks for image processing operations, for example image convolution and the Hadamard product achieve better latency and throughput than previously reported algorithms, as well as reduce the area cost. Reducing the area and constraining the execution to a single row can achieve massive parallelism in the memristive crossbar, which will be discussed later on in this chapter.

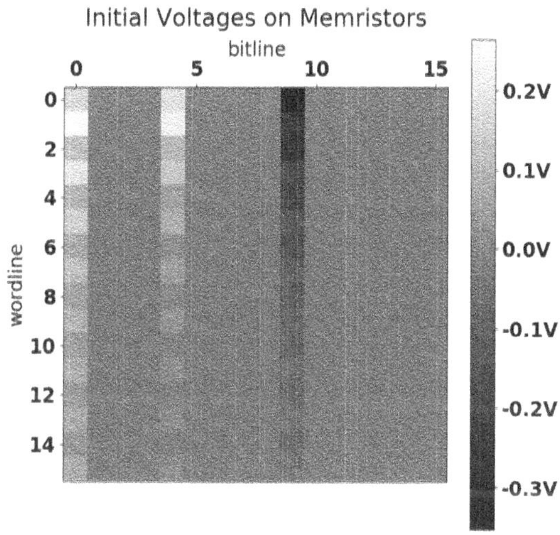

Figure 6.2. *Evaluation tool for MAGIC within crossbar arrays. The initial voltage drop on different cells in a* 16×16 *1T1R memristive crossbar array. A* $V_G = 2V$ *voltage was applied on columns 0 and 4, and a zero voltage was applied on column 9; other bitlines and wordlines were floating. All the memristors were initialized with a* $R_{ON} = 100\Omega$, *except for the memristors in column 0 of the even rows, which were initialized with* $R_{OFF} = 10M\Omega$. *Negative voltages are shaded with darker colors. Negative voltage drop indicates that the memristor is changing its state to* R_{OFF}

6.4. Synthesis and in-memory mapping methods

Manual latency- or area-optimized mapping of an algorithm onto the memristive crossbar is not trivial, since it requires the identification of the execution cycle and the location in the crossbar for each gate in the logic. In addition, there are constraints imposed by the MAGIC gate structure, which should be considered in the mapping process. Since the problem specification is complex and the search space is large, algorithm mapping is not a simple task, as exemplified by the complexity of the aforementioned algorithms.

In this section, we elaborate on different automatic logic mapping methods, which define the logic execution sequence by specifying the execution cycle and location of each gate in the logic. The methods we discuss optimize different execution parameters. One of the methods we present optimizes the latency, i.e. the number of clock cycles needed for execution, while another improves the throughput, i.e. the number of operations performed at a certain time. When a logic function is mapped onto a single row in the memristive array, it can be performed in parallel for different data in the other rows of the array, thereby achieving massive parallelism and high throughput. Other execution parameter optimizations, for example area, are also discussed in this section.

6.4.1. *SIMPLE*

The SIMPLE MAGIC (Synthesis and In-memory MaPping of Logic Execution for Memristor-Aided loGIC) tool (Ben Hur *et al.* 2017) was developed to automatically detect the execution sequence that yields the lowest latency. Its flow is shown in Figure 6.3. The system has two inputs: a .blif file, which describes the logic function to be implemented, and a customized standard cell library, which includes the gates used in the synthesis process. In our case, only NOR and NOT are used, since these gates can be easily implemented using the MAGIC mechanism.

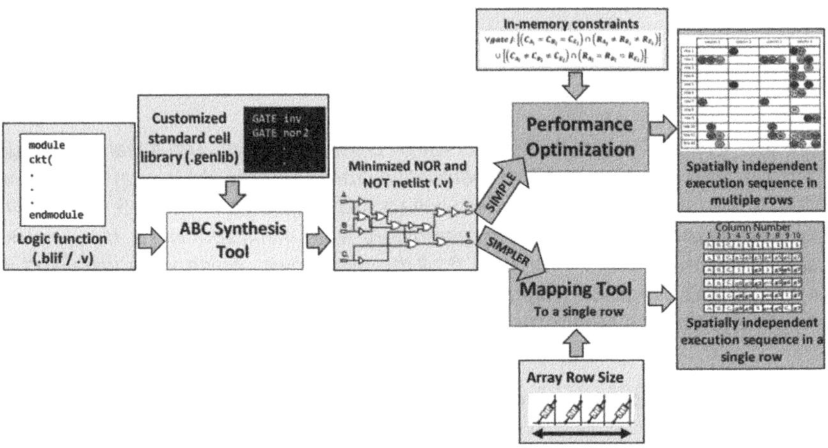

Figure 6.3. *The SIMPLE and SIMPLER flows. In both flows, the logic is synthesized to a NOT and NOR netlist, using the ABC tool. Then, the flows are split into different mapping methods. The SIMPLE flow optimizes the execution latency and maps onto several rows in the memristive array. The SIMPLER flow optimizes throughput and maps onto a single row in the memristive array*

The input logic is synthesized using the ABC tool (Mishchenko 2012) with the input library. The ABC tool outputs a NOR and NOT netlist in a Verilog format. Then, a Python script processes the netlist and generates a constraint file. This file includes all the logic-related (e.g. all the logic gates in the netlist have to operate at one of the execution clock cycles) and memristive crossbar-related (e.g. logic gates can operate in parallel only if the rows of the inputs and outputs are aligned) conditions that the solution must fulfill. The constraint file then serves as the input for an optimization tool, Z3 (De Moura and Bjørner 2008), which solves the mapping optimization problem. The Z3 solver finds the optimum of a defined function. The present work demonstrates latency optimization (i.e. the optimized function is the number of clock cycles of the execution sequence), but other parameters, for example area, can also be optimized. In order to optimize the area, the optimization function can be defined as the product of the number of used rows and number of used columns.

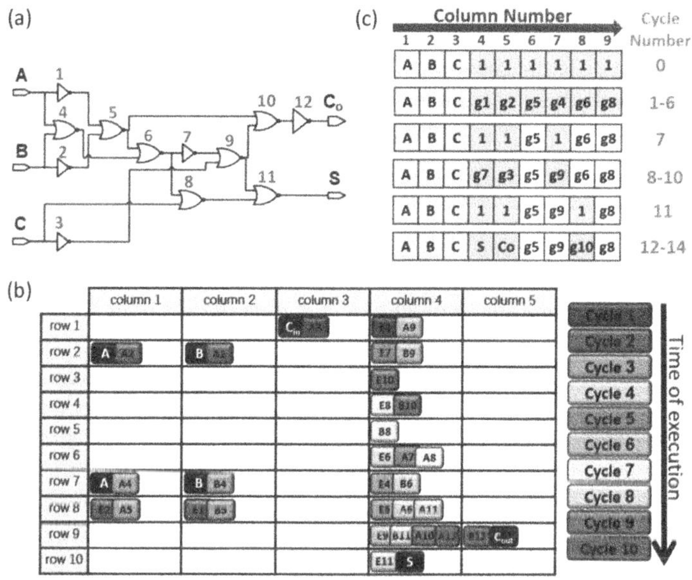

Figure 6.4. *A 1-bit full adder implementation using SIMPLER. (a) A 1-bit full adder NOT and NOR netlist, generated by the ABC tool. (b) SIMPLE execution sequence. Each clock cycle is shown in a different color. The logic gates that operate in a clock cycle are shown with the clock cycle color. (c) SIMPLER execution sequence. Green cells are initialized, and orange cells are written with a new gate output value*

Figure 6.4 illustrates a 1-bit full adder logic, using the latency-optimizing SIMPLE execution sequence. The netlist in Figure 6.4(a) shows the NOT and NOR implementation of the logic (the ABC synthesis tool output). Figure 6.4(b) shows the execution sequence of SIMPLE, based on this netlist. In some cycles, several gates operate in parallel, for example gates 4 and 5 in the third clock cycle.

At the end of the process, the output of SIMPLE describes the execution sequence of the input logic within the memristive crossbar array. The inputs and outputs of each gate are mapped onto a memristive cell in the crossbar and are executed in a defined clock cycle.

Flaws in the SIMPLE mechanism include the complexity of the optimization problem it defines, resulting in a very long runtime for medium-sized logic benchmarks (hundreds of gates). This issue can be partially addressed by eliminating one degree of freedom and specifying the exact latency instead of finding the minimum latency. The runtime can sometimes be shortened by running different processes in parallel, each of which solves a different optimization problem with a specified latency value. In addition, the SIMPLE mechanism does not limit the utilized area. Since the main advantage of the memristive crossbar is its parallelism, i.e. execution of the same logic function in different rows or columns, the level of SIMD parallelism that can be achieved is limited, thereby reducing the throughput.

6.4.2. *SIMPLER*

SIMPLER MAGIC, a tool for synthesis and mapping of in-memory logic executed in a single row to improve throughput (Ben Hur *et al.* 2019), uses a SIMD-oriented mapping technique. Its basic method is founded on the assumption that if a logic is mapped onto a single row in the memristive crossbar, a high level of parallelism can be achieved by performing the same calculation for data in different rows. Hence, SIMPLER attempts to improve the throughput and not the latency. Since the typical size of the crossbar row is 512 cells, execution of a logic using a single row is not always possible and initialization of the cells is often needed in order to enable cell reuse.

The SIMPLER flow has some similar steps to the SIMPLE flow as well as some new steps, as illustrated in Figure 6.3. SIMPLER creates a directed acyclic graph (DAG), which describes the dependencies between the different gates in the netlist. Each gate is represented by a vertex in the graph, and each wire is represented by an edge. In addition, each vertex receives two values: a fan out (FO) value, which is the number of vertices the gate output is connected to (the number of parents of that vertex), and a cell usage (CU) value, which is an estimation of the number of cells needed to execute the sub-tree, starting from this vertex. Using the graph and vertex FO and CU values, the SIMPLER algorithm determines the execution sequence and the initialization cycles of cells that are not needed and can be reused.

Figure 6.4(c) shows the execution sequence of SIMPLER, based on the netlist in Figure 6.4(a). As can be deduced from the execution sequence, the gates are executed until there are no more available cells for execution (cycle 6). Then, one cycle is dedicated to the initialization of cells which contain outputs that are no longer needed for the computation. These cells are used for the execution of other gates, and so on. As shown in Figure 6.4(c), unlike SIMPLE, SIMPLER uses only a single row for mapping. Logic execution requires 14 cycles, which is higher than the SIMPLE execution time (10 cycles). However, parallel execution of 1-bit full adder logic on different rows in the memristive crossbar achieves higher throughput than the use of several instances of the SIMPLE execution sequence in the memristive array.

Both SIMPLE and SIMPLER use existing synthesis tools to create a netlist. These tools are designed to synthesize CMOS logic and are ill-suited for memristor logic for various reasons: for example, while synthesis tools try to optimize the longest path in terms of latency, the longest path is not relevant for the MAGIC implementation, since the signals do not propagate and one NOR gate is executed per cycle. Future work will require development of a synthesis tool that is adapted to MAGIC execution.

6.5. Designing the memory controller

Using SIMPLER, macro-instructions sent from the processor to the mMPU can be split into a sequence of micro-instructions, i.e. a NOR sequence, and executed in the memory. An mMPU controller was designed to enable the execution of these sequences (Ben Hur and Kvatinsky 2016). The mMPU structure is described in Figure 6.5. To execute a command within the mMPU, the memory controller that resides in the processor sends a transaction to the mMPU controller, which splits the command into a sequence of MAGIC NOR operations and applies voltages on the columns and rows of the memristive crossbar to execute this sequence.

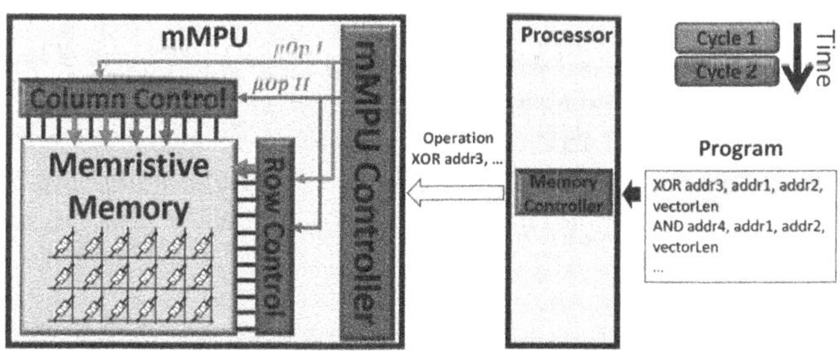

Figure 6.5. *High-level description of the mMPU architecture. A program is executed by the processor. The memory controller within the processor sends a command to the mMPU controller (yellow arrow), which splits the command into micro-operations (MAGIC NOR execution sequence). Then, in each clock cycle, the mMPU controller executes a micro-operation by applying appropriate voltages on the bitlines and wordlines of the memristive crossbar array (green and brown arrows). The command vector length field indicates the number of wordlines participating in the command calculation and operating in parallel*

The internal structure of the mMPU controller is depicted in Figure 6.6. It first interprets the instruction and, according to its type (arithmetic, SET/RESET, read or write), sends it to a suitable block for further interpretation, which determines the control signals that are applied on the crossbar wordlines and bitlines. There are four different blocks: a read block, a write block, an initialization block and an arithmetic block. The first three blocks apply voltages on the wordlines and bitlines of the crossbar array in the clock cycles in which the command operates. However, the arithmetic block operates during more than a single clock cycle, with the specific latency depending on the macro-instruction type. The arithmetic block includes a finite state machine (FSM), which handles different macro-instructions and determines the suitable control signals.

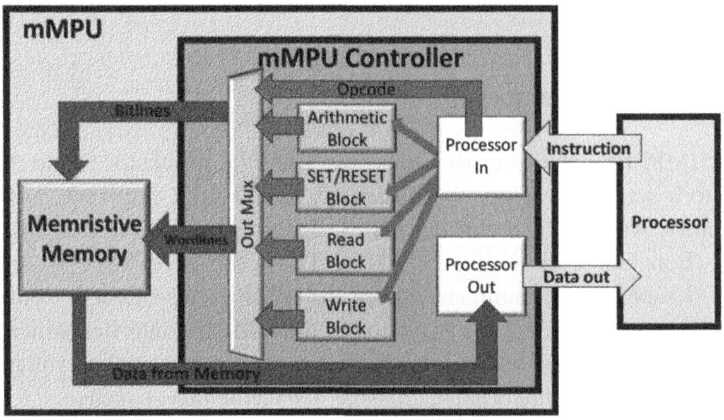

Figure 6.6. *The internal structure of the mMPU controller. First, an instruction is sent from the processor to the mMPU controller and interpreted in the processor in block. Then, it is sent for further interpretation in a suitable block (arithmetic, SET/RESET/read or write), according to the instruction type. The output of these blocks are the control signals, which are propagated to the memristive array bitlines and wordlines. The output is then transferred to the processor out unit, which sends it back to the processor*

The goal of the memory controller is to schedule the mMPU requests in an optimized way, considering the memory technology constraints. Existing memory controllers, which are designed for dynamic random access memory (DRAM), are ill-suited for memristor-based architectures since they cannot fully exploit the potential of these architectures and obey their constraints. For example, the open-page policy applied in DRAM protocols is not suitable for RRAM, which multiplexes multiple bitlines to a single sense amplifier, rendering the entire row of data unavailable in the row buffer. Furthermore, there are operations that are supported in

DRAM controllers, but which are not needed in memristive technologies, for example refresh and restore, and there are operations which are not supported in DRAM, but which are needed in memristive technologies, for example PIM operations. The use of an unsuitable memory controller can cause significant degradation in performance and energy efficiency. In Talati *et al.* (2019), the R-DDR protocol was proposed to optimize these parameters in a memristor-based architecture. The instruction-set architecture (ISA) was extended to support both standard memory operations and processing-in-memory (PIM) operations, based on MAGIC. More addresses and therefore more bits are needed to implement a PIM operation (e.g. a MAGIC NOR gate operates on three cells and therefore requires three addresses). The data field, which is not used in a PIM operation, is used for this purpose. Moreover, the row and column addresses are sent in a multiplexed fashion to reduce the pin count. The R-DDR protocol is responsible for parsing the requests queued in the controller and serving them at the appropriate time, according to the timing constraints of the used technology. For example, a read operation latency is calculated by the sum of the address decoding latency, the resistive-capacitive (RC) delay of the wires, the reading latency and the latency of the data transfer from the sense amplifier to the input/output ports. However, a MAGIC NOR operation has a different delay, which comprises the decoding and RC delays, the logic delay and the delay of pre-charging the array to the idle state in order to prepare for the next operation.

There are some parts of the mMPU architecture that still have not been implemented and are left for future work. One example is the programming model, which is a broad subject that includes several challenges. First, a method should be formulated to determine which parts of the code are executed using PIM and which parts are executed using the processor. The part of the program execution that is run in-memory can be both defined by the programmer and detected automatically by the compiler. We aspire to design a profiling mechanism that will both reduce burden on the programmer and better identify the parts of the program that can benefit from in-memory execution. Second, a compiler should be developed to support the PIM opcodes, in addition to the regular processor opcodes. Other issues, such as the cache coherency model, should also be taken into account.

6.6. Conclusion

In this chapter, the mMPU architecture was presented. The architecture was described in a bottom-up fashion, starting from the manner in which logic is executed within memory, to the design of the memory controller, which manages the memory and calculation transactions. Future work will plan and implement a programming model, which will enable the transparent use of PIM capabilities from the software code, thereby reducing the programming burden.

6.7. References

Angel Lastras-Montaño, M. and Cheng, K.-T. (2018). Resistive random-access memory based on ratioed memristors. *Nature Electronics*, 1, 466–472.

Ben Hur, R. and Kvatinsky, S. (2016). Memristive memory processing unit (MPU) controller for in-memory processing. *IEEE International Conference on the Science of Electrical Engineering (ICSEE)*, 1–5.

Ben Hur, R., Wald, N., Talati, N., and Kvatinsky, S. (2017). Simple magic: Synthesis and in-memory Mapping of logic execution for memristor-aided logic. *IEEE/ACM International Conference on Computer-Aided Design (ICCAD)*, 225–232.

Ben Hur, R., Ronen, R., Haj-Ali, A., Bhattacharjee, D., Eliahu, A., Peled, N., and Kvatinsky, S. (2019). SIMPLER MAGIC: Synthesis and mapping of in-memory logic executed in a single row to improve throughput. *IEEE Transactions on Computer-Aided Design of Integrated Circuits and Systems*, 1–1.

Chua, L. (1971). Memristor – The missing circuit element. *IEEE Transactions on Circuit Theory*, 18(5), 507–519.

David, C.A. and Feldman, B. (1968). High-speed fixed memories using large-scale integrated resistor matrices. *IEEE Transactions on Computers*, C-17(8), 721–728.

De Moura, L. and Bjørner, N. (2008). Z3: An efficient SMT solver. *Proceedings of the Theory and Practice of Software, 14th International Conference on Tools and Algorithms for the Construction and Analysis of Systems, TACAS'08/ETAPS'08*, Springer-Verlag, Berlin, Heidelberg, 337–340. Available: http://dl.acm.org/citation.cfm?id=1792734.1792766.

Haj-Ali, A., Ben Hur, R., Wald, N., Ronen, R., and Kvatinsky, S. (2018). IMAGING: In-Memory AlGorithms for Image processiNG. *IEEE Transactions on Circuits and Systems I: Regular Papers*, 65(12), 4258–4271.

Kaplan, R., Yavits, L., Ginosar, R., and Weiser, U. (2017). A resistive CAM processing-in-storage architecture for DNA sequence alignment. *IEEE Micro*, 37(4), 20–28.

Kvatinsky, S., Belousov, D., Liman, S., Satat, G., Wald, N., Friedman, E.G., Kolodny, A., and Weiser, U.C. (2014). MAGIC–memristor-aided logic. *IEEE Transactions on Circuits and Systems II: Express Briefs*, 61(11), 895–899.

Mishchenko, A. (2012). ABC: A system for sequential synthesis and verification. *Berkeley Logic Synthesis and Verification Group* [Online]. Available: http://www.eecs.berkeley.edu_alanmi/abc/.

Reuben, J., Ben Hur, R., Wald, N., Talati, N., Ali, A.H., Gaillardon, P., and Kvatinsky, S. (2017). Memristive logic: A framework for evaluation and comparison. *27th International Symposium on Power and Timing Modeling, Optimization and Simulation (PATMOS)* ,1–8.

Shafiee, A., Nag, A., Muralimanohar, N., Balasubramonian, R., Strachan, J.P., Hu, M., Williams, R.S., and Srikumar, V. (2016). Isaac: A convolutional neural network accelerator with in-situ analog arithmetic in crossbars. In *2016 ACM/IEEE 43rd Annual International Symposium on Computer Architecture (ISCA)*, 14–26.

Song, L., Zhuo, Y., Qian, X., Li, H., and Chen, Y. (2017). GraphR: Accelerating graph processing using ReRAM. *CoRR*, abs/1708.06248 [Online]. Available: http://arxiv.org/abs/1708.06248.

Strukov, D.B., Snider, G.S., Stewart, D.R., and Williams, R.S. (2008). The missing memristor found. *Nature*, 453(7191), 80–83.

Talati, N., Gupta, S., Mane, P., and Kvatinsky, S. (2016). Logic design within memristive memories using memristor-aided loGIC (MAGIC). *IEEE Transactions on Nanotechnology*, 15(4), 635–650.

Talati, N., Ha, H., Perach, B., Ronen, R., and Kvatinsky, S. (2019). CONCEPT: A column-oriented memory controller for efficient memory and PIM operations in RRAM. *IEEE Micro*, 39, 33–43.

Wong, H.P., Raoux, S., Kim, S., Liang, J., Reifenberg, J.P., Rajendran, B., Asheghi, M., and Goodson, K.E. (2010). Phase change memory. *Proceedings of the IEEE*, 98(12), 2201–2227.

7

Removing Load/Store Helpers in Dynamic Binary Translation

Antoine FARAVELON[1], Olivier GRUBER[2] and Frédéric PÉTROT[1]

[1] *Université Grenoble Alpes, CNRS, Grenoble INP, TIMA, 38000 Grenoble, France*
[2] *Université Grenoble Alpes, CNRS, Grenoble INP, LIG, 38000 Grenoble, France*

During dynamic binary translation, guest memory accesses need to be translated from guest virtual memory addresses to virtual host memory addresses, a translation that is time-consuming and greatly impacts the performance of the overall emulation. In this chapter, we propose translating a guest load/store instruction into a host load/store instruction, leveraging the host MMU to perform the address translation at native speed. This requires that the emulator maps the guest virtual address space in a region of its own address space, using a Linux kernel module to control the host MMU translation in that region. On a large spectrum of emulated code, our prototype performance ranges within 0.59–5.9 of an unmodified QEMU performance. On average, our experiments show a speedup of 67% for x86 guest and 42% for ARM guests. Besides our primary goal for better performance, we were careful to engineer our solution for minimal modifications to QEMU. While we believe our current approach shows that hardware-assisted memory emulation has great potential for near-native speed, we also feel that reaching the full potential of the approach will require an in-depth redesign of the engineering of memory emulation in emulators such as QEMU.

For a color version of all figures in this book, see www.iste.co.uk/andrade/multi1.zip.

Multi-Processor System-on-Chip 1 – Architectures,
coordinated by Liliana ANDRADE and Frédéric ROUSSEAU.

7.1. Introduction

The emulation of one instruction-set architecture (ISA) on another is required in many situations. For hardware designers, they can simulate a new ISA, evaluate its performances and its correctness, before the actual hardware is produced. For embedded software developers, using an emulator is much easier than using the actual hardware, as illustrated with the success of the Android emulator based on QEMU (Bellard 2005). The challenge is the performance of the emulation. Major techniques are instruction interpretation (Wilkes 1969), threaded code (Bell 1973), dynamic binary translation (Deutsch and Schiffman 1984) and compiled emulation (Mills *et al.* 1991). Dynamic binary translation (DBT) provides the fastest emulation for general-purpose code (Cmelik and Keppel 1994; Witchel and Rosenblum 1996; Altman *et al.* 2001), handling both regular application code and operating system code. Nevertheless, its performance can be improved, reducing the overhead of emulating memory accesses, an overhead than can take a significant portion of the overall emulation time, as shown in (Tong *et al.* 2015) and confirmed by our own results, discussed in section 7.5.1.

Emulating memory accesses is expensive for two reasons. The first is quite simple: memory accesses are many in the emulated code, sometimes as many as one memory access to every other set of instructions, according to (John *et al.* 1995), and confirmed by our own experiments, as we will see later. The second is the cost of the emulation of the address translation mechanism, done in software rather than in hardware. Indeed, the emulated instruction-set architecture contains a memory management unit (MMU) in charge of address translation, a translation that has now to be emulated in software. Our ultimate goal is to remove this overhead entirely by emulating the guest address translation with the host hardware.

In the context of DBT, this address translation requires both a code translation and some runtime support. First, the guest code that respects the *guest ISA* must be translated to host code that respects the *host ISA*. Therefore, before executing any guest instruction, the emulator translates each guest instruction into a snippet of one or more host instructions. These snippets of code will be executed natively on the host hardware, emulating the operational semantics of their corresponding guest instruction. For memory accesses, the translation relies on *helper functions*. Every single memory access, either a read or a write, is translated into a call to a corresponding helper function. The helper function will emulate the address translation and then execute the read or write at the proper address in the emulated guest memory within the architected state maintained by the emulator.

Our proposal is to translate guest memory access instructions into native host memory access instructions, for instance, translate guest ARM load (`ldr`) and store (`str`) instructions into Intel x86_64 `mov` instructions. This proposal requires the emulator to coerce the host MMU to emulate the correct guest address translation. For the first step, we need to carefully architect the overall memory layout of the emulator process, making sure that we can map the guest address space as a single contiguous region within that process. In Linux, this is done by a call to `mmap`, using our own Linux kernel module to back up the mapped region, thereby having full access to the host MMU from within the Linux kernel. Thus, we can set up and maintain the host MMU address translation for that region as we see fit.

For the second step, in order to know which translation is necessary, we need to properly emulate any guest modification of the guest MMU setting. This is done by all emulators, but, in our case, we also need to funnel this information down to our module in the Linux kernel and accordingly update the host MMU translation. To do this, the emulator communicates with our kernel module via *ioctls*. To percolate memory-related traps up to the emulator, our kernel module installs its own trap handlers and manipulates the return-to-user sequence in order to invoke specific trap handlers in the emulator code in user mode.

We have implemented our proposal in QEMU for the Linux on x86_64 host and both x86 and ARM guests. Our prototype performance ranges within 0.59–5.9 of the unmodified QEMU performance on a large spectrum of emulated code. On average, our experiments show a speedup of 67% for x86 guest and 42% for ARM guests. The significant increases in speed are explained by three gains. First, address translation is done in hardware rather than in software. Second, memory accesses remain just memory accesses, executing no other instructions and therefore without any negative impacts on the instruction cache locality. Third, the emulation fully benefits from the host hardware data caches.

The reductions in speed are explained by a higher MMU management overhead in some workloads, with the communication overhead between the emulator and our kernel module counter-balancing the performance gain of relying on native address translation. This communication overhead is largely due to our goal of leaving QEMU mostly untouched. We flush our mappings too often and too much, because we could not hook QEMU memory emulation properly for minimal and timely flushes. Also, the original design was to handle page faults by percolating them up to QEMU, but the overhead is high, especially if we flush too much and too often. Here we report on a proof of concept that shows the potential of complete memory emulation within our kernel module, but we could not coerce the QEMU codebase to do what was necessary and were severely limited in what we could do. Nevertheless, the results show that it is a valid path towards near-native speed memory emulation if new versions of QEMU would re-engineer the memory emulation subsystem accordingly.

Our approach has also a few limitations in its current implementation that may limit its performance in specific settings. First, when the guest and host have different endianness, we need larger translated snippets of code, but this is true for all emulator technologies. Second, if the guest and host have radically different MMU features, coercing the host MMU might prove more difficult. Third, the emulation of hardware devices may prove costly since every load or store to memory-mapped hardware registers require a trap to the emulator code in order to carry the proper emulation of the device. We have great hope that these limitations can be addressed with para-virtualized guests, as we will discuss in our conclusion. We believe this work, combined with our earlier work (Faravelon *et al.* 2017), shows a new research path towards near-native speed memory emulation for dynamic binary translation.

This chapter is organized as follows. Section 7.2 gives an overview on what is necessary, in a DBT-based emulator, to emulate memory accesses. Section 7.3 details the design of our solution, and section 7.4 details its implementation in QEMU/Linux. Then, section 7.5 evaluates the performance of our solution. Finally, section 7.6 presents correlated works and section 7.7 concludes, discussing promising research perspectives.

7.2. Emulating memory accesses

Emulating memory accesses in the context of DBT is complex, so in this section we have summarized the key elements that are necessary to understand our proposal. For the sake of clarity, we will focus on a 32-bit ARM processor with a standard architected MMU for the guest ISA and on the Intel x86_64 machine for the host ISA. Both the guest and the host software stacks are based on ArchLinux with Linux kernel version 4.14.15.

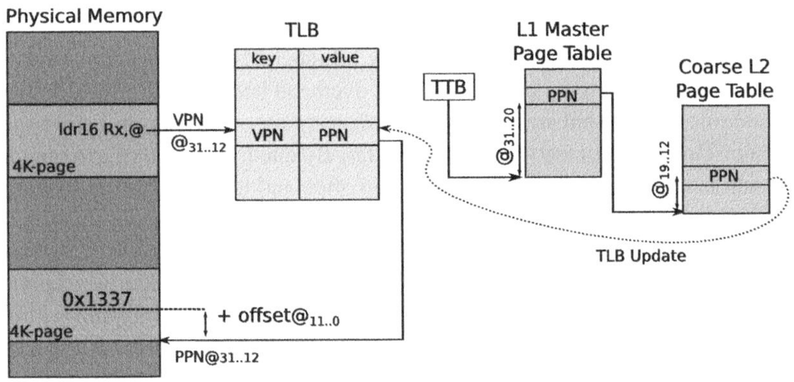

Figure 7.1. *Address translation for the ARMv7 architecture*

From the guest perspective, depicted in Figure 7.1, everything happens as if the emulator was not there. Guest assembly instructions are executed as expected on the guest hardware. Regarding memory accesses, the ARMv7 `ldr` and `str` instructions are executed as expected, manipulating either physical or virtual address. If virtual, this means that the guest code has set up and enabled the guest MMU, with a *translation lookaside buffer* (TLB) and hierarchical page table with 4K pages (the smallest granularity for ARMv7). For clarity, we represented the page table as its own data structure on the right-hand side of the figure, but it is, of course, composed of regular pages within the physical memory of the guest. When accessing memory, the guest MMU first extracts the virtual page number (VPN), given by bits 31 to 12 of the virtual address and the offset, given by bits 11 to 0. Then, via the two-level page table, it translates the VPN into a physical page number (PPN).

Figure 7.1 also shows the execution of a guest `ldr` instruction. The execution starts with the processor fetching, decoding and issuing the instruction for a page in the physical memory. The address translation first accesses the *TLB*, a small fully associative memory that caches the most recent virtual-to-physical translations, to obtain the PPN. If the address translation is not cached, a TLB miss occurs and the hardware traverses the page table in memory. The page table is identified by the translation table base (TTB) register and set up by the guest operating system (OS). If the translation is found, then the TLB is updated and the instruction finishes. If not, a page fault exception is raised and it must be handled by the guest OS. Finally, the memory access happens, reading the 16-bit value (`0x1337`) from the corresponding page in the physical memory.

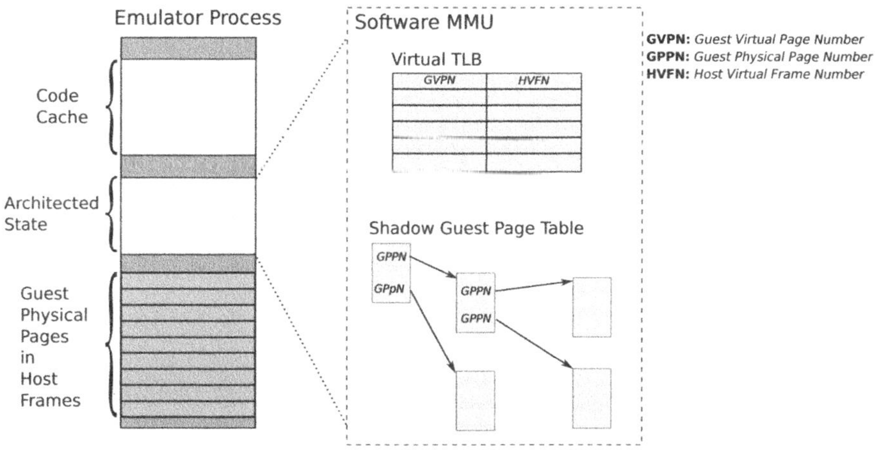

Figure 7.2. *Host view of the architected state of the guest*

In contrast, Figure 7.2 shows the host perspective, with the emulator being a regular Linux process with its own virtual memory, visible on the left-hand side of the figure. In this Linux process, the emulator manages three main data structures: the *architected state* of the emulated guest, the guest physical frames and the code cache with the translated guest code blocks. The architected state represents the state of the emulated guest, such as the guest processor state, the register values or the different device states. In particular, the architected state includes the software MMU that emulates the guest MMU, mainly composed of the virtual TLB and the shadow page table, both detailed on the right-hand side of the figure.

Let us consider the emulation of the same `ldr` guest instruction, discussed just above. The instruction is stored in a page in the physical memory of the guest, but it is not executed there. The instruction is translated into a snippet of x86_64 instructions emulating the operational semantics of the instruction when executed by the host. The code snippet is stored in the code cache. The host only executes instructions from the code cache or from the emulator code. In particular, code snippets often rely on calling *helper functions* that are part of the emulator code and perform various emulation tasks. The helper function for a `ldr` instruction walks through the page table of the software MMU in order to translate the address of the `ldr` instruction. If the guest MMU is enabled, the guest address is a guest virtual address that needs to be translated first to a guest physical address and then translated to a host virtual address to access the correct guest physical frame allocated within the emulator process. The code shown in Figure 7.3 gives an idea of the helper for the emulation of a `ldr` instruction.

```
1  uint32_t ldr_helper(uint32_t addr) {
2      uint32_t hvpn;
3      uint32_t gppn;
4      // extract the guest virtual page number and offset
5      uint32_t gvpn = guest_virtual_page_number(addr);
6      uint32_t offset = guest_virtual_offset(addr);
7      // access the virtual TLB, update if necessary
8      uint32 index = hash(gvpn);
9      if (virtual_tlb[index].key != gvpn) {
10         gppn = shadow_page_table_walk(cpu->ttb, gvpn);
11         hvfn = guest_physical_to_host_virtual(gppn);
12         HT[index].entry = hvfn;
13     } else
14         hvfn = HT[index].entry;
15     // do the guest memory access, reading the 32-bit value
16     return *(uint32_t*)(hvfn + offset);
17 }
```

Figure 7.3. *Pseudo-code of the helper for the* `ldr` *instruction*

The helper first extracts, from the guest address, both the guest virtual page number (GVPN) and the offset in that page. If then it attempts to find the translation in a small hash table (the Virtual TLB) that caches the most recent translations. This means computing the index and checking whether the translation at this index corresponds to a GVPN. If not, the helper walks through the shadow guest page table (line 10) to get the guest physical page number (GPPN), then translates it (line 11) to the host virtual frame number (HVFN) and finally updates the TLB (line 12). The host virtual frames are the page-size and page-aligned chunks of host virtual memory that the emulator allocates to hold the contents of guest physical pages. In most emulators, the host virtual frames correspond to virtual pages in the emulator Linux process that were allocated at startup. Finally, line 16, the helper does the memory access.

So far we have overlooked one important role of the software MMU, the emulation of memory-related traps that may occur during the address translation or the actual memory access. Considering the guest MMU is turned on, the translation of guest virtual addresses by the guest MMU may fail due to a missing mapping or an invalid access right. Furthermore, the resulting load and store operation in the guest physical memory may fail because of an invalid physical address in the guest *memory map*. Of course, load and store operations on an invalid address may also occur when the guest MMU is turned off. Overall, this means that the emulator must correctly emulate the guest MMU and system bus, emulating the appropriate traps and the corresponding branching through the guest vector of guest trap handlers.

This complex emulation of the guest MMU is the traditional approach and works well in practice, but it represents a major part of the emulation overhead, from 60% to 80% according to our own evaluations (Faravelon 2018). First, the overhead introduced by helper calls is significant. To reduce that overhead, modern emulators optimize the code translation with a *software TLB*, inlining a fast path and only making helper calls on the slow path.

Figure 7.4 illustrates this process. The inlined fast path, on the left-hand side of the figure, looks up into the Virtual TLB. The Virtual TLB improves performance, but it is not ideal. First, the fast path is several instructions long with several memory accesses and a branch, which is many more instructions than the original single guest load or store instruction. Furthermore, whenever the fast path fails, the slow path, given on the right-hand side of the figure, is still a full call to the helper function, with the full address translation done in software. Even with this optimization in place, according to (Tong *et al.* 2015), the emulation of the virtual-to-physical address translation still represents 38% of the execution time on a set of representative benchmarks.

Let us discuss the fast and slow paths of Figure 7.4 in more detail. The fast path starts with computing the accessed memory address (lines 0 and 1) and places it in the register `%ebp`. Among those, the *operation index* constant (noted `$oi`) at line 2 of the slow path trampoline code is a constant generated at translation time that indicates if the access is a read or a write, its endianness, signedness and size in number of bytes. The helper needs this information to check if the memory access is valid and to actually perform the memory access in the emulated physical memory of the guest. The execution cost of the helper is significant in itself as well as because it pollutes both the data and instruction caches of the host hardware. The host data cache is polluted because of the page table walk. The host instruction cache is polluted because the code of the software MMU is fairly large.

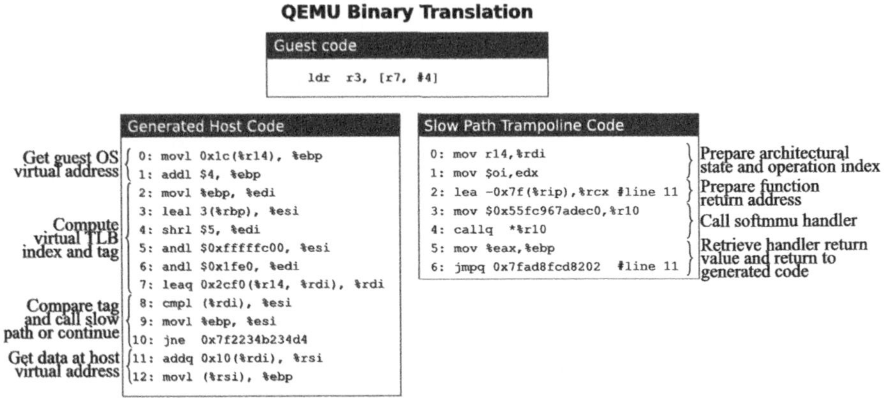

Figure 7.4. QEMU*-generated code to perform a load instruction*

7.3. Design of our solution

The goal of our solution is to strive toward native performance for memory accesses, translating guest memory accesses into host memory accesses. With an ARM guest and an Intel x86_64 host, this means translating a `ldr` or a `str` instruction into a single `mov` instruction. Our design relies on two key points: (1) mapping the guest address space in a region of the virtual address space of the emulator process and (2) control the mapping of that region via the host MMU so that it correctly emulates the guest address translation at all times.

Figure 7.5 shows how we extended QEMU with our new mapping of the guest address space within the Linux process of the emulator. In Figure 7.5(a), we see, in green, the usual QEMU translation that is still available. Let us consider the guest virtual page at guest virtual address 0x2000, mapped to the guest physical page 0x5000 by the guest MMU and stored by QEMU in the host virtual frame at 0x106000 (visible as the green-colored host virtual frame and labeled *Page-2*). This means that QEMU will translate any guest memory access to the guest virtual page [0x2000+`offset`], using its Virtual TLB and Soft MMU, to a host memory access at the address [0x106000+`offset`]. This green-colored translation remains the one in use by the various helpers used by QEMU for the emulation. In red, we see the new mapping of the guest address space at the *mapping base address* (MBA), as a 4GB contiguous region within the virtual address space of the emulator process. This means that the guest virtual page 0x2000 will be mapped at host virtual address `MBA`+0x2000. In the figure, the MBA is at 0x10000000, which means the guest virtual page 0x2000 will be mapped at 0x10002000. With this mapping in place, guest memory accesses can now be translated to native host memory accesses, just like this:

ldr r1, [r0] → mov MBA(r0), r1
str r1, [r0] → mov r1, MBA(r0)

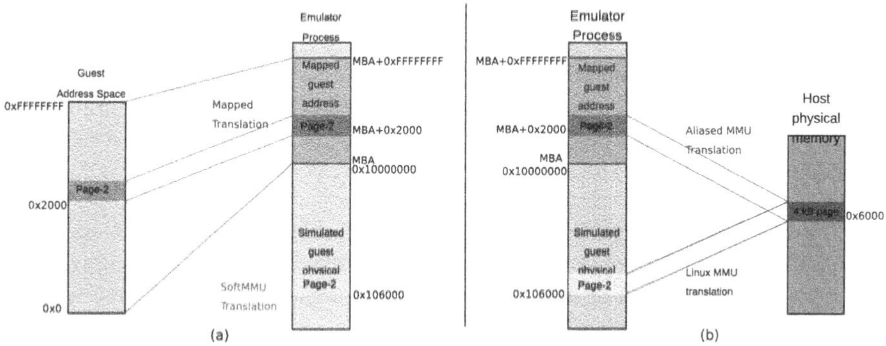

Figure 7.5. *Embedding guest address space*

Figure 7.5(b) describes the host MMU perspective of the emulator process. The host virtual frame at the virtual address 0x106000 is mapped by the host MMU to the host physical page 0x6000, something that is entirely managed by the Linux kernel since the emulator runs on a regular Linux process. However, we alias the host physical frame 0x6000 from both the host virtual address 0x106000 and 0x10002000. How do we know this aliasing? First, we simply ask the QEMU Soft MMU to translate from guest virtual 0x2000 to guest physical 0x5000 and then to host virtual 0x106000. Then, we set up the host MMU to alias 0x106000 and MBA+0x2000, which is 0x10002000. This means that the host MMU will translate the virtual address 0x106000 and 0x10002000 to the same host physical address 0x6000. Hence, reading or writing either [0x106000+offset] or [0x1002000+offset] is actually performing the same memory access, emulating the guest memory access at [0x5000+offset] (guest physical address) or [0x2000+offset] (guest virtual address).

Now that the translated guest code directly uses host memory accesses, within our mapped region, we must pay special attention to access rights. It is now the responsibility of the host MMU to check the validity of emulated memory accesses. Following up on our example, this means that we need to set the access rights to the host virtual page MBA+0x2000 in accordance with the emulated access rights on the guest virtual page 0x2000. Failure to do so may permit illegal guest memory accesses to succeed where they should be trapped. Fortunately, most of the available MMUs are fairly similar and using host MMU access rights to emulate guest MMU access rights is fairly straightforward. It is, however, true that arcane guest MMU may be challenging; we will discuss some of the specifics later in this chapter.

One design point deserves more clarity: the effects of having the guest MMU enabled or disabled. In fact, our design is fairly oblivious to whether the guest MMU is enabled or not. First, our aliasing mechanism is totally oblivious to the type of guest address space in use. If the guest MMU is enabled, it works exactly as explained above, with the guest address `0x2000` being a virtual address. If the guest MMU is disabled, then the only difference is that the guest address `0x2000` would be a physical address. It would therefore be the guest physical page `0x2000` that sits in the host virtual frame `0x106000`. In either case, the translated guest code would read and write at `[MBA+0x2000+offset]` and experience near-native performance.

However appealing our design seems, its practicality still raises a few technical questions. First, the control of the host MMU is privileged on most operating systems, putting it out of the reach of the emulator code. Fortunately, most operating systems are also extensible with user-defined kernel modules and kernel modules have full access to the host MMU; more details on this are presented in section 7.4. Second, the respect of the guest semantics may sometimes be tricky. Most of the time, however, our solution works as described above, with only a single `mov` instruction per translated `ldr` or `str` instruction. But, depending on the guest and host processors, the translated snippet of code might require more instructions than just a single `mov` instruction. For example, simulating different access size or sign extension might be tricky, but it can usually be done by choosing the right host instructions. Also, if the guest and host do not share the same endianness, we will need a few more instructions to shuffle bytes. In some cases, the host processor has an instruction for that, such as the Intel PSHUFB (packed shuffle bytes).

Handling the tricky cases above is not specific to our solution; all DBT techniques face a similar extra overhead in order to handle these cases. However, our approach requires us to pay special attention in order to handle processor modes and the related traps. Indeed, we cannot rely on the host MMU to do the checks related to the guest processor mode (kernel or user), since the emulator always executes in user mode (being a regular Linux process). Thus, we cannot set the privilege protection bit for guest physical pages holding guest kernel pages; if we did, the emulator code could no longer access these pages without triggering an access-right violation. But if we do not set the privilege protection bit, guest pages that should be protected as being privileged would not be protected from unprivileged guest accesses. This seems like a lose–lose situation. But this is not all. The emulation of hardware devices poses a performance problem. The problem comes from memory-mapped input–output hardware registers, commonly called MMIO registers. This means that device drivers are reading and writing such MMIO registers and every such memory access must be emulated, individually, based on each emulated hardware device. This means that these read and write operations are not memory reads and writes, which means that the host MMU would need to trap upon each read or write instruction regarding an MMIO register, percolating the page fault up to the emulator for proper emulation.

Fortunately, there is a solution: the idea is to optimize memory accesses when it is efficient to do so and to use the standard approach otherwise. In other words, we will gain when we can and lose nothing otherwise. The problem is that it is a decision that we need to take at translation time. Indeed, the idea would be for the emulator to translate a memory access as it would normally, if that memory access is expected to be trapping. Otherwise, the emulator would generate code with a single-host memory access. Technically, it is quite easy to translate a code block either one way or the other, or even consider each memory access individually. The hard part is to know, at translation time, when to translate one way or the other. In general, the necessary knowledge is not available statically at translation time, but it so happens that we can leverage the processor state to decide. Indeed, MMIO registers are read or written solely when the processor is in kernel mode, since all device drivers execute in privileged mode. This also means that when the emulator translates a code block, it knows if the processor is in user mode or kernel mode, which happens to be a simple and efficient way to decide how to translate memory accesses. In kernel mode, the translation will be done as usual. In user mode, the translation will rely on our approach.

7.4. Implementation

This section describes our prototype based on Linux and QEMU, on x86_64 processors for the host. Overall, we were able to leave QEMU mostly untouched, with only a few modifications to the dynamic binary translation. Of course, we added our mapping of the guest address space. Regarding Linux, we added our own Linux kernel module to control the host MMU and to handle page faults in our mapped region.

7.4.1. *Kernel module*

At load time, our kernel module creates a virtual memory region to map the guest address space. To create the virtual memory region, our module relies on the Linux mechanism called *virtual memory area* (VMA). A VMA is a contiguous range of virtual memory addresses in a process for which we may define a specific exception handler as part of a kernel module. That exception handler is invoked by Linux whenever an MMU-related exception occurs as a result of a memory access within the address range of the VMA. In our implementation, our exception handler will percolate the exception up to our user-land QEMU *handler*.

It is important to note that Linux invokes the exception handler on the thread that attempted the memory access that induced the hardware MMU-related exception.

In other words, the exception handler executes on the QEMU thread that executes the translated guest code. To percolate up the exception, our exception handler manipulates the Linux thread context so that the *iret instruction* (interrupt return) will not resume the execution of the guest code at the instruction that caused the exception, but at the start address of our user-land handler that we added to QEMU. That QEMU *handler* will interact with the rest of QEMU to determine what needs to be done about this hardware MMU-related exception.

There are two situations that our QEMU handler must consider, as depicted in Figure 7.6. Either the MMU-related exception is a page fault that the emulator must resolve internally or it is a page fault that concerns the guest operating system. With internal page faults, the QEMU handler needs to manipulate the host MMU via `ioctls`. For example, it could be setting up a missing aliasing or changing access rights via `ioctls`. Once our QEMU handler has resolved the MMU-related exception, it will retry the emulation of guest instruction that raised that exception. However, if the page fault was not an internal issue, then our QEMU handler cannot and should not resolve the MMU-related exception, and it must delegate the entire emulation of the corresponding guest trap to QEMU.

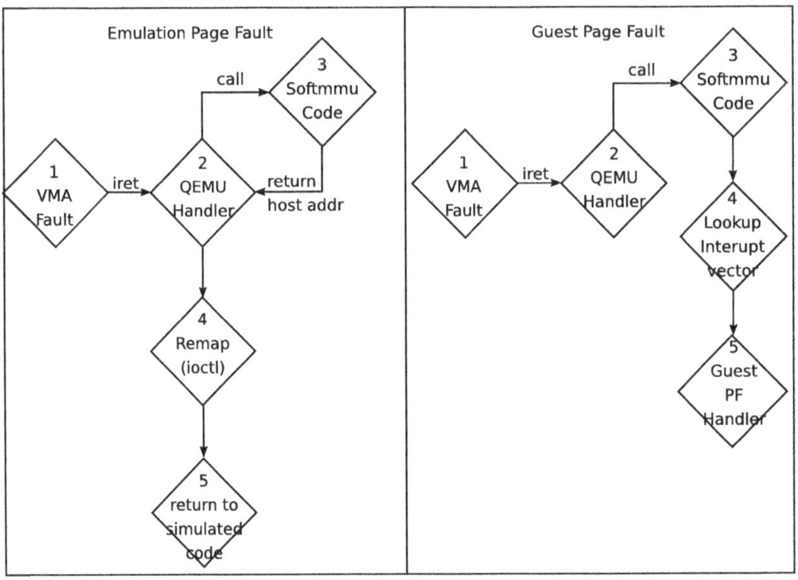

Figure 7.6. *Overview of the implementation*

For this whole process to work, our QEMU handler must be aware of the current state of the guest MMU, which includes the knowledge of the page table. To have this knowledge, our handler must accurately track the emulation of the following guest operations:

– guest enabling and disabling the MMU;

– guest applying changes to the current page table;

– guest switching page tables.

Fortunately, QEMU already tracks these guest operations accurately, so the question becomes where should we hook in QEMU to update the mapping of our VMA region? The answer is, in fact, quite simple: whenever QEMU flushes its Virtual TLB, since it is essentially the same problem. Since QEMU flushes its Virtual TLB quite often, we really need the corresponding flush of our VMA mapping to be fast. Unfortunately, the invalidation of the entire mapping of a 4GB VMA region by Linux is quite expensive, because it is essentially a loop over all the page entries of the host page table. To avoid this entire scan and reduce the overhead of our flush operation, our module maintains its own list of page entries that are currently mapped within our VMA region. Now, flushing our VMA region only requires scanning that shorter list, asking Linux to invalidate individual pages. Since we invalidate often, at least at each guest context switch, the list remains small and the flush operation is considerably faster.

7.4.2. *Dynamic binary translation*

With the mapping of the guest address space in place, the dynamic binary translation can now be changed regarding memory accesses. Figure 7.7 contrasts the original binary translation in QEMU and ours for a guest `ldr` instruction. Note that considering an `str` instruction would be essentially identical.

The left-hand side of Figure 7.7 reproduces the code of the fast path and slow path that we already detailed Figure 7.4. On the right-hand side of Figure 7.7, we show the code that our approach generates for the same memory access when the guest processor is in user mode. As we can see, we have not fully reached our goal that was to translate one `ldr` or `str` instruction to a single `mov` instruction, but we are quite close. We have eliminated both the fast and slow paths, keeping only two extra instructions. The first extra instruction, line 3, is the loading of the `MBA` offset in a register, so that we can compute the memory access address as explained in section 7.3. The other instruction, line 2, loads the operation index that concerns this specific memory access. We do this in case the host MMU raises an exception. That way, the exception handler in our kernel module can find the operation index in the host register values saved by the Linux trap mechanism. This is necessary because we need to percolate that operation index up to QEMU.

In fact, our kernel module passes the same arguments to our QEMU handler as the arguments passed to the regular QEMU slow-path helper. This makes sense since our QEMU handler interacts with the rest of QEMU in pretty much the same way as the original slow path. This is one of the main design decisions that allows our approach to be minimally intrusive with respect to QEMU. Indeed, our handler reuses the unmodified Soft MMU from QEMU. This means, in particular, that the load or store operation that caused the page fault will be actually performed by the Soft MMU, as it would normally. The only difference is that our QEMU handler does not update the Virtual TLB, since our translation of guest user-mode code no longer relies on the Virtual TLB. Instead, our QEMU handler communicates back with our kernel module, via a `remap ioctl`, to update the host MMU translation for the translated address. Our QEMU handler finally resumes the emulation at the next instruction in the translated guest code.

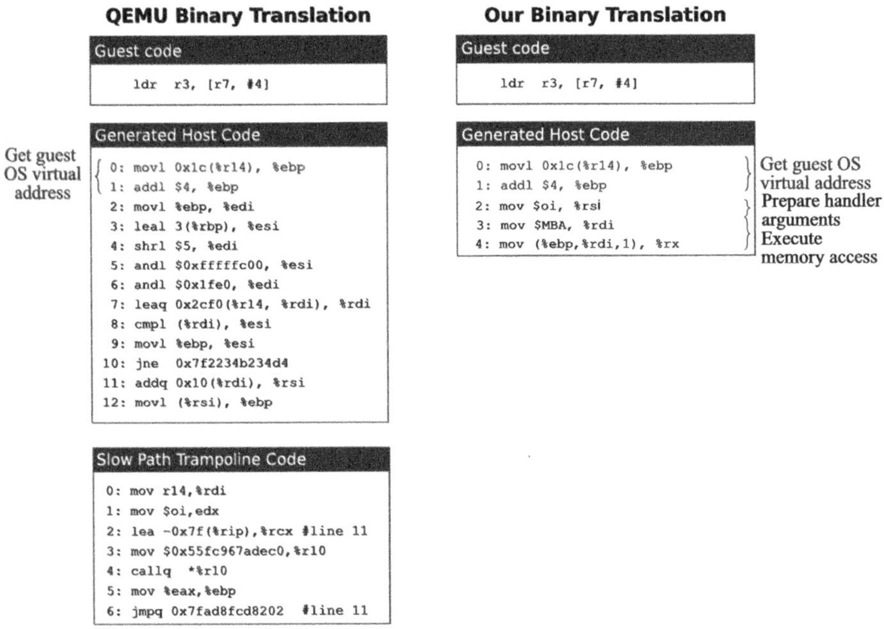

Figure 7.7. *Contrasting memory access binary translations*

This last point is a bit challenging on CISC hosts such as Intel x86_64 processors. Indeed, QEMU generates different `mov` instructions with different instruction length in bytes. Therefore, computing the address of the next instructions to resume the execution at would normally require decoding the `mov` instruction that was trapped. We decided against decoding, and we opted for padding all mov instructions to 8 bytes, generating no-ops for any extra byte. This does not affect performance in practice because QEMU generates `mov` instructions that are usually between 6 and 8 bytes long.

7.4.3. *Optimizing our slow path*

Since our implementation flushes the guest memory mapping when QEMU flushes its Virtual TLB, every first memory access in any page after each Virtual TLB flush goes through a host page fault, a page fault that must be percolated up to QEMU, which is relatively costly. Therefore, we aimed to reduce the cost of page faults, pursuing the idea that we could avoid percolating them up to QEMU and we could resolve them entirely within our kernel module. As it turned out, we could only partially reach our goal, but as our evaluation will show, the performance gain is still significant. To go further would require much deeper changes to QEMU, which would be interesting future research: seeking to adapt the design of memory emulation in order to rely more on leveraging the host MMU.

As a first step, we were only able to optimize page faults induced by memory read operations, but not entirely those induced by memory write operations. This is purely an engineering issue with QEMU. By design, QEMU requires us to trap for the first write on any page after a Virtual TLB flush, even if a read access has already happened. For this reason, we mark pages as read-only upon page faults induced by a read access; that way, the host MMU will trap them again on the first write access, a trap that will be percolated up to QEMU. After that, we obey QEMU with respect to the setting of access rights on mapped guest pages. Therefore, we still sustain quite a high number of unnecessary page faults that we have to percolate up to QEMU. Hence, we feel this line of optimization has potential for even better speed-ups.

But let us backtrack a bit, we have not discussed how we could resolve page faults entirely within our kernel module. Indeed, to resolve page faults internally, our kernel module must be able to do the full address translation, from a guest virtual address to the corresponding guest physical page and then to the corresponding host virtual frame. To achieve this translation, it needs the following two translation tables:

– a *shadow guest page table* that maps guest virtual pages to guest physical pages;

– a QEMU *frame table* that maps guest physical pages to host virtual frames.

The shadow page table is the one managed by QEMU that our kernel module directly accesses using the Linux support for accessing user data from kernel code: the functions `getuser` and `copyfromuser`. Fortunately, QEMU emulates the guest page table, using a fairly standard tree of guest physical pages, resembling page tables architected for hardware MMUs. Therefore, it is fairly easy for our kernel module to walk through that shadow guest page table, but the only delicate point is that the shadow guest page table is managed within guest physical pages and thus linked via guest physical addresses. This is where our QEMU frame table comes in to play, translating guest physical addresses to host virtual addresses, which is compatible with using the Linux user access functions mentioned above.

The QEMU frame table is implemented as a contiguous array, indexed by the guest physical page number that provides the host virtual address where that particular guest physical page can be found in the QEMU process. Therefore, for instance, let us assume that QEMU allocated the guest physical page 0x00000000 in the host virtual frame at the address 0x2F003000. Since the guest physical page index is zero in this case, the array contains the address 0x2F003000 at index 0. The QEMU frame table is allocated once and for all, when our kernel module is loaded. Its content is set up by hooking QEMU code where it allocates a host virtual frame to hold a guest physical page, which happens only once, at startup. Indeed, with 32-bit targets, QEMU allocates all guest physical pages when initializing the guest architected state at startup. The frame table size is therefore 8MB when emulating 32-bit guests with 4KiB pages (2^{20} entries of 8 bytes), which is reasonable[1].

```
   Data: pf: page fault information, including register values
   Result: update embedded guest mapping or forward fault to simulator
1  if isReadAccess() then
2      guest_addr ← pf.addr − MBA
3      guest_phys_addr ← guest_page_table(guest_addr)
4      if is_valid(guest_phys_addr) then
5          host_virtual_frame ← qemu_frame_table(guest_phys_addr)
6          if host_virtual_frame != NULL then
7              setup_host_mmu_alias(pf.addr, host_virtual_frame)
8              return
9          end
10     end
11 end
12 pf.reg[pc] ← qemu_handler
13 return
```

Figure 7.8. *Kernel module page fault handler*

With these two tables in place, our host-MMU-exception handler in our kernel module can service page faults, as described in the algorithm shown in Figure 7.8. First, line 1, the handler determines if the memory access that provoked the page fault was a read or write operation; the handler can easily determine this since the translated guest code preloads a description of the operation in one of the processor registers – the operation index explained earlier for Figure 7.7. If the access is a write access, we percolate it up to QEMU for proper emulation, as explained in section 7.4.1. This happens, line 12, by setting up the address of our QEMU handler as the address that Linux will use to resume the execution in user mode.

1. This design would not scale for the emulation of 64-bit guests, but a design using a hash map would.

If it is a read access, the handler tries to resolve the page fault internally. Line 2, it translates the address that caused the MMU exception to the corresponding guest address, subtracting the mapping base address (MBA) discussed earlier. Line 3, it uses the shadow guest page table in order to translate the guest address to a guest physical address. Line 4, it verifies that the guest physical address is valid because the requested address translation may be missing in the guest page table or read access on the page may not be granted. Line 5, the valid guest physical address is translated to the address of the corresponding host virtual page, using the QEMU frame table. This translation may also fail if the guest physical address is not a valid address in the guest memory map or it is a page holding memory-mapped hardware registers for which QEMU does not allocate host virtual pages. In case of any translation failure, the read access must be percolated up to QEMU for proper emulation. However, if the address translation succeeds, our handler has the host virtual address corresponding to the guest physical page. From there, it is easy to establish the needed aliasing in the host MMU (line 7). At this point, our handler is done – the page fault has been handled internally, without percolating the page fault up to QEMU. As explained earlier, the page is left write-protected in the host MMU so that the first write access after a flush can be percolated up to QEMU. Nevertheless, we expected this optimization to pay off and our evaluation confirms it.

7.5. Evaluation

The evaluation is done with QEMU, running on Intel x86_64 host. We evaluated two Linux guests, one for ARM and one for Intel x86 (i386), both 32-bit guests, since our prototype currently only handles 32-bit guests over a 64-bit Intel host. The configuration gcc details of the host machine are given in Table 7.1.

CPU	Intel Xeon E3 1240V2 @3.4GHz
RAM	16G @ 1600Mhz
OS	Archlinux (kernel 4.14.15)
QEMU version	git January 24, 2018

Table 7.1. *Setup details*

The Linux guests execute entirely in memory (*initramfs*) in order to focus the evaluation on the emulation of memory accesses and not the emulation of disk operations. Above our Linux guests, we execute both computational and system benchmarks. On the one hand, the Polybench suite (Pouchet 2012) provides mostly *computational* programs. On the other hand, we used three home-grown *system* benchmarks. The first, called `compile`, compiles the Polybench suite using the GNU compiler (`gcc`) with full optimization (`-O3`) and link time optimization (LTO) enabled. The other compresses the `/usr` directory, using the utility `tar` followed by compression using `bzip2` (called `tarbz`).

Results concerning execution times and cache metrics were acquired with Linux's `perf` tool (De Melo 2010). Guest execution times are obtained using the Linux `time` command, with a precision of one-hundredth of a second, since QEMU virtual time is kept aligned with host time. This is also true for the ARM Linux guest since reading the guest cycle count (register `ccnt` on ARM) returns the value of the host cycle count (register `tsc` on x86_64).

7.5.1. QEMU *emulation performance analysis*

In this section, we discuss the overall performance of *vanilla* QEMU, focusing on understanding the costs of emulating memory accesses. We will see that the number of memory accesses is indeed significant and that their emulation is indeed costly. We will confirm the importance of an inlined fast path to enable a faster emulation of guests exhibiting a good memory locality. We will also confirm the importance of an optimized slow path when emulating guests with a poorer locality.

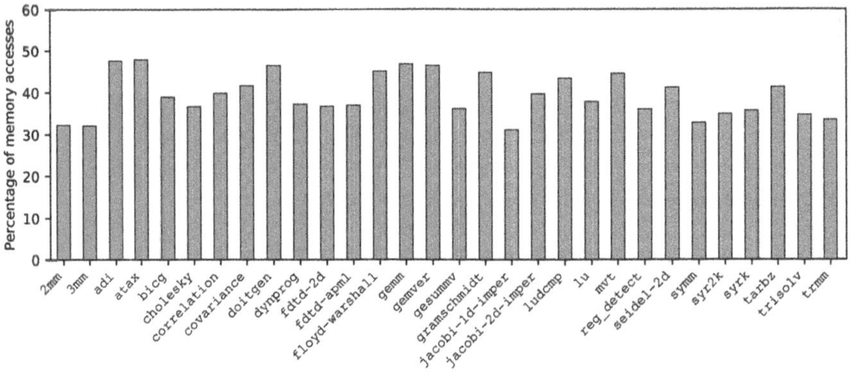

Figure 7.9. *Percentage of memory accesses (with Linux)*

Figure 7.9 shows the percentage of memory access within all emulated instruction, across all our benchmarks. Overall, it confirms the importance of memory accesses, with memory accesses being within the range 30–50% of the total number of emulated instructions. This total number of emulated instructions includes both the emulated benchmark and the emulated Linux. In other words, this is the total number of emulated instructions from booting Linux up to the termination of the benchmark.

Figure 7.10 shows the time spent in the Soft MMU. We clearly see two groups of programs: one with percentages in the range of 50–75%, and the other with percentages in the range of 5–20%. This is not about more or fewer memory accesses, since we know that all benchmarks have more or less the same percentage of memory accesses (30–50%). Therefore, this is the result of memory locality. Some benchmarks have a better hit ratio in the Virtual TLB and therefore spend less time in the Soft MMU, whereas other benchmarks have a poorer hit ratio in the Virtual TLB, resulting in more time spent in the Soft MMU.

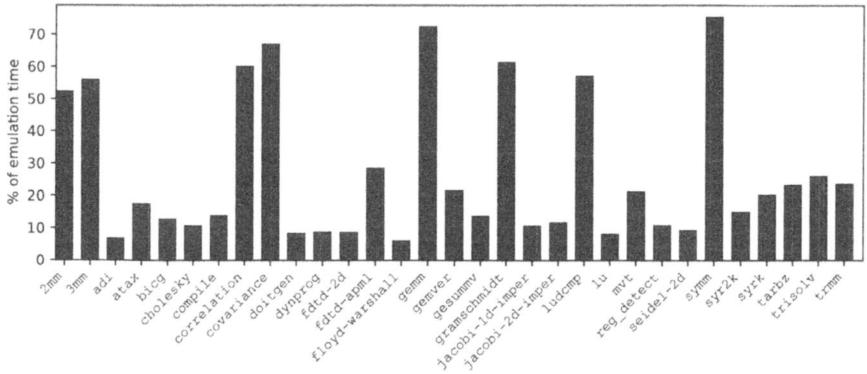

Figure 7.10. *Time spent in the Soft MMU*

Overall, we can conclude that emulating memory accesses is indeed a major part of the emulation time, given the high number of memory accesses. We can further say that having an efficient fast path is important, as it enables us to improve performance when the emulated code has a good memory access locality. However, the search for an efficient fast path should not hide the importance of improving the performance of the slow path, since its overhead can dominate the emulation time when memory access locality drops. The evaluation of our proposed solution will also agree with these statements, with a fast path with near-native performance and a more expensive slow path that requires special optimization.

7.5.2. *Our performance overview*

Figure 7.11 gives an overview of our performance, when compared to the unmodified QEMU as a baseline. We immediately recognize the same two groups of benchmarks identified earlier in Figure 7.10. Unsurprisingly, benchmarks with a good memory locality are emulated much faster with our solution. This is because our fast path is faster than the Virtual TLB of QEMU– we have near-native performance since we are relying on the host hardware MMU. For the second group, we are very happy with the performance since, overall, we are able to maintain a similar performance to QEMU, although our slow path is slower than the slow path of QEMU.

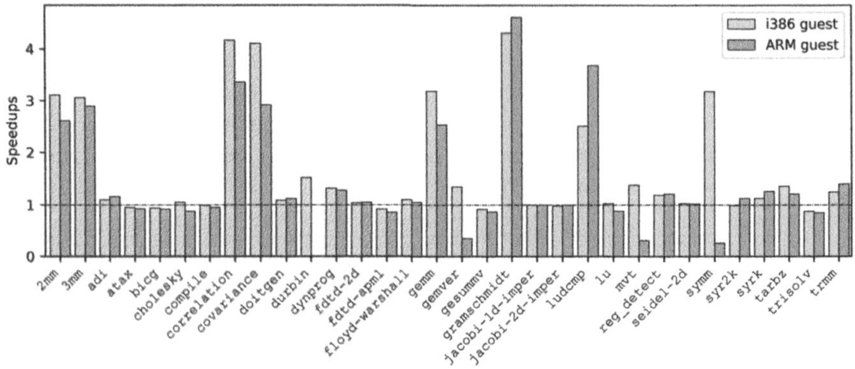

Figure 7.11. *Benchmark speed-ups: our solution versus vanilla* QEMU

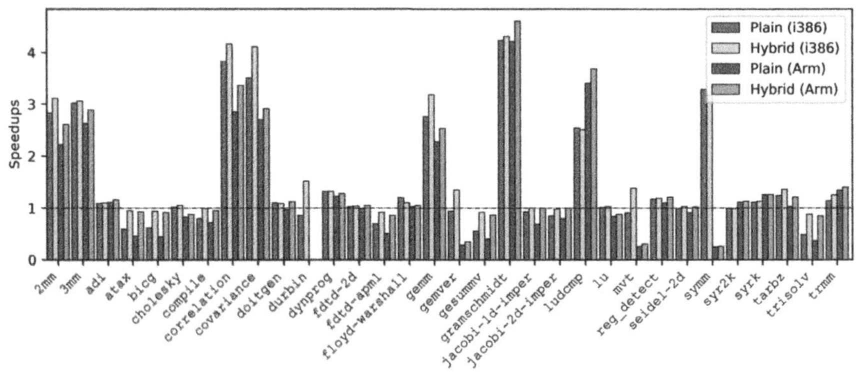

Figure 7.12. *Plain/hybrid speed-ups versus vanilla* QEMU

We explain such good results through the distinction that our solution makes between kernel and user code. As explained before, our DBT only optimizes guest user code via our memory mapping, while translating guest kernel code exactly as vanilla QEMU does. This approach, which we call *hybrid*, needs to be contrasted with a *pure* proposal that translates all memory accesses to use our memory mapping. Figure 7.12 shows the importance of the hybrid design over a pure design, for both ARM and Intel guests. Overall, we can see that our *hybrid* approach always wins over a *pure* approach. But more importantly, our hybrid approach does not degrade performance significantly, while a pure approach shows some slowdowns below 0.5. As explained earlier, with kernel code, the higher cost of our slow path is not counter-balanced by our faster fast path.

Let us discuss how we are able to produce better speed-ups with our hybrid design. The obvious first answer is the near-native performance when emulating memory accesses for pages already mapped. When compared with the Virtual TLB in QEMU, not only are we faster, but we do not suffer from the size limitation of the Virtual TLB. Indeed, to be efficient, the Virtual TLB needs to be small. In comparison, we use the host MMU to map the entire guest address space. This means that, in principle, all guest pages could end up being mapped, with all guest memory accesses being emulated near-native performance. Unfortunately, this does not happen in practice since our mapping is flushed quite often, whenever QEMU flushes its Virtual TLB. At the very least, a flush happens for each guest context that switches between guest processes. Nevertheless, Figure 7.13 shows that our hybrid design exhibits orders of magnitude fewer calls to the slow path than QEMU.

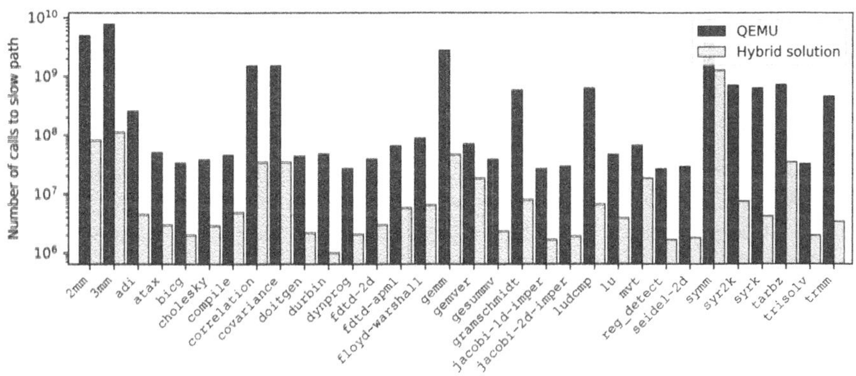

Figure 7.13. *Number of calls to slow path during program execution (i386 and ARM summed)*

7.5.3. *Optimized slow path*

We just discussed that our approach does drastically reduce the number of calls to our slow path, but we need to discuss how to optimize our slow path, since its overhead is still significant when percolating all page faults to QEMU. In section 7.4.3, we proposed an alternative design that could reduce the need to percolate page faults up to QEMU; the details are given in the algorithm shown in Figure 7.8. Figure 7.14 shows the performance potential of striving to handle page faults entirely within our kernel module. We see that certain benchmarks benefit a lot, with an increase in speed-ups from 3 to 6. We see that some benchmarks that were slowed down by our hybrid proposal are now running faster, close to the performance of vanilla QEMU. We can see that one benchmark (symm) still suffers from our proposal and is emulated faster by vanilla QEMU. Nevertheless, these numbers are encouraging for pursuing research in the direction of performing more memory emulation within the kernel module. Indeed, let us remember that, because of the engineering problem with the QEMU codebase, this proof-of-concept prototype still percolates up many page faults induced by write accesses.

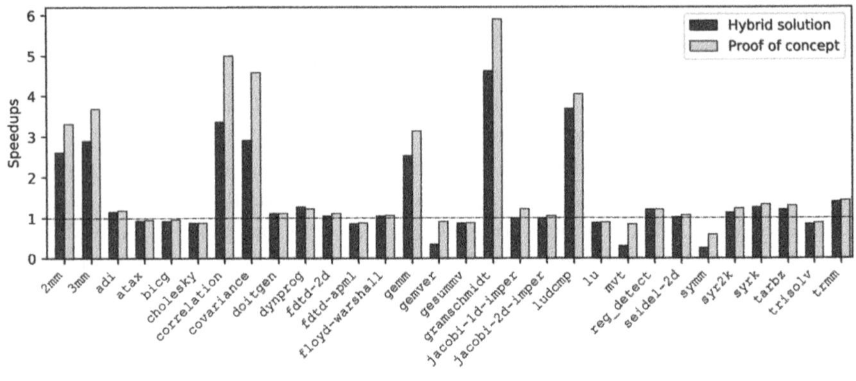

Figure 7.14. *Page fault optimization speed-ups (ARM guest)*

Figure 7.15 illustrates this very point, showing how many page faults are handled in our kernel module and how many are percolated up. The figure gives the total number of page faults on our VMA region as a basis, then it shows how many of those page faults were percolated up to our QEMU handler, as induced by either a read or write access. We clearly see that many page faults are still percolated up to QEMU, however, mostly page faults induced by write accesses. We percolate those up, not because we could not resolve them internally, but because otherwise QEMU fails to correctly emulate the guest program if we do not. Figure 7.15 clearly shows the remaining potential of entirely emulating the guest memory in a kernel module, only percolating up page faults that must be handled by the guest.

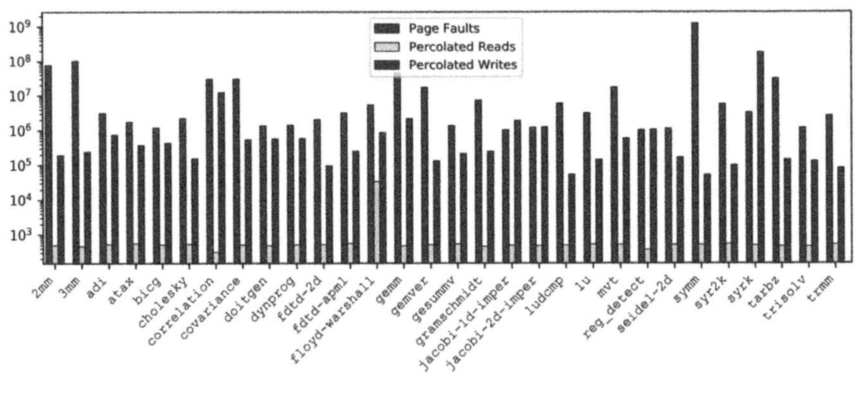

Figure 7.15. *Page fault handling – internal versus percolated (note the logarithmic scale on the y-axis)*

7.6. Related works

Accelerating the process of dynamic binary translation has been the subject of many related works. However, among those, only a handful specifically target accelerating memory accesses. To avoid a lengthy literature overview, we stick to these relatively few proposals in this state of the art. A first group focuses, in a cross-ISA context, on reusing hardware-assisted virtualization (HAV), originally introduced in Creasy (1981) and now available in all general-purpose processors.

In Spink *et al.* (2016), the authors rely on HAV to a large extent to implement their solution. The idea is that the execution of the generated code, as well as code generation, is offloaded on a virtual CPU. Basically, the generated code memory is mapped on the lower part of the virtual address space, while the simulator and generated code lies on the higher part. This allows fairly efficient emulation of memory accesses, with the goal of one host instruction per guest memory access. However, this approach suffers from a few catches. A major catch is that an x86_64 kernel has to be developed, which the authors call the *execution engine*. Indeed, the emulated code and code generator are not able to run bare-bones on the virtual CPU because they need to have some kind of environment to run on. As a result, the approach will require a lot of re-engineering to run on other host architectures. Furthermore, this work was carried out in an *ad hoc* emulator, called CAPTIVE, which does not support as many guest/host couples as an industrial grade emulator, such as QEMU. Finally, the emulator does not fully run in the HAV context; device emulation and interrupt management remains in a user context process. A handshake mechanism has been devised to communicate between those two contexts, but a similar work from Sarrazin *et al.* (2015) has shown that these solutions are impractical when the number of guest CPUs increases. In addition, this work also shows that the cost of switching from host mode to guest mode for device access remains an important performance bottleneck.

The work (Faravelon *et al.* 2017) has taken an approach that is close to the one of CAPTIVE. However, it has important differences regarding implementation. First, instead of creating a full OS, it uses the *Dune* framework (Belay *et al.* 2012). This framework allows us to run an unmodified Linux process on top of a virtual CPU that is virtualized thanks to HAV. It then runs QEMU on top of this framework, with it being both the code generator and device emulator, and as such not needing to go back and forth between virtual CPUs and the host environment. As with our current work, it does not require any heavy modification to QEMU. However, it is tied to both Intel CPUs and the *Dune* framework for back-end. It also suffers from multiple limitations due to the fact that *Dune* is a research proof-of-concept and not a fully stable framework.

Given the complexity of using HAV, another work (Wang *et al.* 2015) suggests using the `mmap` syscall to manage the embedded guest virtual address space and

SIGSEGVs to handle both internal faults and guest faults. While the work on the embedded page table itself is fairly interesting, notably to support multiple guest processes efficiently, the principle itself suffers from constraints of the SIGSEGV and `mmap` interfaces. With our approach, we can hope to handle internal faults at kernel level, while this is not possible with their solution. In addition, they assume that only one host thread exists for emulation. In our work, mappings are made per guest CPU, thus supporting the multi-threaded implementation of QEMU (Rigo *et al.* 2016) for emulating multi-cores in a much more natural way. Finally, they rely on core identifiers (CID), an information they assume is available in the guest TLB. While this can bring a performance advantage, this makes the implementation much more guest-dependent. If this feature is not present in the guest ISA, their approach cannot work. And even if they are present in the guest hardware, CIDs may not be used by the guest OS, which, on Intel, for example, may just choose to perform a full TLB flush at process context switch. Indeed, this is still the case for Linux on AMD CPUs and on Intel CPUs if kernel page table isolation is not used. Even worse, the use that Linux is making of CIDs on Intel is not compatible with this approach (Newbies 2017) as PCID is *not* used to identify a process, but only as a cache for user side mappings to avoid flushing them when making a syscall.

The work (Chang *et al.* 2014) uses a Linux kernel module, like we do, to make the memory aliasing and embed shadow page tables in the host process. However, it suffers from the same limits as (Wang *et al.* 2015) concerning CID and single-thread emulation. It also does not use the module in any way that differs from an `mmap` operation. Faults are handled through SIGSEGV signals, and the module basically only does what the `mmap` does, i.e. map the embedded shadow memory map onto a QEMU vision of the guest physical frames. This makes performances slightly lower than that of our hybrid solution, quite probably due to slower handling of page faults. The other important point for both the works Chang *et al.* (2014) and Wang *et al.* (2015) is that, in both cases, the problem of system code emulation was not discussed. This is probably due to the lack of benchmarks with pathological behaviors such as `symm`, `mvt` or `gemver`. Parts of the performance issues have also probably been masked by the use of CID-based optimization. But, overall, benchmarking very short programs tends to expose setup times or focus on micro-benchmarking, which, in our experience, was not enough, and benchmarking large realistic guests played a crucial role in our quest for optimized solutions.

The work (Hong *et al.* 2016) represents another line of work with a focus on enhancing QEMU Virtual TLB rather than accelerating memory accesses themselves. To that aim, they use a dynamically sized Virtual TLB. Depending on application profile, the size will either increase or decrease so as to reach a sweet spot for TLB fills versus flush cost. This approach, which is fairly elegant, has multiple advantages against all previously cited works, the most important being that it integrates well in the existing codebase of QEMU and probably other emulators. Its main disadvantage is that it only partially solves the problem. It only marginally reduces the number of flushes, and memory accesses are still far from native performance (one host memory access for one guest memory access). Nevertheless, it is interesting that its average speed-up is higher than ours, while its peak speed-up is lower. While interesting, the comparison must be taken with a grain of salt because our benchmarks are not the same and we do not use the same combo QEMU version and host system. Future work might reveal that combining the approaches might prove efficient, especially in improving the emulation of system code, which we do not optimize with our solution.

7.7. Conclusion

In this work, we have shown a path to evolve dynamic binary translation for much faster memory emulation. The starting point was to emulate guest memory accesses with host memory accesses, with near-native performance. This was done by mapping the whole guest address space into the emulator process and translating guest memory access instructions at a given address with their host-equivalent memory access instructions at that same address, but used as an offset within the mapped guest address space. The result is an average speed-up of 67%, with some benchmarks being up to four to six times faster than the unmodified QEMU. We believe that this research path should be continued.

From the start, we wanted to engineer a solution that would require as few modifications to QEMU as possible. We succeeded, and we believe our ideas and techniques could be integrated into most emulators based on dynamic binary translation. However, the engineering challenges were not small, since memory emulation is a central performance-crucial subsystem of any emulator and QEMU is no different in that regard. Knowing when to flush our mapping was one of the main challenges and we chose to flush our mapping whenever QEMU flushes its Virtual TLB. It worked, but we undoubtedly flush too often and too much this way. Decoupling memory emulation so that it can be done entirely within our kernel module also proved quite difficult. We managed a proof-of-concept that limited percolating page faults from the Linux kernel up to our QEMU handler, showing great potential for near-native memory emulation.

However, overall, the engineering challenges and the limitations raise an open question: how should emulators be designed to benefit fully from hardware-assisted memory emulation? We are confident that memory emulation can be divided between the user-land emulator and a kernel module much more efficiently, leading to near-native performance across many, if not all, workloads. Let us note that this work was done without using any special hardware to assist memory emulation, just a regular MMU was necessary. With an earlier work (Faravelon *et al.* 2017), we successfully explored leveraging the ability to control the mechanisms for hardware-assisted virtualization that are available on modern processors, even though the attempt was hindered by the fact that the *Dune* framework (Belay *et al.* 2012) was only a research proof-of-concept. Nevertheless, we believe that if Linux integrated *Dune*-like capabilities and QEMU-like emulators to use them, then dynamic binary translation could deliver outstanding cross-ISA performance.

We are convinced that this work is a promising research path, even though our current prototypes do suffer from several limitations. We only support 32-bit guests, but with adequate hardware support in the host MMU, leveraging existing mechanisms from hardware-assisted virtualization, this would no longer be an issue. We currently require that the guest and host MMUs are sufficiently similar so that the host MMU can reasonably be used to emulate the guest MMU. This may be seen as a real limitation for hardware simulation, but para-virtualization offers a great opportunity for general-purpose emulation. Indeed, all operating systems do wrap the underlying platform with a hardware abstraction layer (HAL), where the emulator can expose an easy-to-emulate MMU to guests.

Para-virtualization will also help with the emulation of devices, which is a challenge for all emulators alike. Our approach is no exception – the performance penalty when guest mmio hardware registers are manipulated by guest device drivers can be quite high. Indeed, the corresponding guest load and store operations are not memory operations; they must lead to an accurate emulation of the corresponding guest hardware devices. With our design, this means that each such access must trap and the trap must be percolated up to our QEMU handler so that QEMU is given the opportunity to emulate the corresponding device. By leveraging para-virtualization with split drivers as in Xen (Barham *et al.* 2003), there is no need for guest drivers to access mmio registers – guest drivers send and receive messages with host drivers through a message-oriented bus called the Xen bus. QEMU already supports the emulation of the Xen bus, and our next prototype ought to evaluate this support towards faster emulation of operating system code.

Overall, our research opens up new opportunities for dynamic binary translation as a viable option for high-performance cross-ISA emulation. Our current understanding clearly suggests that a clean design bridging dynamic binary translation, hardware-assisted virtualization driven from user processes and para-virtualized guests would be a potent combination. The only shadow is the substantial engineering effort that it would require across the QEMU and Linux communities.

7.8. References

Altman, E., Ebcioğlu, K., Gschwind, M., and Sathaye, S. (2001). Advances and future challenges in binary translation and optimization. *Proceedings of the IEEE*, 89(11), 1710–1722.

Barham, P., Dragovic, B., Fraser, K., Hand, S., Harris, T., Ho, A., Neugebauer, R., Pratt, I., and Warfield, A. (2003). Xen and the art of virtualization. *ACM SIGOPS Operating Systems Review*, ACM, 37, 164–177.

Belay, A., Bittau, A., Mashtizadeh, A. J., Terei, D., Mazières, D., and Kozyrakis, C. (2012). Dune: Safe user-level access to privileged CPU features. *10th USENIX Symposium on Operating Systems Design and Implementation (OSDI)*, 335–348.

Bell, J.R. (1973). Threaded code. *Communication of the ACM*, 16(6), 370–372.

Bellard, F. (2005). QEMU, a fast and portable dynamic translator. *USENIX Annual Technical Conference*, FREENIX Track, 41–46.

Chang, C.-J., Wu, J.-J., Hsu, W.-C., Liu, P., and Yew, P.-C. (2014). Efficient memory virtualization for Cross-ISA system mode emulation. In *Proceedings of 10th ACM SIGPLAN/SIGOPS International Conference on Virtual Execution Environments*. ACM, 117–128.

Cmelik, B. and Keppel, D. (1994). Shade: A fast instruction-set simulator for execution profiling. *Proceedings of the 1994 ACM SIGMETRICS Conference on Measurement and Modeling of Computer Systems*, ACM, 128–137.

Creasy, R.J. (1981). The origin of the VM/370 time-sharing system. *IBM Journal of Research and Development*, 25(5), 483–490.

De Melo, A.C. (2010). The new linux "perf" tools. *Slides from Linux Kongress*, 18 [Online]. Available at: http://vger.kernel.org/~acme/perf/lk2010-perf-acme.pdf.

Deutsch, L.P. and Schiffman, A.M. (1984). Efficient implementation of the smalltalk-80 system. *Proceedings of the 11th ACM SIGACT-SIGPLAN Symposium on Principles of Programming Languages*, 297–302.

Faravelon, A. (2018). Acceleration of memory accesses in dynamic binary translation, PhD thesis, Grenoble Alpes University.

Faravelon, A., Gruber, O., and Pétrot, F. (2017). Optimizing memory access performance using hardware assisted virtualization in retargetable dynamic binary translation. *2017 Euromicro Conference on Digital System Design*, IEEE, 40–46.

Hong, D.-Y., Hsu, C.-C., Chou, C.-Y., Hsu, W.-C., Liu, P., and Wu, J. J. (2016). Optimizing control transfer and memory virtualization in full system emulators. *ACM Transactions on Architecture and Code Optimization (TACO)*, 12(4), 47.

John, L.K., Reddy, V., Hulina, P.T., and Coraor, L.D. (1995). Program balance and its impact on high performance RISC architectures. *First IEEE Symposium on High-Performance Computer Architecture*, IEEE, 370–379.

Mills, C., Ahalt, S.C., and Fowler, J. (1991). Compiled instruction set simulation. *Software: Practice and Experience*, 21(8), 877–889.

Newbies, K. (2017). Longer-lived TLB Entries with PCID. Available at: https://kernelnewbies.org/Linux_4.14#Longer-lived_TLB_Entries_with_PCID.

Pouchet, L.-N. (2012). PolyBench/C 3.2 [Online]. Available at: http://web.cs.ucla.edu/~pouchet/software/polybench/.

Rigo, A., Spyridakis, A., and Raho, D. (2016). Atomic instruction translation towards a multi-threaded QEMU. *30th International ECMS Conference on Modelling and Simulation*, 587–595.

Sarrazin, G., Fournel, N., Gerin, P., and Pétrot, F. (2015). Simulation native basée sur le support matériel à la virtualisation : cas des systèmes many-cœurs spécifiques. *Technique et Science Informatiques*, 34(1–2), 153–173.

Spink, T., Wagstaff, H., and Franke, B. (2016). Hardware-accelerated cross-architecture full-system virtualization. *ACM Transactions on Architecture and Code Optimization (TACO)*, 13(4), 36.

Tong, X., Koju, T., Kawahito, M., and Moshovos, A. (2015). Optimizing memory translation emulation in full system emulators. *ACM Transactions on Architecture and Code Optimization (TACO)*, 11(4).

Wang, Z., Li, J., Wu, C., Yang, D., Wang, Z., Hsu, W.-C., Li, B., and Guan, Y. (2015). HSPT: Practical implementation and efficient management of embedded shadow page tables for cross-ISA system virtual machines. *ACM SIGPLAN Notices*, 50(7), 53–64.

Wilkes, M.V. (1969). The growth of interest in microprogramming: A literature survey. *ACM Computing Survey*, 1(3), 139–145.

Witchel, E. and Rosenblum, M. (1996). Embra: Fast and flexible machine simulation. *ACM SIGMETRICS Performance Evaluation Review*, 24(1), 68–79.

8

Study and Comparison of Hardware Methods for Distributing Memory Bank Accesses in Many-core Architectures

Arthur VIANES[1] and Frédéric ROUSSEAU[2]

[1] *Kalray S.A., Grenoble, France*

[2] *Université Grenoble Alpes, CNRS, Grenoble INP, TIMA, 38000 Grenoble, France*

Multi-core and many-core architectures have evolved towards a set of clusters, each cluster being composed of a set of cores communicating with each other through networks-on-chip. Each cluster integrates cache memory and a local memory shared by all the cores of the cluster. The local memory bandwidth should be able to support all the accesses requested by the cores to gain maximum performance. One solution to reach this objective is to split the local memory into several banks, which are accessible in parallel. If some cores get access to the same bank simultaneously, then this leads to memory bank conflict. A conflict increases the latency for the other cores, decreasing overall bandwidth.

In this chapter, we first present, in detail, when collisions occur and their consequences on the running application. We then focus on hardware solutions to improve the performances of multi-bank accesses, from hash-function address coding

For a color version of all figures in this book, see www.iste.co.uk/andrade/multi1.zip.

to additional memory banks. All of these solutions have been simulated on the Kalray MPPA processor, and we provide a comparison and discussion of the results obtained, in order, mainly, to highlight that there is no universal solution!

8.1. Introduction

8.1.1. *Context*

The increasing demand for computing power implies a huge pressure on the memory architecture. Indeed, processing elements require very high bandwidth and low latency to/from the memory. With such constraints, multi-processor architectures no longer use only one huge memory component for the processing elements and with a high latency.

Multi-core or many-core architectures are now used to run high-performance computing applications. As with the MPPA of Kalray, the cores of modern MPSoCs are often grouped into clusters that communicate with one another through networks-on-chip (NoCs). Each cluster integrates cache memory and a local memory shared by all the cores of the cluster, and is thus close to the cores to benefit from low latency. The difference between cache and local memory is the ability of the programmer to control the content of the local memory, as there is no hidden mechanism to maintain or update its content, as with the existing ones for cache memory (i.e. cache coherence, for instance). In this chapter, we will examine optimizations of the local memory architecture, however, the solutions or optimizations discussed can also be applied to the cache memory.

The local memory bandwidth should be able to support all the accesses requested by the cores to gain maximum performance. One solution to reach this objective is to split the local memory into several banks, which are accessible in parallel. In this case, the overall bandwidth of the memory can be close to the overall bandwidth of cores.

If some cores get access to the same bank simultaneously, then this leads to memory bank conflict, as only one access is possible, delaying the other accesses. A conflict increases the latency for the other cores, decreasing the overall bandwidth. This could happen even if cores try to access separate resources or data if they are mapped to the same memory bank. Such a problem can be solved by using an efficient spread of resources or data in the memory banks.

Software solutions exist to spread resources in different memory banks, but they are expensive in terms of memory space and not well suited for dynamic data size, or, they are complex to set up for some specific applications. Hardware solutions also exist which are based on the assumption that memory accesses follow regular and well-known patterns. This makes it possible to smartly spread resources or data in memory banks, in order to decrease memory bank access conflicts as much as possible.

In this chapter, we first identify pathological access patterns that produce a great deal of memory bank access conflicts. Focusing on hardware solutions, we study and compare previous work and optimizations, such as prime modulus indexing or pseudo-randomly interleaved memory. We then perform simulations to evaluate several different approaches to reduce conflicts in the local memory of a cluster of the Kalray MPPA processor.

8.1.2. *MPSoC architecture*

This study focuses on hardware solutions to reduce collisions in memory banks within a many-core processor, dedicated to high-performance computing. A many-core processor is an architecture made up of a large number of cores, forcing several computing units to be clustered together around a local memory, shared by the cores of the cluster.

For instance, the MPPA of Kalray (2016) (Figure 8.1) is composed of clusters of 16 cores (PEx for Processing Element plus one, called RM for Resources Management).

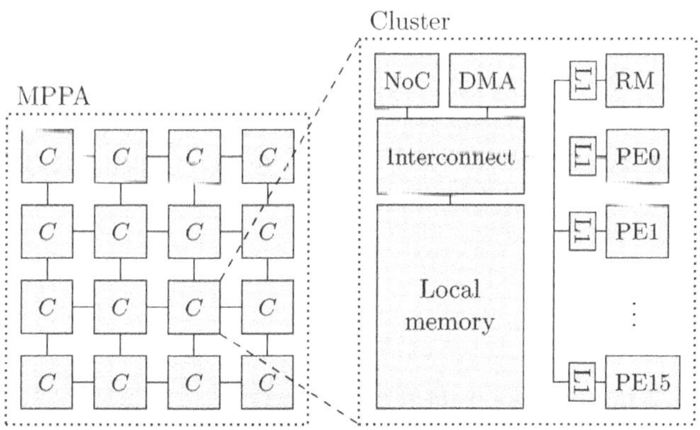

Figure 8.1. *Kalray MPPA overall architecture*

All of these 16 computing units are cores based on a VLIW architecture, which are capable of generating one memory access per cycle while performing computations simultaneously. This implies that the local memory must be able to handle 16 parallel accesses per cycle. To meet this requirement, the local memory is split into 16 memory banks which can be accessed independently (Figure 8.2). All memory banks provide a complete addressing space, with an interleaving of 32 bytes (8×4 bytes), meaning the selection of the memory bank for a given address is obtained by the following formula: $bank = ((address \div 32) \bmod 16)$. We describe banked memory and memory bank interleaving in more detail in section 8.2. All the cores are connected to the memory through a local interconnect and a global interconnect allows the clusters to communicate with one another.

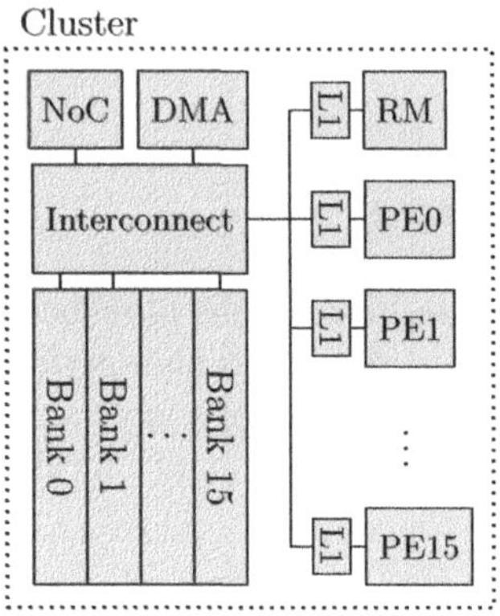

Figure 8.2. *Memory banks and local interconnect in a Kalray cluster*

In the context of high-performance computing, a perfect scenario supposes, within each cluster, that the local interconnect will be saturated with access to memory and that the memory will also be saturated. At each cycle, the memory bank accesses are perfectly distributed and collision-free, which means that an access to the same memory bank at the same cycle never happens. In case of collision, one access per memory bank is performed, delaying all the other accesses and therefore the next steps of computation.

8.1.3. *Interconnect*

Interconnects are essential parts of the architecture since all memory accesses are routed through them. Choosing to split the processor into clusters as the architecture implies that the processor must integrate at least two levels of interconnects: a local interconnect, inside each cluster that connects each element (cores, memory banks, etc.) of the cluster to one another, and a global interconnect, which connects the different clusters together inside the processor. In this chapter, the global interconnect is not considered as we do not discuss communication between clusters. We assume that each core has access to all the data it needs to carry out its computations within the local memory of its cluster – through the local interconnect – and therefore we are interested in the optimization of these internal memory accesses within the cluster.

Since the local interconnect is in charge of handling all core accesses to the local memory, it must then be able to handle the full workload when all processors are using the memory; otherwise, it is impossible to use the full computing capability of the cluster.

8.2. Basics on banked memory

8.2.1. *Banked memory*

The efficiency of high-performance systems relies mainly on two factors: the computational throughput – their ability to deliver a large amount of processing – and the memory throughput – their ability to supply the processing units with data to process. Over the last few decades, we have continuously produced more and more efficient processing units. Nowadays, in terms of computational throughput, the processing units are generally used in parallel which makes the system even more data-intensive. Regarding the memory throughput, the evolution is slow, which leads us today to the point where the memories are a bottleneck for processing units. This explains why we are conducting this study on optimizing memory use.

To understand how to optimize memory throughput, we need a global picture of the memory implementation. A memory can be seen as a large matrix containing tiny memory elements, and each with a position in the matrix and an address assigned to it. The memory also includes a memory controller which is the only one that can manipulate the memory matrix. The memory controller also exposes read/write ports. In order to access the memory, we should use these ports. When an access is made, it must be routed to a port of the memory controller that decodes the access address to determine which memory element is being accessed and therefore its position in the matrix. Then, the memory controller turns on the lines corresponding to the memory element addressed, to read or write its contents (depending on the port used), and the result of the operation is sent back through the memory port. Since memories are large, the realization of all these steps can take several cycles depending on the size of the memory and on the distance from the memory to the computing unit. We use the term latency to refer to the number of cycles needed to perform all of these steps, and the term throughout to refer to the amount of data transferred per cycle. We are able to produce memory controllers with more than one read port and one write port, but the number of ports is restricted by hardware constraints. Therefore, in this study, we will consider only one read port and one write port. Thus, the throughput of one memory controller will only be constrained by the port size. While increasing the port size to increase the memory bandwidth is theoretically efficient, this is not the case in all contexts. If several accesses are performed at different locations – by different computing units in a many-core architecture – the increase in the port size is not sufficient. The commonly used solution is then to split the memory into several smaller memories called memory banks. Each memory bank contains a memory controller with its ports, and a memory matrix. Thus, if several accesses are made, they can all be distributed across a different memory bank and thus benefit from a maximized memory throughput. Then, a mechanism is necessary to route the memory access to the corresponding memory bank, which is the task of the interconnect between processing units and memory.

An address range will be assigned to each memory bank, meaning that all accesses occurring in this range will be handled by this memory bank. Therefore, if each processing unit produces accesses in a different memory bank, the memory throughput will be fully exploited. However, if two or more processing units access the same memory bank – i.e. with addresses mapped to the same bank – then the accesses cannot be handled in parallel, so we will not benefit from memory banking. This is called a collision, and this chapter will focus on how to prevent collisions in order to make more efficient use of the memory bandwidth.

8.2.2. *Memory bank conflict and granularity*

In the previous section, we explained that splitting the memory into memory banks is essential to achieve the bandwidth required to feed a many-core system. However, it is not enough to improve performance in all circumstances, as we must be able to use these memory banks efficiently; in other words, we must be able to spread memory accesses of all the actors through all the memory banks in order to handle all the accesses in parallel. Otherwise, if two or more actors access the same memory bank during the same cycle, a collision occurs, also referred to as memory bank conflict. A collision does not require the actors to access the same address; it is sufficient that the addresses are mapped to the same memory bank. Since a memory bank is only capable of handling one memory operation per cycle, it is impossible for it to perform all accesses: during the cycle, only one access is performed and the others are delayed. We therefore lose all benefits from banking when collisions occur; hence, we try to reduce the number of collisions. The way memory words are distributed in memory banks has an impact on the amount of collisions and therefore on memory performance. The chosen distribution of words in memory banks is called the memory bank mapping or the interleaving scheme. We define memory interleaving in section 8.2.3.

Let us assume the following configuration as an example to illustrate collision issues: a 32-bit memory with a total capacity of 4 kB, which is thus addressed on 12 bits. To meet the memory access requirement of four cores, we will split this memory into four memory banks (1 kB each). For real cases, the memories that require banking are much larger than 4 kB, but the consequences of banking will be the same. The naive way to map these memory banks is to map them contiguously one after the other – bank 0 handles addresses 0 to 1023, bank 1 handles addresses 1024 to 2047, and so on. Figure 8.3 describes this memory layout, showing the address limit of the banks.

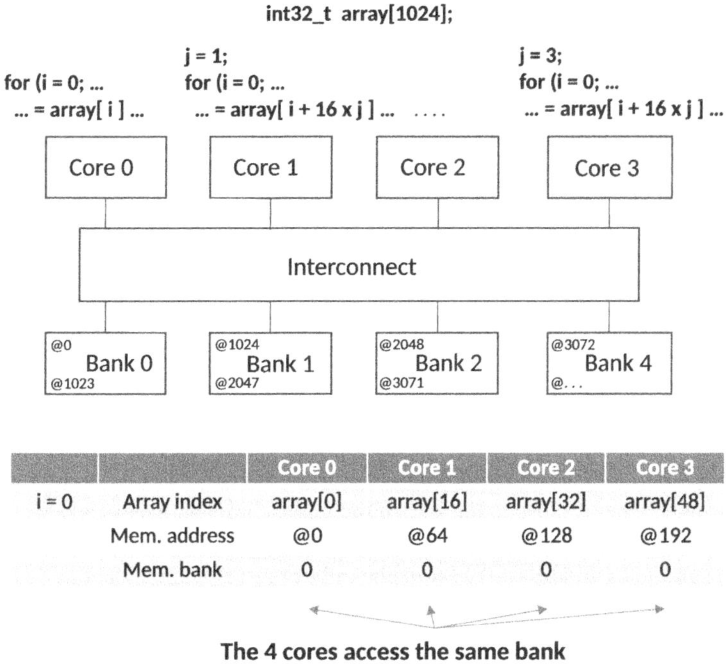

Figure 8.3. *Example of collisions with four accesses to the same memory bank*

Figure 8.4 shows the decoding of the address to achieve this layout. The two least significant bits select the byte in the 32-bit word, the following eight bits give the address of the word in the memory bank, and the two most significant bits (bits 10 and 11) select the appropriate memory bank (one of the four banks).

Figure 8.4. *Description of memory address bits and their use for a 32-bit memory with four banks of 1 kB each*

This naive approach is inexpensive and intuitive, but it is also inefficient if we take into account the spatial locality of memory accesses. Indeed, under the assumption of the spatial locality, the majority of accesses will be addressed close to each other. With contiguous address mapping, this results in all accesses being routed to the same memory bank. As an example, in Figure 8.3, all programs executed by cores access different elements of the array, but all these array elements are mapped to the same memory bank, leading to collisions. This example illustrates that if several processors use the same resource in memory, then we will observe traffic congestion (due to many collisions) of accessing the memory bank where the resource is located and the memory bank will be unable to process all the accesses in parallel. The effective bandwidth shared by all the cores will be the bandwidth of a single bank instead of four banks; thus, in this situation, we do not benefit from banking.

8.2.3. *Efficient use of memory banks: interleaving*

In order to realize an efficient memory mapping, the spatial location of the memory accesses must be taken into account. This requires a memory mapping that distributes accesses close to each other in terms of address across all the memory banks. This is done with interleaving, which means slicing up the address range of the memory and then distributing it among the memory banks. However, it is counterproductive to cut the memory too finely since this would result in cross-bank accesses. It is desirable to choose a "slice" that is close to the size of the largest memory accesses that a core can perform at once; for example, it is advisable to choose the same size as the L1 cache line to ensure that the refill of the L1 cache only accesses a single memory bank. The slice size is called the interleaving granularity.

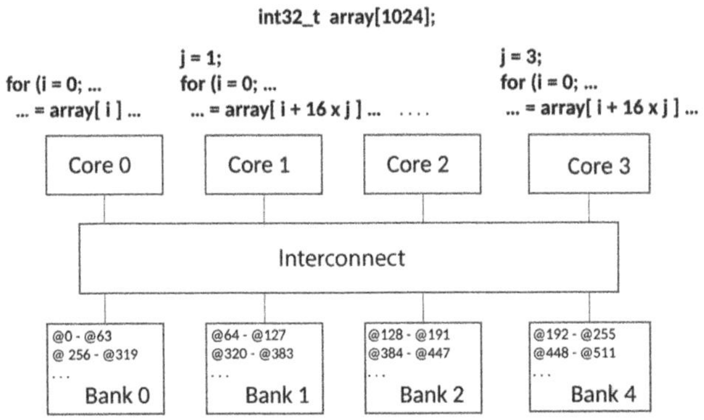

		Core 0	Core 1	Core 2	Core 3
i = 0	Array index	array[0]	array[16]	array[32]	array[48]
	Mem. address	@0	@64	@128	@192
	Mem. bank	0	1	2	3

The 4 cores access different banks

Figure 8.5. *Example of interleaving with four accesses to different memory banks*

Figure 8.5 describes this memory layout. The interleaving in the four memory banks provides better data distribution. All programs executed by cores access different elements of the array, and all of these array elements are mapped to different memory banks, avoiding collisions. Figure 8.6 shows the decoding of the address to achieve this layout. The two least significant bits select the byte in the 32-bit word, the following two bits (bits 2 and 3) select the appropriate memory bank, and the eight most significant bits give the address of the word in the memory bank.

11 … 4	3 2	1 0
Bank Offset	Bank Index	Word Offset

Figure 8.6. *Description of memory address bits and their use for a 32-bit memory with four banks of 1 kB each with interleaving*

8.2.3.1. *Access patterns*

To study the efficiency of a memory system, we need to develop hypotheses about the distribution of memory accesses generated by the processors. Based on these hypotheses, we can both develop theoretical models to find optimal solutions and model the behavior of the processors with simulators, to evaluate the performance of a memory system. In the latter case, it is also possible to use memory access traces of real applications.

Access patterns can be seen from a spatial or temporal point of view. A temporal access pattern corresponds to all the accesses that a single core will produce during its compute over time. While a spatial access pattern represents the accesses produced during a cycle by a set of processors. Here we are interested in how the cores share memory banks within a cluster and during a cycle, so our focus is on the space access model.

In the following sections, we will distinguish three types of access patterns: random access pattern, sequential access pattern and stride access pattern.

8.2.3.1.1. Random access pattern

The random access pattern describes the behavior of processors when they produce accesses to unpredictable addresses. In practice, this turns out to be wrong because the access patterns are not really random, but they follow a logic that is encoded in the program being executed. However, this pattern is useful for modeling the behavior of accesses produced under certain conditions where the distribution of accesses is very irregular. For example, if several processors scatter/gather in local memory at disordered addresses, then we obtain a result close to random behavior. Since this kind of access pattern does not allow us to make strong assumptions about their distributions, and since their behavior is close to random, we allow this assumption to be made. In addition, if cores are randomly delayed from one another, then even if individually their access pattern is temporally organized, globally their access patterns are likely to be random. This time delay can be caused by collisions or a DMA that uses the memory. The cores need to be perfectly synchronized or temporally shifted in a regular way for the access pattern to be regular.

8.2.3.1.2. Sequential access pattern

The sequential access pattern corresponds to accesses to a contiguous memory area; in other words, from an address, each memory cell is accessed one by one. This pattern is very common and can be found in almost every application, which is therefore important to take into account.

8.2.3.1.3. Stride access pattern

The stride access pattern corresponds to the access whose addresses are spaced by a constant amount. Such a pattern typically appears in programs that handle large matrices or large amounts of regular data. This makes it particularly interesting to study as it is used in a vast majority of high-performance computing scenarios. Note that the sequential access pattern is a special case of the stride access pattern, which is also called the unit stride access pattern.

In the following section, we therefore have two distinct approaches: one approach to deal with stride access patterns and another approach to deal with random access patterns. For stride access patterns, we will focus on the interleaving scheme (section 8.4) while for random access patterns we will have no choice but to study the variations in the architecture of the local interconnect, which we will briefly discuss in section 8.5.4.4.

8.3. Overview of software approaches

The literature offers several approaches to address the problems of memory bank conflict, as outlined in section 8.2.2. In this section, we will first go through software solutions, and then focus on hardware solutions that seem more appropriate in coping with memory constraints and dynamic data size.

8.3.1. *Padding*

To identify which type of program produces collisions, we will focus on programs that produce stride access patterns – the type of pattern that produces most of the collisions. The primary source of the stride access pattern is operations on multidimensional objects, such as operations with vectors or matrices.

Under these circumstances, the stride size of the accesses is directly related to the vector or matrix size. This implies that the designer is able to control the stride size by adjusting the dimensions of its objects. When the dimension size of the objects has common factors with the number of memory banks, we can expect to find collisions.

For example, let us consider a four-bank memory system and a 4×16 matrix containing 1-byte words. The matrix is stored in row-major order, which means that cells of a row are contiguous in memory, while cells of a column will be spaced the size of a row in memory. Then, since the size of a row is 16 bytes, a column access in this matrix produces a 16-byte stride access pattern. This configuration corresponds to Figure 8.7 in which it can be observed that a 16-byte stride access pattern, in a system with four banks, produces accesses only to the same bank, therefore leading only to collisions.

		Bank			
		0	1	2	3
Address	0	•			
	16	•			
	32	•			
	48	•			
	total	4	0	0	0

Figure 8.7. *A four-bank architecture, 1-byte words, with a 16-byte stride access pattern*

If we change the size of the matrix to 4×17, then column accesses produce a 17-byte stride accesses pattern, as shown in Figure 8.8, which produces no collisions. As the number 17 has no common factor with 4, we obtain a perfect distribution of the accesses to the four banks.

		Bank			
		0	1	2	3
Address	0	•			
	17		•		
	34			•	
	51				•
	total	1	1	1	1

Figure 8.8. *A four-bank architecture, 1-byte words, with a 17-byte stride*

To solve the collision problem, we added an unused column to the matrix, which increases from 4×16 to 4×17: this method is called padding.

Padding has the advantage of being extremely simple and efficient, but it has the disadvantage of wasting part of the memory since the added column is not used. This column can still be used to store other information, such as metadata about the matrix, but this requires special management of this part of the memory.

8.3.2. *Static scheduling of memory accesses*

In the case where there are several addressing modes for the memory banks, if there is an addressing mode without interleaving memory banks, we can take advantage of it. This allows interleaving to be disabled to statically assign to each core a memory bank that is the only one to use. Thus, no collisions can occur. This kind of method can be used, for example, in real-time applications to meet determinism constraints. However, this makes programming more complex since the memory bank assignment needs to be considered. It also makes it more complex to share a resource across multiple cores.

In the case of a memory system with an interleaved addressing mode, it is possible to statically allocate memory banks to the cores. For example, there are methods such as Static Address Generation Easing (SAGE) (Chavet *et al.* 2010) which allow us to statically shuffle the memory cells in a matrix in order to place memory cells, used at the same time, in different memory banks. This technique requires an additional matrix at runtime that contains the mapping; however, the use of small tiles can keep this matrix small enough. This method also requires a pre-processing of the data in order to shuffle it.

8.3.3. *The need for hardware approaches*

All of these methods require programmers to respect rather strict constraints to improve performance: specific array size choice, static memory area assignment, strong assumptions on the number of banks, etc. By exploring hardware approaches, as in section 8.4, we will try to relax these constraints. In other words, removing constraints or making constraints easier will enable easier and more sophisticated development. For example, we can attempt to use hardware to encourage 2^n stride access patterns that would be pathological with a classical architecture of interleaved banked memory system.

8.4. Hardware approaches

8.4.1. *Prime modulus indexing*

In section 8.3, we discussed the padding method that suggests reducing the number of collisions by padding the array being processed with unused cells. The effectiveness of this approach comes from the fact that when the number of banks and the stride size are two numbers that do not have common factors, there is no collision. However, collisions are common when these two numbers share many common factors, such as the powers of two. The padding approach reduces the number of common factors between these two numbers by modifying the only one that can be modified from the software: the stride size. The hardware approach presented here consists of modifying the second number that is specific to the hardware: the number of memory banks. Choosing a number of banks that is co-prime with stride size will reduce the number of collisions. Therefore, we can choose a prime number of banks to avoid collisions with most stride sizes.

Regarding the implementation, to select the bank and the position of the word in a memory bank according to the address, division is used to compute the bank, and the rest of the division (mod operation) is used to compute the index (position of the word within the bank) as in [8.1]:

$$\begin{cases} Bank = \lfloor Addr \div S_{interleave} \rfloor \mod N_{bank} \\ Index = \lfloor Addr \div S_{interleave} \rfloor \div N_{bank} \end{cases} \quad [8.1]$$

where N_{bank} is the number of memory banks and $S_{interleave}$ the interleaving granularity.

When N_{bank} is a prime number, there are several implications. First of all, as explained above, the memory system will not produce any collisions for stride access patterns with a conventional stride size (2^n bytes or $K \times 2^n$ bytes with K co-prime with N_{bank}).

As an example, let us consider a five-bank memory system with 1 byte words. Figure 8.9 shows the distribution of the addresses of 20 words in the five-bank system. Bank and word indexes can be determined using the above formula: the access to address 16 is in bank 1 ($\lfloor 16 \div 1 \rfloor \mod 5 = 1$) and at index 3 ($\lfloor 16 \div 1 \rfloor \div 5 = 3$). With a stride of 16 bytes, there is no collision for the four accesses, which will be the same as long as the stride size is co-prime with 5.

Bank 0	1	2	3	4
0	1	2	3	4
5	6	7	8	9
10	11	12	13	14
15	16	17	18	19
		⋮		

		Bank 0	1	2	3	4
Address	0	•				
	16		•			
	32			•		
	48				•	
	total	1	1	1	1	0

Figure 8.9. *Left: the distribution of addresses within a 5-bank memory system. Right: a 16-byte stride access pattern in this memory system*

Another implication of a prime number of memory banks is that this means a hardware division of prime numbers must be implemented. However, as shown in Seznec's 2015 paper, it is not necessary to implement a division of prime number to obtain the bank index. When the bank size is 2^c cells and the number of banks is odd, we can use the Chinese theorem to show that the following distribution does not map two addresses in the same cell, as in [8.2]:

$$\begin{cases} Bank = \lfloor Addr \div S_{interleave} \rfloor \mod N_{bank} \\ Index = \lfloor Addr \div S_{interleave} \rfloor \mod 2^c \end{cases} \quad [8.2]$$

where 2^c is the memory bank size. It is assumed that the memory bank size is a power of two because it is a very common and more practical design choice.

For example, Figure 8.10 shows this distribution in a system with five memory banks, where each bank is 4 bytes. The access to address 15 is in bank 1 ($\lfloor 15 \div 1 \rfloor$ mod $5 = 0$) and at index 3 ($\lfloor 15 \div 1 \rfloor \div 4 = 3$).

Bank				
0	1	2	3	4
0	16	12	8	4
5	1	17	13	9
10	6	2	18	14
15	11	7	3	19

Figure 8.10. *Distribution of addresses across five memory banks with* $Index = \lfloor Addr \div S_{interleave} \rfloor \mod 2^2$

The remaining implication of this choice of bank number is that it requires the use of prime number modulo that are more complex and less efficient operations than the 2^n modulo when they are implemented in hardware with binary numbers. However, it is possible to build these operations in an optimized way for some specific numbers, as described in Dupont de Dinechin (1991) and Diamond *et al.* (2014).

With this approach, we allow developers to choose the memory layout of their data with more freedom, aligned to powers of two, for example. Thus, it is the responsibility of the memory system to provide the minimum number of collisions. Therefore, the selection of the number of banks is decisive: the programmer has to choose a number of banks that has the least amount of common factors with the size of the arrays.

8.4.2. *Interleaving schemes using hash functions*

The methods suggested in section 8.4.1, using a prime number of banks, require us to change the number of memory banks in our system and to use a hardware implementation of the prime number modulo, which is quite complex. In order to keep 2^n memory banks (which keeps the architecture simple), other solutions have emerged based on bitwise operations. These solutions can be observed globally as hash functions applied to the address provided by the core; the hash result is used as a bank index.

Figure 8.11 shows how the hash function ($h(\cdot)$) should be applied to the address. The least significant bits of the address that correspond to the byte index in the memory word should be ignored, as they are not usable for bank selection. All other bits of the address must be included and used by the hash function. These bits, shown in the figure, are called the effective address since it is the part of the address that will really be used to select the memory bank. If some of these bits are ignored, then this would produce collisions when these bits are affected by an access pattern. The bank selection is therefore done using the following formula: $b = h(a)$, with the result on m-bits and where the input a corresponds to the effective address on n-bits.

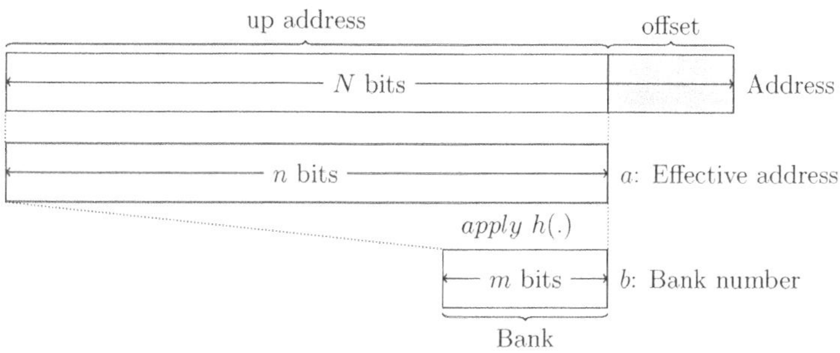

Figure 8.11. *Using a hash function for memory bank selection. N is the address size, n is the address size without the offset byte, m is the memory bank number size and h is the hash function used*

8.4.2.1. *XOR scheme*

As described in Sohi (1993), a XOR-based hash function is a hash function that combines input bits (in our case, the address of an access) with XOR gates, to generate an output number (in our case, the index of the bank). This hash function is similar to the Hamming matrix, which can be written in the same way as a matrix-vector product using the following expression: $B = H \times A$, which encodes the XOR operations with a matrix-vector product. A and B are vectors that represent the input address and output index, respectively. These vectors encode the address and the bank index. For this purpose, each bit of these numbers is placed within the corresponding vector fields. Therefore, if we consider the address $a = 11001_2$, it is translated into the vector $A = (1, 0, 0, 1, 1)$, and we proceed in the same way to encode the bank indexes in vector B. The H matrix is also composed of 0 and 1 that define the behavior of the hash function. To perform this matrix-vector product, the "inner product" $\times$ is replaced by a logical AND, while the "additive operator" $+$ is changed to XOR ($\oplus$). The same result can also be obtained by computing the matrix product in the classical way and then applying a modulo 2 on the output vector, but writing this matrix-vector product with logical operations highlights the hardware implementations of the hash function.

It is convenient to represent the H matrix using a grid, where black cells correspond to a 1 (which means a XOR input) and white cells correspond to a 0 (which means unconnected). Figure 8.12 gives a minimalist example of an H matrix of size 2×3, with a 3-bit address input and a 2-bit index output. This example is unrealistic, but it allows us to explain, simply, how this hash function works.

$$H = \begin{pmatrix} 1 & 0 & 1 \\ 0 & 1 & 1 \end{pmatrix}$$

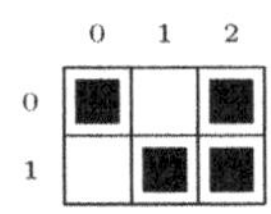

Figure 8.12. *Left: an example of an H matrix of size 2×3. Right: the same H matrix displayed as a grid*

As the H matrix will be a constant, all the AND operators will disappear, leaving only a number of XOR. To demonstrate this, we can expand $B = H \times A$ with the H matrix of Figure 8.12, as in equation [8.3]:

$$\begin{cases} B_0 = (H_{00} \cdot A_0) \oplus (H_{01} \cdot A_1) \oplus (H_{02} \cdot A_2) \\ B_1 = (H_{10} \cdot A_0) \oplus (H_{11} \cdot A_1) \oplus (H_{12} \cdot A_2) \end{cases} \qquad [8.3]$$

Then, when the values of the H matrix are substituted into equation [8.3], we end up with two XORs, as shown in [8.4]:

$$\begin{cases} B_0 = (1 \cdot A_0) \oplus (0 \cdot A_1) \oplus (1 \cdot A_2) = A_0 \oplus A_2 \\ B_1 = (0 \cdot A_0) \oplus (1 \cdot A_1) \oplus (1 \cdot A_2) = A_0 \oplus A_1 \end{cases} \qquad [8.4]$$

With an input address $a = 5 = 101_2 = (A_2 A_1 A_0)_2$, we are able to compute the output index b, as calculated in [8.5]:

$$\begin{cases} B_0 = A_0 \oplus A_2 = 1 \oplus 1 = 0 \\ B_1 = A_0 \oplus A_1 = 1 \oplus 0 = 1 \end{cases} \Rightarrow b = 10_2 = 2 \qquad [8.5]$$

It is important to note that each row and each column must contain at least 1; otherwise it would mean that some input or output bits are ignored. If there is no 1 on a line, this indicates that the corresponding bit of the output value will not be used, which halves the dynamic range of the output values. When a column is found without 1, it implies that the corresponding input bit will be ignored, which will be collision-prone in the case of a stride of the form 2^n.

In section 8.4.2.2, we show how to generate an H matrix that favors strided access patterns.

8.4.2.2. *PRIM – pseudo-randomly interleaved memory*

Rau (1991) suggested a solution called pseudo-randomly interleaved memory (PRIM) to the problem of memory bank collisions, based on a XOR scheme, as described in section 8.4.2.1. Unlike the classical XOR scheme, the PRIM hash function allows us to prove some properties concerning stride access patterns and to define conditions to reach a perfect distribution. More specifically, it is designed to produce perfect memory bank distributions for 2^n stride access patterns. Figure 8.13 gives an example of a PRIM allocation. We can observe that for four accesses with a 4-byte stride access pattern, all data are in different banks, leading to high performance.

Bank			
0	1	2	3
0	1	2	3
7	6	5	4
9	8	11	10
14	15	12	13
	⋮		

		Bank			
		0	1	2	3
Address	0	•			
	4				•
	8		•		
	12			•	
	total	1	1	1	1

Figure 8.13. *Example of PRIM allocation in a four-bank architecture, and four memory accesses with a 4-byte stride access pattern*

For a 2^n stride access pattern, there is an interesting property to note: only the consecutive bits from the position n in the address will be incremented, in order to take all possible values. Therefore, in a 2^b bank system, a hash function that would generate the bank indexes from the consecutive b bits at the position n in the address could then produce a perfect distribution of accesses in the memory banks. This can be achieved with a XOR-based hash function. Figure 8.14 shows an H matrix that behaves as just described. Such a solution has good properties; however, we can find in Rau (1991) that some stride patterns may cause some collisions.

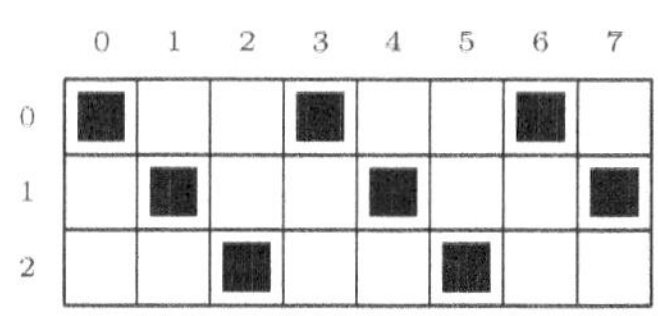

Figure 8.14. *H matrix for the PRIM solution*

PRIM is a generalization of this method but eliminates this kind of problem. The advantage of PRIM is to keep the hardware cost of the XOR scheme, but with the guarantee that stride access patterns use all the memory banks.

8.4.2.2.1. How it works

In the first approach, we can intuitively reveal that the PRIM method is based on a different algebra to that of the integer. In other words, it is based on an algebra where $+$ and $\times$ operations behave in a different way from the integer algebra operators, which is used to compute the addresses of stride access patterns. Thus, if the algebra used to compute a sequence of addresses is different from the one used to compute the bank index, then they are unlikely to share common patterns, which would be pathological patterns. This will result in a pseudo-random distribution (with no pathological patterns) of the memory banks during stride access patterns. However, PRIM goes further than this first intuition by using good mathematical properties of polynomials to achieve efficient distribution of memory banks for stride access patterns. This is what we will briefly describe in this section.

In section 8.4.2.1 on the XOR scheme, we saw that the input addresses and output indexes were encoded as a vector in order to use them for calculations. For PRIM, we will use polynomials rather than vectors to encode these numbers.

Thus, previously, we encoded number $a = 24 = 11000_2$ in vector $A = (0, 0, 0, 1, 1)$, and here we encode in polynomial $A(X) = 1.X^4 + 1.X^3 + 0.X^2 + 0.X^1 + 0.X^0 = X^4 + X^3$. Note that if we replace X by 2, we obtain the number a – it always works – but this is not the purpose of this transformation. In the following, we switch from one representation to another: implicitly when the number is lower case, we refer to its value (e.g. b); when it is upper case, we refer to the vector representation (e.g. B); and when it is parameterized with X, we refer to the associated polynomial (e.g. $B(X)$). We also label the polynomials by their associated number, for example polynomial 24 ($= 11000_2$) corresponds to $X^4 + X^3$.

Similar to the vectors mentioned earlier, these polynomials have coefficients that only take the value 0 or 1. Let us consider two polynomials $C(X) = X + 1$ and $D(X) = X$. Thus, as with vectors, the $+$ operation corresponds to a bitwise XOR, for example $C(X) + D(X) = (X + 1) + X = 1$. Regarding the $\times$ operator, which does not exist for vectors, this operator has the potential to shift coefficients, for example $C(X) \times D(X) = (X + 1) \times X = X^2 + X$.

As the product between polynomials is defined, it allows us to use the division and modulo of polynomials. PRIM is indeed based on the idea of replacing the integer modulo by the polynomial modulo. The polynomial modulo is written $R(X) = A(X) \bmod P(X)$, where $R(X)$ is the remainder of the polynomial division and $P(X)$ is the divisor. As in the classical Euclidean division, the polynomial division must satisfy the following equations: $A(X) = Q(X) \times P(X) + R(X)$ and $Deg(R(X)) < Deg(P(X))$.

These operations seem very complex to implement with hardware. However, when we expand all the calculations, we observe that, for a constant $P(X)$ polynomial, all the operations are simplified and only XORs will remain.

By directly substituting $A(X)$ into the equation $R(X) = A(X) \mod P(X)$, we obtain equation [8.6]:

$$R(X) = (a_{n-1} \cdot X^{n-1} + \cdots + a_0 \cdot X^0) \mod P(X) \qquad [8.6]$$

Then, by applying the modulo distributivity to [8.6], we obtain equation [8.7]:

$$R(X) = a_{n-1}(X^{n-1} \bmod P(X)) + \cdots + a_0(X^0 \bmod P(X)) \bmod P(X) \qquad [8.7]$$

As the XOR operation cannot increase the degree of polynomials, the last *mod* operation is not required to satisfy the constraint $Deg(R(X)) < Deg(P(X))$, so we can then remove it and obtain [8.8]:

$$R(X) = a_{n-1} \cdot (X^{n-1} \mod P(X)) + \cdots + a_0 \cdot (X^0 \mod P(X)) \qquad [8.8]$$

Let $H_i(X) = X^i \mod P(X)$, by replacing this in the previous equation, we obtain a simplified equation [8.9]:

$$R(X) = a_{n-1} \cdot H_{n-1}(X) + \cdots + a_0 \cdot H_0(X) \qquad [8.9]$$

Polynomials $H_i(X)$ are only dependent on the polynomial $P(X)$, which is itself a constant, so they are also constants that can be calculated in advance. We end up with a structure very close to the XOR scheme that is only made up of XORs. We can even show that [8.10] is equivalent to a XOR. Equation [8.10] shows this transformation to a matrix-vector product scheme using the vector representation of polynomials $H_i(X)$:

$$R = a_{n-1} \cdot H_{n-1} + \cdots + a_0 \cdot H_0 = H \times A \qquad [8.10]$$

Through this demonstration of a few lines, it is shown that it is possible to transform a complex calculation of polynomial division, the first formula, into a computation that is very easy to implement in hardware, the last result which consists only of logical gates XOR (the operation +).

As an example, we can determine the bank of the address $a = 13$ from Figure 8.13, which uses PRIM 7 (i.e. PRIM with polynomial 7). If $p = 7 = 111_2$, then $P(X) = X^2 + X + 1$. However, the address $a = 13 = 1101_2$ corresponds to the polynomial $A(X) = X^3 + X^2 + 1$ or the vector $A = (1, 0, 1, 1)$. There are two ways to compute this bank number, with a polynomial division: $A(X) \div P(X)$; or with a matrix-vector product: $H \times A$.

Equation [8.11] shows how to compute the bank number using the Euclidean division of the polynomial:

$$\begin{array}{r|l} X^3 + X^2 \qquad + 1 & X^2 + X + 1 \\ \cline{2-2} X^3 + X^2 + X \qquad & X \\ \cline{1-1} R(X) = \qquad X + 1 & \end{array} \qquad [8.11]$$

If $R(X) = X + 1$, then the bank is $11_2 = 3$, as shown in Figure 8.13.

Equation [8.12] shows how to compute the bank number using the matrix-vector product $B = H \times A$. The H matrix is generated by PRIM using polynomial 7:

$$B = H \times A = \begin{bmatrix} 1\,0\,1\,1 \\ 0\,1\,1\,0 \end{bmatrix} \times \begin{bmatrix} 1 \\ 0 \\ 1 \\ 1 \end{bmatrix} = \begin{bmatrix} 1 \\ 1 \end{bmatrix} \qquad [8.12]$$

If $B = (1, 1)$, then the bank is $11_2 = 3$, as shown in Figure 8.13. These two methods are equivalent since the H matrix is generated for this purpose.

In his 1991 paper, Rau showed several good properties of this hash function that allow us to choose $P(X)$ polynomials that produce better distributions.

8.4.2.3. *PRIM example – Intel LLC Complex Addressing*

This section briefly explains the principle of Intel's "Complex Addressing", used in the L3 cache – Last-Level Cache (LLC) – of Intel processors. We can show how techniques such as PRIM can be exploited in mainstream circuits to improve performance. This example is based on reverse engineering work on the hash function of Intel's Last-Level Cache (Maurice *et al.* 2015; Yarom *et al.* 2015).

The LLC of Intel processors is made up of slices; the concept of the cache slice is similar to that of the memory bank as it consists of partitioning the cache memory into memory slices which are independently controlled. The number of slices is then chosen according to the number of processor cores, and each slice must provide the bandwidth equivalent to one core. To determine in which slice each address should be mapped, Intel uses an undocumented hash function that is called Complex Addressing. This mapping must produce a distribution of addresses as uniform as possible in the cache slices, to avoid contention in the cache and therefore a reduction in performance.

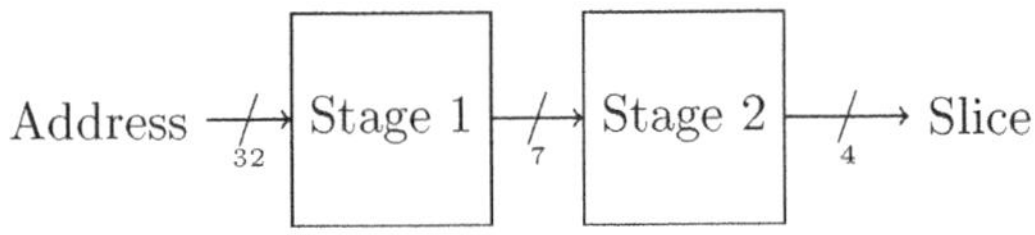

Figure 8.15. *Complex Addressing circuit overview*

Reverse engineering works show that Complex Addressing is made up of two stages, as illustrated in Figure 8.15. The first stage inputs a 32-bit physical address and outputs a 7-bit index, then the second stage reduces this index from 7 bits to a number that allows slice selection – a number between 0 and the slice count minus 1. This two-stage structure makes it possible to keep nearly the same hash function regardless of the number of cores in the processor; only the last stage of the hash function must be adjusted to the number of cores/slices.

The analysis shows that the first stage consists of a XOR combination of the input bits that is equivalent to a XOR scheme, as described in section 8.4.2.1. The H matrix extracted from the first stage of the Complex Addressing hash function is reproduced at the top of Figure 8.16.

Interestingly, the structure of the Complex Addressing H matrix looks similar to the type of pattern that can be generated by the PRIM algorithm.

The second matrix shown in Figure 8.16 corresponds to an H matrix generated with PRIM and the polynomial 67. The two red spots highlight the differences between the two H matrices. The H matrix extracted from Complex Addressing also includes an extra line that is not highlighted in red to improve readability. We can then note that PRIM 67 matches the hash function extracted by the researchers almost perfectly. Therefore, it seems reasonable to conclude that the construction of this Intel XOR scheme is based on PRIM or a similar method.

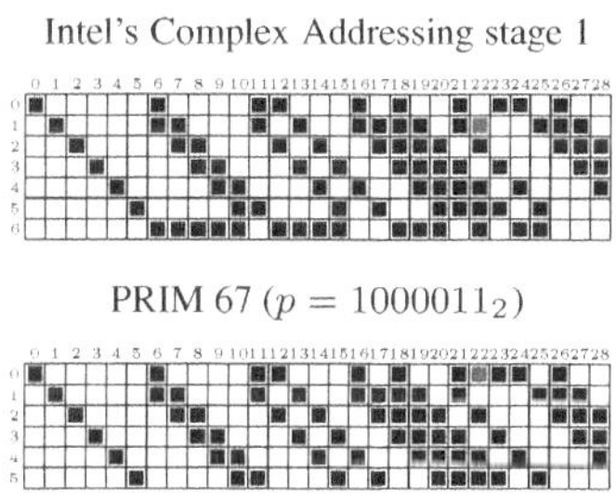

Figure 8.16. *Intel Complex Addressing stage 1 and PRIM 67*

Using a hash function based on PRIM is a reasonable choice from Intel as it makes it possible to massively eliminate collisions produced by stride access patterns that can be generated with a DMA, or from the AVX extensions and their gather/scatter instructions.

8.5. Modeling and experimenting

In sections 8.4.2 and 8.4.1, we presented several solutions to deal with collision issues. Evaluating and comparing these methods with only theoretical considerations is quite complex and not effective enough as they are built and based on different theoretical aspects. The best way to compare them is to define a set of tests and use simulation models of the memory architecture. Thus, we can inject data traffic in the memory architecture to compare performance.

8.5.1. *Simulator implementation*

In order to evaluate the performance, we need to build a realistic model of the memory system. The model will thus be a complete simulation of how an interconnect behaves. However, we focus on the arbitration model of memory accesses in the architecture. This means that our simulation model does not include the computational part of applications but only the path to or from the memory elements, as well as the different patterns of memory accesses.

In addition, the simulation model should be modular and evolutive, to allow the exploration of different memory architectures and different hash functions for memory bank interleaving. The simulation model is built as a set of several simulation components: memory banks, computational cores, FIFOs and arbitration units. Each of these simulation components is able to transfer memory accesses to the other components.

Accesses may be represented by a single structure containing the origin, the destination and the date of access. The inclusion of the date allows latency measurements. Accesses are generated and then injected by the cores into the interconnect; they will then flow through the interconnect and be routed by the arbitration components to their destination memory bank. As each component is able to compute statistics on all memory accesses that pass through it, it is possible to track which component is used more than the others.

Computing units (mainly cores) are modeled as simple memory access sources. They generate predefined memory access patterns or replay a memory access record extracted from programs or applications. The model presented in the following section does not consider software dependencies of memory accesses. Thus, if some access date depends on the latency of previous accesses, this dependence will be ignored. But in high-performance applications, these dependencies are kept as low as possible. However, if the latency becomes too high, the measured performance will be overestimated.

8.5.2. *Implementation of the Kalray MPPA cluster interconnect*

To use a real system as a comparison, we have implemented the interconnect architecture of a Kalray MPPA cluster. Figure 8.17 shows a simplified cluster containing only four cores (instead of 16). This figure gives an overview of the overall memory architecture with cores C_n and memory banks B_n connected together by the interconnect.

Figure 8.18 represents the simplified architecture with four banks and four cores of the interconnect with its crossbar and arbitration components. Here the architecture is a crossbar with four masters and four slaves, with each master having a direct link with the slaves. Each master can use only one link at the same time, so it could be linked with only one slave. Such an architecture prevents internal collisions inside the interconnect as there are no shared links by several masters. According to the figure, each core is connected to an scr_n component that is in charge of directing the accesses to the corresponding bank, and that scrambles the accesses following the interleaving scheme. Thus, it is in these scr_n components that the hash functions are implemented. Finally, the outputs of these components are linked to arbiters arb_n that have the function – in case of collision – to decide which access will be performed.

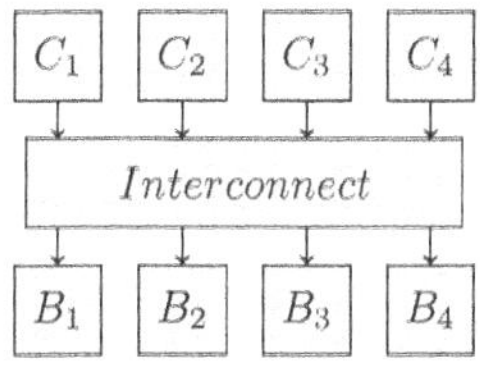

Figure 8.17. *Overview of the Kalray MPPA simplified local memory architecture*

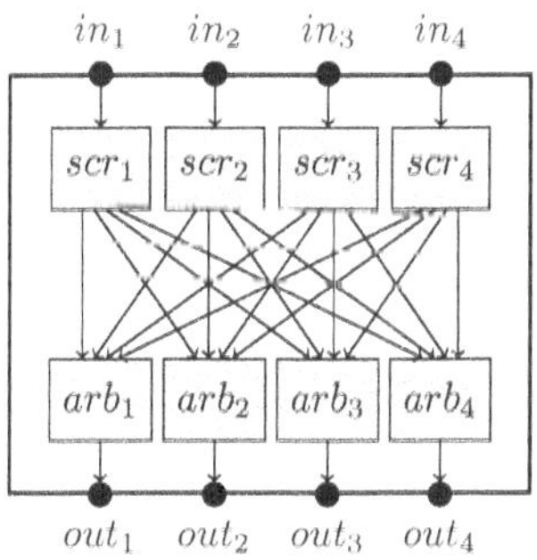

Figure 8.18. *Kalray MPPA simplified crossbar internal architecture*

In our simulator, each component of this interconnect is modeled to reproduce its real behavior. Since the model is split into simple components, it is easy to modify any architectural detail, allowing architectural exploration. For example, if we want to add a 17th bank, then we have to modify the scr_n to add an output, change their hash function and add an arb_17.

8.5.3. *Objectives and method*

The objective of the experiments is to compare the efficiency of the memory system using the different hash functions, discussed previously, in order to determine which hash functions offer the best memory performance. Memory performance is highly dependent on stride size, so we will measure performance as a function of stride size. We then consider a cluster of 16 cores based on the Kalray MPPA cluster, with 17 memory banks if prime modulo indexing is used and 16 memory banks overwise. Every access is 8 bytes aligned, and it is assumed that, at each access, 8 bytes are read. It is unnecessary to study misaligned access patterns that are not efficient and infrequently used, such as a stride aligned on 7 bytes. Hence, they are excluded from the results to avoid adding unnecessary noise.

The memory banks interleaving granularity is 32 bytes, corresponding to cache line size, but not all the accesses will pass through the cache – they are sent directly to the memory.

In order to visualize performance of the methods described in section 8.4, we use the memory model, as described previously, and inject several types of access patterns generated by the simulation. We obtain information such as the average use of memory banks (in number of accesses performed per cycle) depending on the stride size. In the case of perfect distribution of accesses in the memory system, this measure should be 16 accesses/cycle. This means the maximum bandwidth is used on all the banks, and therefore the memory throughput of the cores is maximized and all cores will be running at full capacity. On the contrary, if we are measuring 15 accesses/cycle, then it means that, on average, one access per cycle cannot be achieved. This corresponds to the bandwidth of one core, and therefore we lose the processing capacity equivalent to one core. This can also be translated as 15 active cores/cycle, meaning that, on average, at each cycle one core has to wait for its access to be performed. A method is considered to be good if it allows high performance for widely used stride sizes – multiples of power of 2 ($K \times 2^n$).

We can use a random access pattern as the reference value. Indeed, the random access pattern provides average performance regardless of the memory bank selection method. If the performances are worse than the random pattern, then it would be more efficient to use a pseudo-random distribution of the banks. This means that the method used is not very efficient.

In addition to the simulation with the real architecture and its variations, we will add a simulation with a modified architecture, where a FIFO has been inserted on all links between the core and the memory bank. This architecture is very costly in terms of hardware; the addition of $N \times M$ FIFOs requires a huge amount of surface area, with N being the number of cores and M being the number of memory banks. However, this model is particularly efficient in absorbing collisions and will be used as an indicator. This is because when sporadic collisions occur, these FIFOs will be used to buffer the access and serve them later, without blocking the other access from the core. This helps to smooth out collisions when they are only occasional, but not frequent. This has the effect of avoiding the accumulation of minor delays. Thus, this simulation will help us to determine whether the loss of performance is due to isolated collisions or whether the problem is recurrent, which indicates that there is a structural problem with the interleaving scheme.

8.5.4. *Results and discussion*

8.5.4.1. *Mapping MOD 16*

Figure 8.19 shows simulation results for the modulo 16 indexing (MOD 16). In other words, a classical modulo is used for bank selection in a 16-bank memory system. In the figure, the abscissa represents the stride size S, i.e. the distance between the addresses generated by the cores. When the 16 cores generate addresses with a stride of S, all the 16 generated addresses are spaced by S.

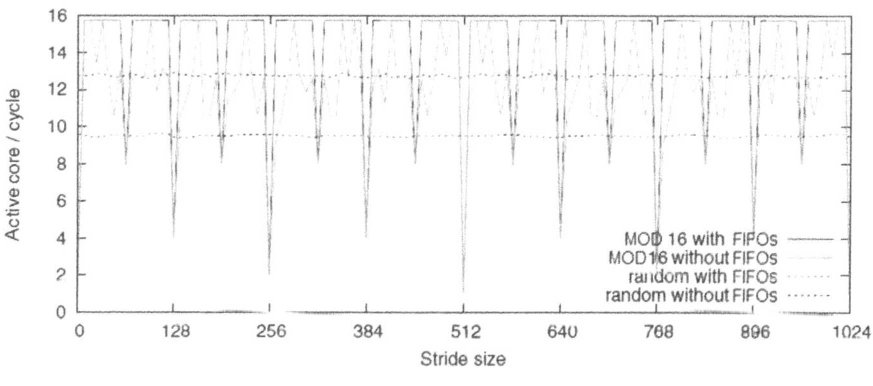

Figure 8.19. *Theoretical performance measure (in accesses per cycle) for stride access with MOD 16*

As a reminder, we use two simulations: one without FIFOs, which measures all collisions (purple curve), and another with FIFOs, which only measures recurrent collisions (blue curve). The graphs of the results with and without FIFOs share a similar peak-down pattern on stride sizes of the powers of two and their multiples. This highlights the structural pathological behaviors of MOD 16 that we would like to avoid.

This pattern can be explained as follows. The stride-by-512 bytes indicates only one access per cycle for the 16 requested accesses. This means that all cores try to access the same memory bank. Indeed, if core 0 generates address 0, it tries to access memory bank 0. If core 1 generates address 512, then it also tries to access memory bank 0 ($\lfloor 512 \div 32 \rfloor \mod 16 = 0$). Core 2 with generated address 1024 also goes to access bank 0 ($\lfloor 1024 \div 32 \rfloor \mod 16 = 0$) and so on, with all the cores going to access bank 0. Something similar happens with stride-by-128 bytes, which allows an average of only four accesses per cycle. Regardless of adding buffers in the memory system, all the accesses are directed to the same memory bank. As a comparison, for a stride of 32, all cores access a different memory bank, so we obtain 16 accesses per cycle, which is the best possible performance.

The curve without FIFO, in addition to the peak-down pattern, which is a structural issue, has other flaws that can also be explained. As an example, let us take stride-by-240 bytes that has a performance of about 10 accesses/cycle. Given that $240 \div 32 = 7.5$, the memory bank selection will behave almost like a stride-by-224 bytes (as $224 \div 32 = 7$) but with a displacement caused by the decimal part of 7.5 (which is 0.5). This will cause occasional collisions as the decimal part will accumulate until it reaches 1, which will increase the index, causing a bank change and occasionally a collision. These collisions are occasional and all banks are used evenly over time, which allows the architecture with FIFOs to absorb them.

This curve shows very good performance, way above the performance of the random pattern, with 16 accesses per cycle for strides that are not powers of two or their multiples (except for small powers of two). However, we prefer to encourage the opposite behavior, with good performance for the powers of two and inferior performance otherwise.

8.5.4.2. *Mapping MOD 17*

Figure 8.20 shows simulation results for modulo 17 indexing (MOD 17). It uses prime modulo indexing with a prime number of bank (17 memory banks).

The graph of MOD 17 shows a structural pathological behavior similar to MOD 16 with a peak-down pattern, but aligned on multiples of $17 * 2^n$. This pattern has the same origin as for MOD 16: the stride-by-544 bytes indicates that only one access per cycle is performed among the 16 requested accesses. All the cores try to access the same memory bank. Indeed, if core 0 generates address 0, it tries to access memory bank 0. If core 1 generates address 544, then core 1 also tries to access memory bank 0 ($\lfloor 544 \div 32 \rfloor \mod 17 = 0$). The same happens for core 2 with generated address 1088 ($\lfloor 1088 \div 32 \rfloor \mod 17 = 0$), and so on, with all the cores going to access bank 0. This problem is structural since it also appears on the curve of the results with FIFO (purple curve). As 17 has factors in common with fewer numbers than 16, the number of pathological accesses is lower in MOD 17 than in MOD 16 (fewer bad performance peaks). In addition, these pathological patterns are on less commonly used stride sizes (stride sizes multiple of 17 are not common), which makes MOD 17 more interesting than MOD 16 in terms of performance.

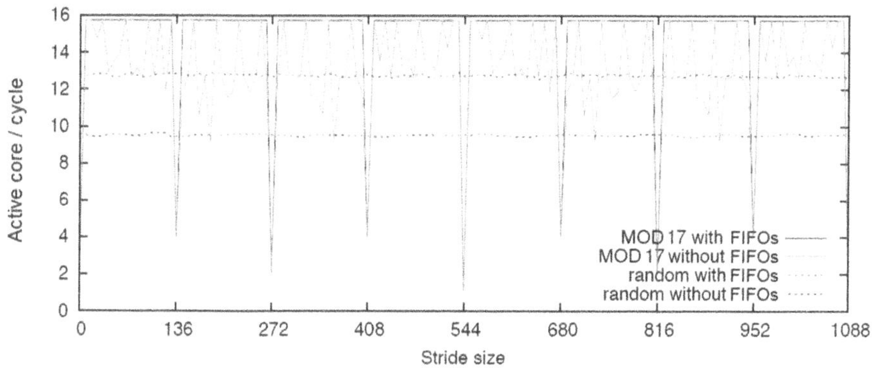

Figure 8.20. *Theoretical performance measure (in accesses per cycle) for stride access with MOD 17*

The curve without FIFO (blue curve) also shows the same types of occasional collisions as MOD 16 and for similar reasons. It is interesting to note that these occasional collisions never happen on stride sizes that are powers of two. The curve is almost always above the performance of the random access pattern, and reaches 16 accesses per cycle for powers of two stride sizes but not necessarily all their multiples.

Broadly speaking, we reduce both pathological patterns and the probability of generating some of them. However, the downside of MOD 17 is that it requires specific hardware components, thereby increasing the size of the circuit and the cost of this solution.

Figure 8.21 shows some differences between MOD 16 and MOD 17 for the same executable code. With a stride of 16 words ($16 \times 8 = 128$ bytes), we can observe that the architecture with 17 banks provides better efficiency for this program, but with a higher complexity for the modulo 17 computation.

8.5.4.3. *Mapping PRIM 47*

Figure 8.22 shows simulation results for the pseudo-randomly-interleaved-memory indexed memory system using the polynomial 47 (PRIM 47). It therefore uses a 16-bank memory system to keep the interconnect and bank selection simple.

The results are generally worse than MOD 16 and MOD 17, and slightly below the random access pattern (because PRIM is not a perfect random number generator). However, it does not suffer from any pathological stride access pattern, as all the stride sizes have approximately the same performance. This is because the accesses remain evenly distributed in the memory banks which avoids pathological patterns. It can be noted that the curve with FIFO is higher because it manages to absorb some collisions, but there are too many collisions to resolve all of them as the FIFOs become saturated.

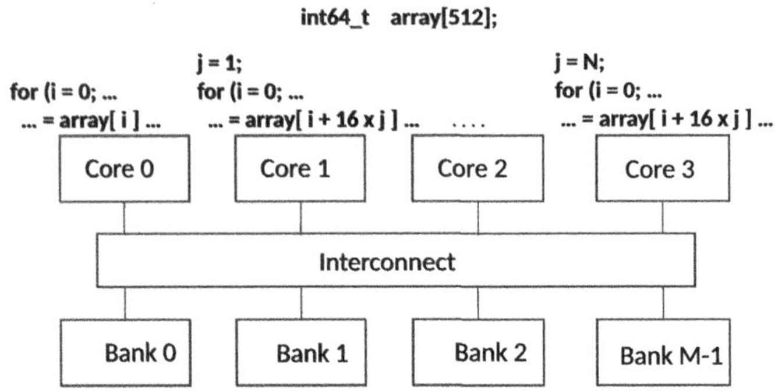

		Core 0	Core 1	Core 2	Core 3
i = 0	Array index	array[0]	array[16]	array[32]	array[48]
	Mem. address	@0	@128	@256	@384
	MOD16 bank	0	0	0	0
	MOD17 bank	0	9	1	10
i = 1	Array index	array[1]	array[17]	array[33]	array[49]
	Mem. address	@8	@136	@264	@392
	MOD16 bank	8	8	8	8
	MOD17 bank	8	0	9	1

Stride of 16 x 8 B = 128 B

The 4 cores access the same bank

The 4 cores access different banks

Figure 8.21. *Comparison between MOD 16 and MOD 17 for the same executable code and the same architecture*

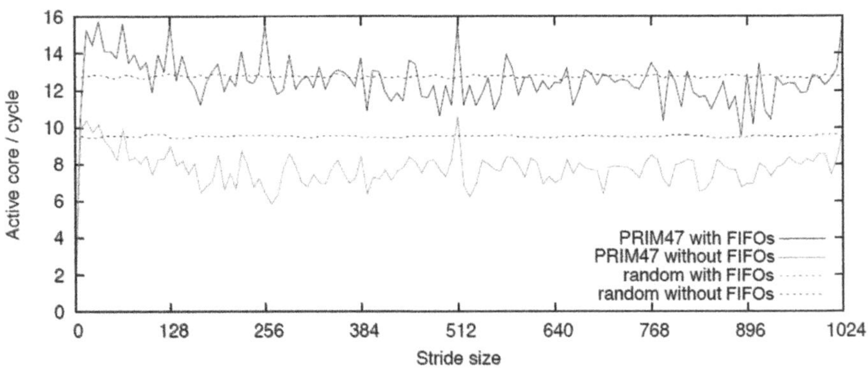

Figure 8.22. *Theoretical performance measure (in accesses per cycle) for stride access with PRIM 47*

We also observe that the power-of-two stride sizes can reach 16 accesses per cycle if we can absorb a few isolated collisions. These isolated collisions probably come from carry propagation (within the address), which changes more bits than PRIM is designed for, and from the desynchronization between execution of cores that the collisions produce.

Thus, PRIM needs to be considered in specific conditions that require a simple implementation, no pathological stride size and/or use the power-of-two stride sizes.

8.5.4.4. *Addition of memory banks*

Previous approaches are based on predictable access patterns, and hence there is a solution to find an optimal memory distribution. However, this does not work for random access patterns as we cannot predict the distribution. It is then interesting to explore other solutions such as additional memory banks and access reordering.

Let us consider the bandwidths of memory banks and cores. If the bandwidth of memory banks is equal to the bandwidth of generated addresses by cores (16 memory banks for 16 cores, for example), when a collision occurs, we accumulate delays of one cycle. Such a delay cannot be absorbed if cores continue to exploit the full bandwidth. Consequently, under these conditions, the access latency can only increase, and thus it will slow down the program and the application. To absorb this latency, one solution is to provide a high memory bandwidth, higher than that which cores can generate. This could be done by adding more memory banks.

With more memory banks, collisions should be reduced if data are still well distributed in these memory banks to prevent all cores from accessing the same memory bank for a stride access. This solution imposes the use of one of the previous hash functions to ensure a correct distribution.

However, adding memory banks has a downside; we should recall that the hardware cost is not negligible. The interconnect is more complex as it has to manage more slave components (memory banks) and includes more memory controllers.

In section 8.5.4.6, we discuss the performance improvement that this allows.

8.5.4.5. *Access reordering*

We note that some accesses are delayed because they are stuck behind another access, which is itself stuck by a collision, even if these two accesses are not directed to the same bank memory. This behavior decreases the overall efficiency of our memory system, regardless of the access pattern used. To solve this problem, the memory system needs to allow accesses that are stuck behind another when they are not directed to the same memory cell. Consequently, the memory accesses are reordered to allow better use of the memory system. Reordering accesses has an impact on memory consistency, since this allows us to observe the accesses of another processor completed in a different order from the original. It is therefore necessary to be attentive to preserve the memory consistency model of the system. The subject of this chapter is not to study different memory consistency models, or different architectures for access reordering, but to consider only the weak consistency model as this is a common memory consistency model for high-performance processors, and to consider a single abstract architecture. The weak consistency model allows the reordering of most of the memory accesses and specific instructions such as "fence" must be used to guarantee the order of the accesses.

To set up the access reordering, we use a simple buffer at the output of each core. When a core generates an access, it fills this buffer and the memory system reads from this buffer the first available entry that does not collide with an access from another core. The buffer size corresponds to the reorder capacity of the memory system. This mechanism prevents reordering the accesses of one core on the same memory bank, but allows accesses to be reordered to different memory banks. More complex and more optimized architectures are realizable, but here we use this mechanism only to show that it allows us to improve performance and to see how it combines with other optimizations.

8.5.4.6. *Combination of access reordering and addition of memory banks*

The two architectural modifications, addition of memory banks and access reordering, do not exclude each other and can be used together. Thanks to our modular simulator, we can measure the impact of the memory bank number on collisions, and their ability to absorb them.

Figure 8.23 shows the performance of our memory system under a random access pattern using these two modifications. The abscissa shows the number of memory banks within the memory system, and in the ordinate, we vary the reordering capacity of the memory system. Both are found to have positive effects on performance, but, more importantly, they have better performance when both are used at the same time.

This figure shows that adding memory banks allows us to improve performance, but requires a large number of additional banks to fully exploit the memory system, due to the dependency that blocks the accesses one after the other. Similarly, the figure shows that increasing the buffer size allows us to improve performance, but the gain is limited because of memory bank collisions. It is more interesting to examine what happens when both are used together. Both effects are combined, reducing collisions and dependencies between accesses at the same time. The best choice is therefore to find a balance between a memory system with more banks and a memory system with relaxed dependencies between accesses. These are architectural ideas to be considered together for applications where a collision-free distribution is not feasible. It should be noted that these architectural modifications are not incompatible with the interleaving scheme discussed previously.

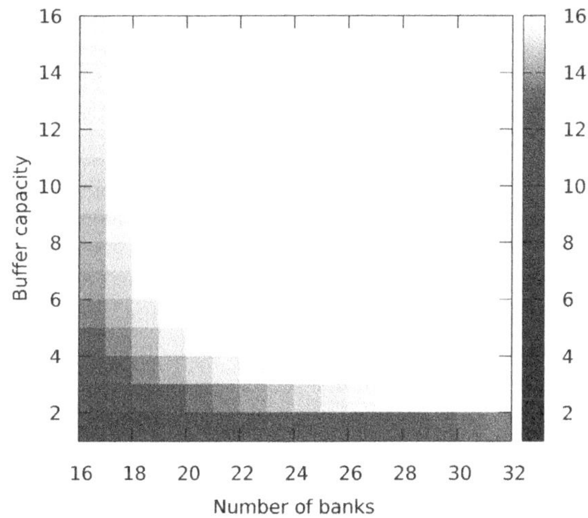

Figure 8.23. *Hotmap of memory access efficiency according to the number of banks and buffer size during random access patterns*

8.5.4.7. *Discussion*

The comparative results of these methods do not show an advantage for a method on all criteria: complexity/effectiveness/usability. There is no universal solution to this problem, i.e. one that is simple, effective and easy to use for all applications. They all have their strengths depending on the application, and this is what we have shown here in these experiments.

The architect's job is precisely to choose among all these options – interleaving scheme, architectural modification or software solutions – and to combine them to find the right compromise to fit the constraints of his or her system. They can also choose to include several modes in their system to suit different applications.

Hash function-based methods such as PRIM can be very effective when used under the conditions for which they are built; however, outside of these specific conditions, they become almost equivalent to random access patterns. Methods based on prime number modulo are often very effective but have a fairly high material cost, compared to PRIM. However, in order to be effective with random access patterns that may arise in complex applications, these methods must be combined with other architectural changes, such as adding more bank memory or reordering memory accesses.

8.6. Conclusion

This chapter dealt with optimizations in memory bank accesses performed within many-core or multi-core architectures. When some cores get access to the same bank simultaneously, this implies a memory bank conflict, delaying the other accesses. Reducing memory bank conflicts could be solved using an efficient spread of resources or data in the memory banks. In this chapter, we focused on hardware solutions that place fewer constraints on the software, rather than on software solutions. Hardware solutions are based on the assumption that memory accesses follow regular and well-known patterns. This makes it possible to smartly spread resources or data in memory banks, in order to decrease memory bank access conflicts as much as possible, while keeping the software architecture suitable.

We first identified pathological access patterns that produce a large number of memory bank access conflicts. We studied and compared previous work and optimizations, such as prime modulus indexing or pseudo-randomly interleaved memory. We then performed simulations to evaluate several different approaches to reduce conflicts in the local memory of a cluster of the Kalray MPPA processor.

Comparative results did not show a universal solution, i.e. simple to implement, effective and easily used by all software applications. Unfortunately, all the methods presented, based on hash functions, have both strengths and weaknesses. There are still some tasks for system architects to evaluate to choose the best solution and the best trade-off between complexity, effectiveness and usability.

8.7. References

Chavet, C., Coussy, P., Martin, E., and Urard, P. (2010). Static address generation easing: A design methodology for parallel interleaver architectures. *Proceedings of the 35th International Conference on Acoustics, Speech, and Signal Processing*, Dallas, USA, 1594–1597 [Online]. Available: https://hal.archives-ouvertes.fr/hal-00455121.

Diamond, J.R., Fussell, D.S., and Keckler, S.W. (2014). Arbitrary modulus indexing. *Proceedings of the 47th Annual IEEE/ACM International Symposium on Microarchitecture*, Washington, DC, USA, 140–152.

Dupont de Dinechin, B. (1991). An ultra fast euclidean division algorithm for prime memory systems. *Proceedings of the ACM/IEEE Conference on Supercomputing*, 56–65.

Kalray (2016) [Online]. Available: http://www.kalrayinc.com/kalray/products.

Maurice, C., Le Scouarnec, N., Neumann, C., Heen, O., and Francillon, A. (2015). Reverse engineering intel last-level cache complex addressing using performance counters. In *Research in Attacks, Intrusions, and Defenses*, Bos, H., Monrose, F., and Blanc, G. (eds). Springer International Publishing, Cham.

Rau, B.R. (1991). Pseudo-randomly interleaved memory. *Proceedings of the 18th Annual International Symposium on Computer Architecture*. Toronto, Canada, 74–83.

Seznec, A. (2015). Bank-interleaved cache or memory indexing does not require euclidean division. *11th Annual Workshop on Duplicating, Deconstructing and Debunking*, Portland, United States [Online]. Available: https://hal.inria.fr/hal-01208356.

Sohi, G.S. (1993). High-bandwidth interleaved memories for vector processors: A simulation study. *IEEE Transactions on Computers*, 42(1), 34–44.

Yarom, Y., Ge, Q., Liu, F., Lee, R.B., and Heiser, G. (2015). Mapping the intel last-level cache. *IACR Cryptology ePrint Archive*, 905.

PART 3

Interconnect and Interfaces

9

Network-on-Chip (NoC): The Technology that Enabled Multi-processor Systems-on-Chip (MPSoCs)

K. Charles JANAC

Arteris IP, Campbell, USA

Over the last two decades, networks-on-chip (NoCs) have advanced from PhD theses to being the key technology enabling the creation of today's production of multi-billion transistor multi-core systems-on-chip (SoCs). This chapter first explains the history of on-chip interconnect technology, from buses to crossbars and finally NoCs, and then explains the key technical components of NoCs and the metrics that make them fit for a particular purpose. Methods of NoC configurability are explained, including component assembly and synthesis approaches. System-level services implemented by NoC technologies are explained, including quality-of-service (QoS)/bandwidth and latency guarantees, voltage domains, frequency domains, security, debug, performance profiling and functional safety. The chapter then explains the use of NoCs to carry more advanced communications semantics and explains how NoCs have led to a new era of heterogeneous cluster cache coherency and inter-die communications. Finally, the chapter speculates on the future of NoC technology including floorplan-guided automatic NoC topology synthesis, on-chip distance spanning technologies, chiplets and integrated diagnostics to optimize system-level runtime performance, safety and security.

For a color version of all figures in this book, see www.iste.co.uk/andrade/multi1.zip.

Multi-Processor System-on-Chip 1 – Architectures,
coordinated by Liliana ANDRADE and Frédéric ROUSSEAU.

9.1. History: transition from buses and crossbars to NoCs

On a system-on-chip (SoC), intellectual property blocks (IPs) communicate over an interconnect system using transactions. A transaction is the operation of reading or writing bytes of data to and from addresses in a memory space. Transactions are composed of a *request*, which goes from an *initiator* to a *target* and contains at least the address and data in the case of a write operation, and a *response* flowing back from the *target* to the *initiator*, with status and data in case of read operation. The physical interfaces between IPs and the interconnect system are called *sockets*, usually synchronous, with a *master* side and a *slave* side. A number of different socket protocols have been defined by the SoC industry (Arm® AMBA® AXI, APB and AHB, OCP, etc.). Sometimes, transactions have to be converted between the initiator socket and the target socket of an interconnect, for instance, if both socket protocols are not identical. Such conversions must preserve the byte transfer semantics, i.e. the functionality of the transaction.

Interconnect systems have gradually evolved from simple buses to more scalable networks-on-chip (NoCs). It could be tempting to consider a NoC as a downscaled TCP/IP or ATM network; it is not. TCP/IP and ATM are connection-based networks, where connections are dynamically opened and closed between producer (often called *initiator* or master) and consumer (also called *target* or slave) ports. History shows that small systems are transaction-based, while large systems are connection-based. Attempts to design small connection-based systems (INMOS transputers), or large transaction-based systems (distributed shared memory PC clustering), have up to now failed. A time may come when SoCs are so huge that a connection-based network is required between transaction-based clusters, but this is not yet the case.

In the 1980s and 1990s, buses were the on-chip interconnect technology implemented in virtually all system-on-chip (SoC) semiconductor designs. Buses were typically handcrafted around either a specific set of features relevant to a narrow target market or support for a specific processor. On-chip buses were essentially implemented as scaled-down versions of board-level interconnects implemented in the 1980s (Benini and De Micheli 2002).

The next evolution of bus interconnects was the hybrid bus, which included a centralized crossbar switch with all the internal IP connections needed for the chip. This presented capacity and routing congestion problems as SoCs became more complex, so multi-layer hybrid buses were the next evolution of the interconnect technology. However, these multi-layer hybrid buses posed latency, frequency, power and area problems as SoC complexity continued to evolve.

Several trends forced evolutions of systems architectures that drove evolutions of required on-chip interconnect technologies. These trends are:

– Application convergence: the mixing of various traffic types in the same SoC design (video, communication, computing, etc.). These traffic types, although very different in nature, for example, from the quality-of-service (QoS) point of view, had to share resources, typically a unified off-chip memory, that were assumed to be "private" and handcrafted to the particular traffic in previous designs.

– Moore's law drove the integration of many IP blocks into a single chip. This is an enabler to application convergence, but it also allowed entirely new approaches (parallel processing on a chip using many small processors) or simply allowed SoCs to process more data streams (such as communication channels).

– IP reuse: the pervasive reuse of existing IP blocks is an absolute necessity. The shapes those to-be-reused IPs can take is quite variable: hard (layout-ready) or soft, in-house or third-party, simple (USB) or complex (complete H.264 decompression engine). The NoC interconnect has to convey transactions between sockets of different protocols (OCP, AXI, proprietary, etc.), each protocol being highly parameterized. This variety has major impacts from a performance perspective. The IPs may have to be clocked at different frequencies, and their sockets may present different data widths, leading to a wide range of peak throughputs, which must be adapted between initiators and targets. Initiators have different traffic patterns, in terms of transaction lengths, address alignments and regularity. Their reactions and resistance to latency or transient backpressure differ as well. Also, meeting the (QoS) required by each initiator despite the presence of the other traffic becomes more challenging as the number of IPs increases.

– Arrival of 40 nm silicon processes that allowed the building of SoCs with a higher level of functional integration. A consequence of silicon process evolution was that gates cost relatively less than wires, both from an area and performance perspective, than with previous process generations. Transistor logic gates scaled as silicon process critical dimensions decreased, but wires did not shrink at the same level because wire resistance parasitics became increasingly important.

– Time-to-market pressures drove most designs to make heavy use and reuse of synthesizable RTL rather than manual layout, in turn restricting the choice of available implementation solutions to fit a bus architecture into a design flow.

These trends drove the evolution of many new interconnect architectures. These include the introduction of split and retry techniques, removal of tri-state buffers and multi-phase clocks in favor of pure synchronous digital design, introduction of pipelining and various attempts to define standard communication sockets, such as VCI (Virtual Component Interface).

However, history has shown that there are conflicting trade-offs between compatibility requirements, driven by IP blocks reuse strategies, and the introduction of the necessary bus evolutions driven by technology changes: in many cases, introducing new features required many changes in the bus implementation, more importantly in the bus interfaces (e.g. the evolution from AMBA ASB to AHB2.0, then AMBA AHB-Lite, then AMBA AXI), with major impacts on IP reusability and new IP design.

Buses do not decouple the activities generally classified as transaction, transport and physical layer behaviors. This is the key reason they cannot adapt to changes in the system architecture or take advantage of the rapid advances in the silicon process technology. In addition, buses require centralized arbiter and response decoding logic, which becomes a performance bottleneck as the number of masters and slaves increases. An example of this is the Arm AHB 2.0 bus shown in Figure 9.1 (ARM 1999).

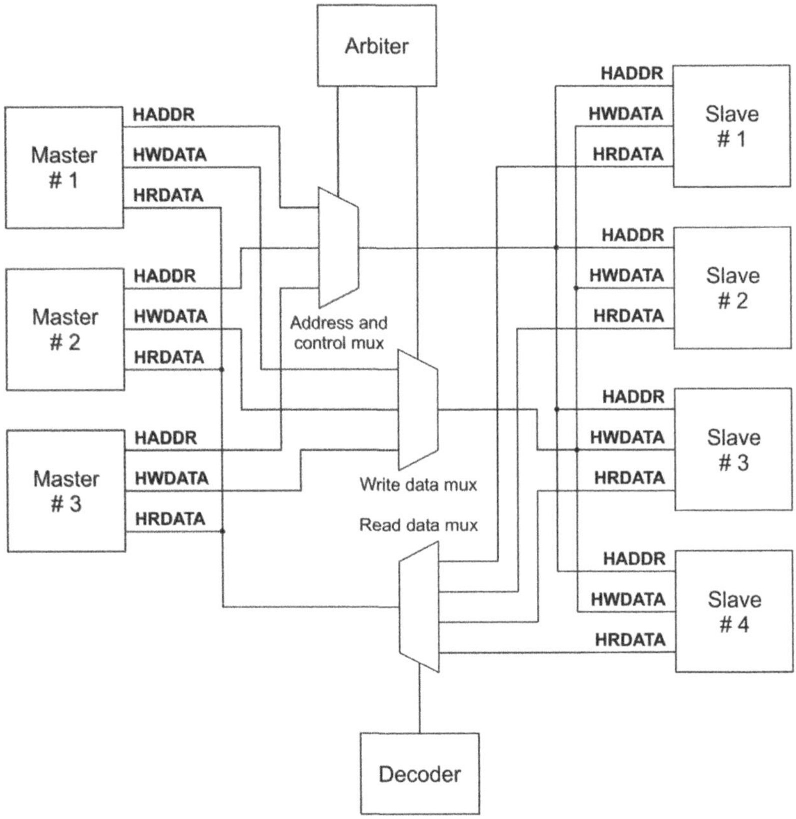

Figure 9.1. *A traditional synchronous bus, in this case the implementation of an AMBA AHB bus, requires centralized arbiter and decoder logic (ARM 1999)*

Consequently, changes to bus physical implementation can have serious ripple effects upon the implementations of higher-level bus behaviors. Replacing tri-state techniques with multiplexers had little effect upon the transaction levels but allowed much higher frequencies. Conversely, the introduction of flexible pipelining to ease timing closure had area and latency effects on all bus architectures up through the transaction level.

Similarly, system architecture changes required new transaction types or transaction characteristics. New transaction types such as exclusive accesses were introduced nearly simultaneously within the OCP 2.0 and AMBA AXI standards. Out-of-order response capability is another example. Unfortunately, such evolutions typically affected the intended bus architectures down to the physical layer, if only by addition of new wires or opcodes. Thus, the bus implementations had to be redesigned.

As a result, bus architectures could not closely follow process evolution nor system architecture evolution. The bus architects always had to make compromises between the various driving forces and resist changes as much as possible.

In the data communications space, LANs and WANs successfully dealt with similar problems by using a layered architecture. By relying on the OSI model, upper- and lower-layer protocols independently evolved in response to advancing transmission technology and transaction level services. The decoupling of communication layers using the OSI model successfully drove commercial network architectures and made it possible for networks to follow very closely both physical-layer evolutions (from the Ethernet multi-master coaxial cable to twisted pairs, ADSL, fiber optics, wireless, etc.) and transaction-level evolutions (TCP, UDP, streaming voice/video data). This produced incredible flexibility at the application level (web browsing, peer-to-peer, secure web commerce, instant messaging, etc.), while maintaining upward compatibility (old-style 10 Mb/s or even 1Mb/s Ethernet devices are still commonly connected to LANs).

Following the same trends, networks started to replace buses in much smaller systems: PCI-Express is a network-on a-board, replacing the PCI board-level bus. Replacement of SoC buses by NoCs followed the same path, driven by economics that proved the use of NoC:

- reduced SoC manufacturing cost;
- increased SoC performance;
- lower power consumption;
- reduced SoC time to market and/or non-recurring engineering expense (NRE);
- reduced SoC time to volume;
- reduced SoC design risk.

In each case, if all the other criteria are equal or better, NoCs are the better interconnect option compared to buses.

The next section describes network-on-chip architecture and explains how it improves SoC-level performance, power consumption and area.

9.1.1. *NoC architecture*

The advanced network-on-chip uses system-level network techniques to solve on-chip dataflow transport and management challenges (De Micheli and Benini 2017). As discussed in the previous section and shown in Figure 9.1, synchronous bus limitations lead to system segmentation and tiered or layered bus architectures.

Contrast this with the NoC approach, first commercialized by Arteris in 2006, illustrated in Figure 9.2. The NoC is a homogeneous, scalable switch fabric network.

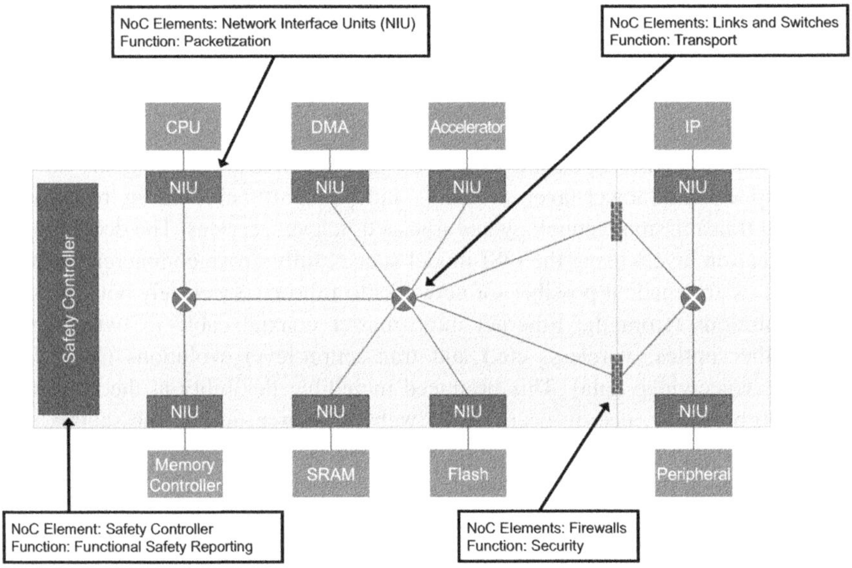

Figure 9.2. *Arteris switch fabric network (Arteris IP 2020)*

This switch fabric forms the core of the NoC technology and transports multipurpose data packets within complex, IP-laden SoCs. Key characteristics of this architecture are:

– layered and scalable architecture;

– flexible and user-defined network topology;

– point-to-point connections and a globally asynchronous locally synchronous (GALS) implementation decouple the IP blocks.

9.1.1.1. *NoC layers*

IP blocks communicate over the NoC using a three-layered communication scheme (see Figure 9.3), referred to as the transaction, transport and physical layers.

The *transaction layer* defines the communication primitives available to interconnected IP blocks. Special NoC interface units (NIUs), located at the NoC periphery, provide transaction-layer services to IP blocks with which they are paired. This is analogous, in data communications networks, to network interface cards that source/sink information to the LAN/WAN media. The transaction layer defines how information is exchanged between NIUs to implement a particular transaction. For example, a NoC transaction is typically made of a request from a master NIU to a slave NIU, and a response from the slave to the master (Lecler and Baillieu 2011, p. 134). However, the transaction layer leaves the implementation details of the exchange to the transport and physical layer. NIUs that bridge the NoC to an external protocol (such as AMBA AXI) translate transactions between the two protocols, tracking transaction state on both sides. For compatibility with existing bus protocols, Arteris NoC implements traditional address-based load/store transactions, with their usual variants including incrementing, streaming, wrapping bursts and so forth. It also implements special transactions that allow sideband communication between IP blocks.

The *transport layer* defines rules that apply as packets are routed through the switch fabric. Very little of the information contained within the packet (typically, within the first cell of the packet, a.k.a., header cell) is needed to actually transport the packet. The packet format is very flexible and easily accommodates changes at the transaction level without impacting the transport level. For example, packets can include byte enables, parity information or user information depending on the actual application requirements, without altering packet transport or physical transport.

A single NoC typically uses a fixed packet format that matches the complete set of application requirements. However, multiple NoCs using different packet formats can be bridged together using translation units.

The transport layer may be optimized to application needs. For example, wormhole packet handling decreases latency and storage but might lead to lower system performance when crossing local throughput boundaries, while store-and-forward handling has the opposite characteristics. The Arteris architecture allows optimizations to be made locally. Wormhole routing is typically used within synchronous domains in order to minimize latency, but some amount of store-and-forward is used when crossing clock domains.

The *physical layer* defines how packets are physically transmitted over an interface, much in the same way that Ethernet defines 10 Mb/s, 1 Gb/s, etc., physical interfaces. As explained above, protocol layering allows multiple physical interface types to coexist without compromising the upper layers. Thus, NoC links between switches can be optimized with respect to bandwidth, cost, data integrity and even off-chip capabilities, without impacting the transport and transaction layers. In addition, Arteris has defined a special physical interface that allows independent hardening of physical cores, and then connection of those cores together, regardless of each core clock speed and physical distance within the cores (within reasonable limits guaranteeing signal integrity). This enables true hierarchical physical design practices.

A summary of the mapping of the protocol layers into NoC design units is illustrated in Figure 9.3.

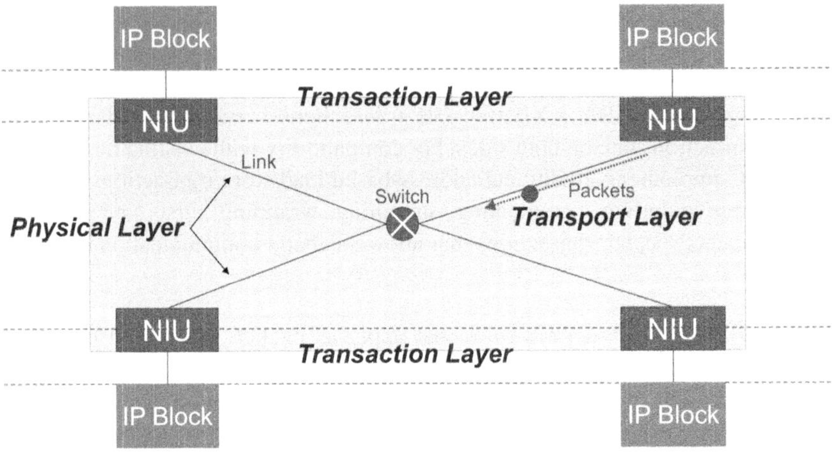

Figure 9.3. *NoC layer mapping summary (Arteris IP 2020)*

9.1.1.1.1. NoC layered approach benefits

The benefits of this layered approach are:

– *Separate optimizations of transaction and physical layers*: the transaction layer is mostly influenced by application requirements, while the physical layer is mostly influenced by silicon process characteristics. Thus, the layered architecture enables independent optimization on both sides. A typical physical optimization used within the NoC is the transport of various types of cells (header and data) over shared wires, thereby minimizing the number of wires and gates.

– *Scalability*: since the switch fabric deals only with packet transport, it can handle an unlimited number of simultaneous outstanding transactions (e.g. requests awaiting responses). Conversely, NIUs deal with transactions and their outstanding transaction capacity must fit the performance requirements of the IP block or subsystems that they service. However, this is a local performance adjustment in each NIU that has no influence on the setup and performance of the switch fabric. In addition, since the complex logic is typically located inside the NIUs, the elements of the transport network (switch, pipelines, etc.) are typically small and fast. This gives much flexibility to scale up or down the switch fabric to match the needed performance and layout constraints at the lowest cost.

– *Aggregate throughput*: throughput can be increased on a particular path by choosing the appropriate physical transport, up to even allocating several physical links for a logical path. Because the switch fabric does not store transaction state, aggregate throughput simply scales with the operating frequency, number and width of switches and links between them, or more generally with the switch fabric topology.

– *Quality-of-service (QoS)*: transport rules allow traffic with specific real-time requirements to be isolated from best-effort traffic. It also allows large data packets to be interrupted by higher-priority packets transparently to the transaction layer.

– *Timing convergence*: the transaction and transport layers' clocking schemes are dependent on the attached IP clocking schemes, while the physical layer has freedom to choose another clocking scheme, typically one that is best suited for distance spanning, low power and optimum routing congestion reduction. Arteris IP's NoC architecture uses a globally asynchronous, locally synchronous (GALS) approach: NoC units are implemented in traditional synchronous design style (a unit being, for example, a switch or an NIU), sets of units can either share a common clock or have independent clocks. In the latter case, special links between clock domains provide clock resynchronization at the physical layer, without impacting transport or transaction layers. This approach enables the NoC to span a SoC containing many IP blocks or groups of blocks with completely independent clock domains, reducing the timing convergence constraints during back-end physical design steps.

– *Easier verification*: layering fits naturally into a divide-and-conquer design and verification strategy. For example, major portions of the verification effort need only concern themselves with transport-level rules since most switch fabric behavior may be verified independent of transaction states. Complex, state-rich verification problems are simplified to the verification of single NIUs; the layered protocol ensures interoperability between the NIUs and transport units.

– *Customizability*: user-specific information can be easily added to packets and transported between NIUs. Custom-designed NoC units may make use of such information. For example, the design of "firewalls" that make use of pre-defined and/or user-added information to shield specific targets from unauthorized transactions. In this case, and many others, such application-specific design would only interact with the transport level and not even require the custom module designer to understand the transaction level.

9.1.1.2. *NoC modularity and hierarchical architecture*

A NoC packetizes transactions into transport packets and sends data from source to destination. The route that the packets take is based on the network topology. A flexible NoC technology enables designers to design the topology based on performance and floorplan requirements (Lecler and Baillieu 2011, p. 145). This allows the user to make trade-offs and optimize individual data paths to best fit design constraints.

To design a NoC that is flexible, all of its different elements or units need to be implemented in a modular fashion. This allows any of the units to be easily inserted or removed, at any arbitrary location. Some examples of NoC units include:

– *Network interface units (NIUs)*: these are the units that perform packetization and de-packetization. They are the entry and exit points of the NoC, converting between the IP's interface protocol (e.g. AXI) and the internal NoC transport packets.

– *Switches*: these are the arbiters and routers. A combination of switches makes up the NoC topology.

– *Links*: placeholders for adding potential pipeline stages.

– *Pipeline stages*: the actual pipeline elements used to break long timing paths.

As the SoC interconnect, all these NoC units are placed at the top level of the chip, connecting different IPs and routes together. As we will see in section 9.3, NoC modularity and hierarchy makes the efficient implementation of a variety of SoC-level "services" possible, such as QoS, safety, security, debug, performance analysis and perhaps some that are yet to be discovered or defined.

9.1.1.3. *NoC trade-offs*

In spite of the obvious advantages, a layered strategy to on-chip communication must not model itself too closely on board- or system-level data communication networks.

In data communication networks, the transport medium (e.g. optical fiber) is much more costly than the transmitter and receiver hardware and often uses "wave pipelining" (i.e. multiple symbols on the same wire in the case of fiber optics or controlled impedance wires). Inside the SoC, the relative cost and performance of wires and gates is different and wave pipelining is too difficult to control. As a result, NoCs will not, at least for the foreseeable future, be able to efficiently serialize data over single on-chip wires but must find an optimal trade-off between clock rate (100 MHz to multi-GHz) and number of data wires (16, 32, 64, etc.) for a given throughput.

Further illustrating the contrast, data communication networks tend to be focused on meeting bandwidth-related quality of service requirements, while SoC applications also focus on latency constraints.

Moreover, a direct on-chip implementation of traditional network architectures would lead to significant area and latency overheads. For example, the packet dropping and retry mechanisms that are part of TCP/IP flow control require significant data storage and complex software control. The resulting latency and area would be prohibitive for most SoCs, and largely useless as the source of errors in the transmission of packets are totally different in a SoC compared to, say, inside a data center.

Designing a NoC architecture that excels in all domains compared to buses requires a constant focus on appropriate trade-offs.

9.1.2. *Extending the bus comparison to crossbars*

In the previous section, the NoC was compared and contrasted with traditional bus structures. We pointed out that system-level throughput and latency may be improved with bus-based architectures by using pipelined crossbars or multilayer buses.

A crossbar has the advantage of connecting all initiators to all targets simultaneously without requiring centralized arbitration logic. However, because traditional crossbars still mix transaction, transport and physical layers in a way similar to traditional busses, they present only partial solutions (Hennessy and Patterson 2019, pp. F31–F32). They continue to suffer in the following areas:

– *Scalability*: to route responses, a traditional crossbar must either store some information about each outstanding transaction or add such information (typically, a return port number) to each request before it reaches the target and relies on the target to send it back. This can severely limit the number of outstanding transactions and inhibit one's ability to cascade crossbars. Conversely, Arteris' switches do not store transaction state, as packet routing information is assigned and managed by the NIUs and is invisible to the IP blocks. This results in a scalable switch fabric able to support an unlimited number of outstanding transactions.

– *IP block reusability*: traditional crossbars handle a single given protocol and do not allow mixing IP blocks with different protocol flavors, data widths or clock rates. Conversely, the Arteris transaction layer supports mixing IP blocks designed to major socket and bus standards (such as AHB, OCP and AXI), while packet-based transport allows mixing data widths and clock rate.

– *Maximum frequency, wire congestion and area*: crossbars do not isolate transaction handling from transport. Crossbar control logic is complex, data paths are heavily loaded and very wide (address, data read, data write, response, etc.) and SoC-level timing convergence is difficult to achieve. These factors limit the maximum operating frequency. Conversely, within the NoC, the packetization step leads to fewer data path wires and simpler transport logic. Together with a globally asynchronous locally synchronous implementation, the result is a smaller and less congested switch fabric running at higher frequency.

Common crossbars also lack additional services that Arteris NoC offers, such as functional safety data protection, error logging and runtime reprogrammable features.

9.1.3. *Bus, crossbar and NoC comparison summary and conclusion*

The major innovation of NoCs is that they separate the physical, transport and transaction layers, which allows the on-chip interconnect to be less tied to the attached

IP blocks and their chosen transaction protocols, and vice versa. This allows for flexible methodologies that support incremental changes to the SoC design.

As discussed in section 9.1.2, hierarchies of crossbars or multilayered buses have characteristics somewhere in between traditional buses and NoCs; however, they fall far short of the NoC with respect to performance and the ability to enable complex SoC designs.

Detailed comparison results necessarily depend on the SoC application, but with increasing SoC complexity and performance, the NoC is clearly the best IP block integration solution for high-end SoC designs today and in the foreseeable future.

9.2. NoC configurability

For a SoC design team, the NoC development or configuration overlaps the whole SoC design project lifetime: from early SoC architecture phase, to final place and route, passing through performance qualification, RTL design, IP integration, verification, debug and bring-up and software integration. We must pay special attention so that a decision made during an early phase, such as design architecture, does not jeopardize the feasibility of a later phase, such as physical layout. Moreover, it is not unusual that marketing requirements for a given SoC change while the chip is being designed, adding or removing some IPs, and thus impacting the NoC specification. As the complexity and cost of brand new architecture increases, organizing projects around reusable platforms becomes more efficient. Small variations of the platform or derivatives may then be brought to the market in shorter cycles. The interconnect system design methodology must make sure that those late specification changes are smoothly integrated in the design and verified.

9.2.1. *Human-guided design flow*

The NoC user's primary challenge is to navigate between the flexibility required to only pay for useful resources required without incurring the cost of overengineering the interconnect. A deliberate human-guided design flow is required to do this.

– In a first phase of NoC design flow, we define the *functional specification* of the NoC: what are the different sockets? The transaction protocols? Subsets of the protocols used? What is the memory map? The objective is to have clear statements about the possible transactions that initiators can request and targets can accept. Remember, some initiators may emit unaligned or wrapping transactions towards targets which do not support them. This implies conversions, and it is important to know whether that underlying logic must be implemented or not. The specification phase leaves completely aside how the functionality is implemented.

– The *performance specification* is the definition of the quality-of-service (QoS) expected from the NoC. We should note that performance of an interconnect is not in any way determined by its internal resources (number of pipeline stages, number of possible pending transactions, etc.), but by the fact that a system using this NoC either behaves correctly or not: will a displayed video image freeze, the loud speaker click or the embedded web-browser be too slow? At this stage, we only define the objectives. Again, the actual implementation is left for a further step. This phase is also a good time to collect data on the actual behavior of the IP blocks around the NoC: what are the transaction patterns? How do they react to transient back pressure? To latency? This behavior of the system around the interconnect is defined in *scenarios* and examined through performance simulation that is built into the NoC configuration tooling.

– In the *architecture* phase, we now allocate hardware resources in the NoC, defining how transaction flows are merged and scattered, tuning wire bundle serialization, buffer size, arbitration schemes, pipeline stages location, clock and power crossings location, etc. The architecture may be simulated against the previously defined performance-oriented scenarios in order to determine, on the one hand, if the scenario performance criteria are met, and on the other hand, what the saturated or under-used resources in the NoC are, and make the corresponding adjustments to the NoC configuration.

– In the *NoC structural* phase, the NoC is translated into hardware blocks. The translation is automated, yet some minor but crucial details may be fine-tuned, such as the presence and/or reset values of configuration registers. At the end of this phase, the register transfer level (RTL) description of the NoC can be exported together with synthesis scripts, ready to be integrated, simulated and synthesized.

– In the *verification* phase, the RTL description is checked against its functional and performance specifications. For the functional verification, a test bench and stimuli are generated automatically out of the specification. The test bench uses third-party protocol checkers and ensures that transactions are correctly translated. Protocol coverage is measured to ensure the stimuli have stressed the interconnect in a number of predefined corner cases. For performance verification, the scenarios are played against the RTL description of the NoC to verify that their performance criteria are met.

9.2.2. *Physical placement awareness and NoC architecture design*

Designing the NoC architecture is one of the most critical steps in achieving a feasible design. At 16 nm and smaller process generations, physical design effects intruded on the SoC architecture phase. It became quite possible to design a valid SoC architecture that was not implementable during the place and route phase of the design. Therefore, modern NoCs have to take into account physical constraints during the architecture and RTL development phases at 16 nm process generations and below. A NoC topology needs to take into account floorplan constraints in order to verify that timing can be closed given the proposed architecture. The timing closure then needs to be verified using physical synthesis prior to turning over the design to the place and route team. Figure 9.4 contrasts two different NoC topologies – the diagram on the left has a NoC that is designed without taking the floorplan into consideration, while the diagram on the right shows a NoC that has a floorplan-friendly topology. The gray rectangles with the letters "SW" represent a switch in the NoC.

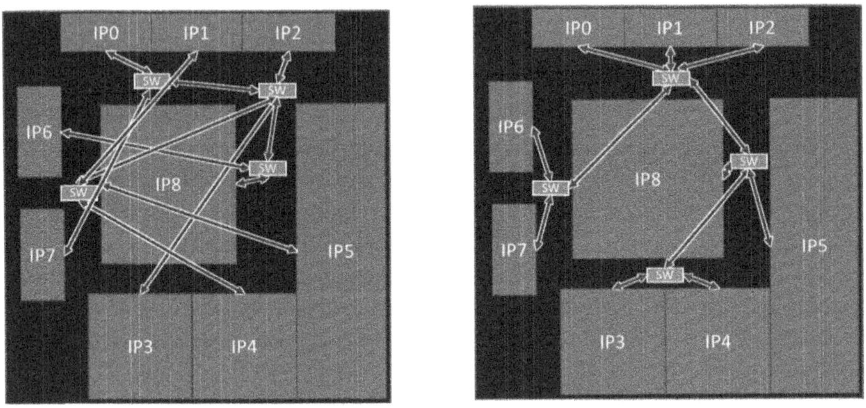

Figure 9.4. *The NoC on the left has a floorplan-unfriendly topology, whereas the NoC on the right has a topology that is floorplan-friendly (Arteris IP 2020)*

The second part of the NoC architecture design is to place pipeline stages at proper locations to allow the NoC paths to span distances. Given a clock period and transport delay of the process node, it is possible to calculate the number of pipeline stages required to span a particular distance.

In the example presented in Figure 9.5:

– if a path is running at a 600 MHz clock frequency, the cycle time is 1.67 ns, of which 1.42 ns is usable;

– in the TSMC 28HPM process, transport delay is approximately 0.644 ns/mm;

– therefore, the pipeline-to-pipeline maximum distance is $1.42/0.644 = 2.2$ mm;

– for a path that is 7 mm in length, at least three pipeline stages are required to span the distance, as illustrated in Figure 9.5.

This calculation needs to be done for every path in the NoC and have pipeline stages enabled accordingly in order for it to close timing. It is very common at this point to:

– fail to enable certain critical pipeline stages resulting in the need for more timing debug and synthesis, place and route (SP&R) iterations;

– over-provision of the number of pipeline stages in anticipation of reducing the number of SP&R runs.

The problem of misconfiguring pipeline stages results in more back-end iterations, and increased latency, and area – therefore power consumption – of the NoC. As a result, automating the process of adding pipeline stages is becoming essential to help address this problem in a systematic manner. This is a prime area of research for future NoC technology development.

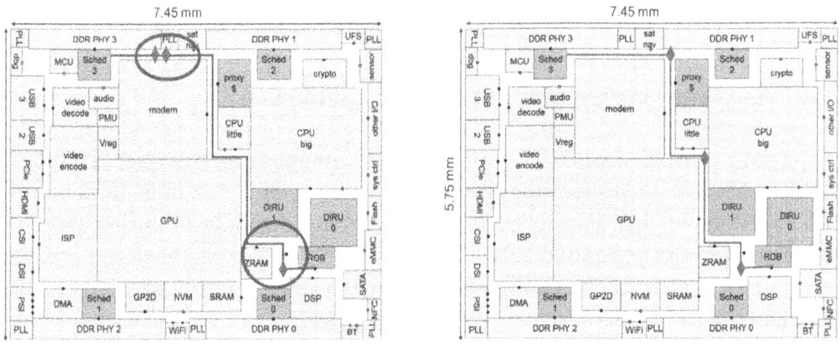

Figure 9.5. *Pipeline stages are required in a path to span a particular distance given the clock period and transport delay (Arteris IP 2020)*

9.3. System-level services

As previously discussed, the modular and hierarchical nature of a NoC allows the creation of system-level "services" that provide value to the SoC designer, the chip end-user or both.

9.3.1. *Quality-of-service (QoS) and arbitration*

Sophisticated QoS and arbitration schemes are system-level services that are made possible thanks to the modular nature of NoC elements. Any type of communications within the NoC can be prioritized at different levels, and these priorities can be changed dynamically at runtime either automatically by the NoC itself (based on information obtained from other NoC services described in section 9.3.2) or at runtime through software interaction with programmable NoC elements using control and status registers (CSRs). Switches and other elements within the NoC can then order the packet communications passing through them using the priorities associated with the packets (Hennessy and Patterson 2019, pp. F21–F23).

Multiple types of NoC hardware mechanisms can be used to implement the optimal ordering within the NoC, with "optimal" being defined as the proper trade-off between bandwidth, latency, area and power consumption required to meet the desired use cases. One example of an ordering mechanism is hardware logic that ensures that higher-priority traffic is prioritized through the NoC before lesser-priority data to ensure system-level latency requirements are met. An example technology to do this is credit-based flow control mechanisms that can ensure prioritized use of links between NoC elements. However, there are other QoS technologies that can perform this task with less area and buffering. The key is to use the proper technology that best helps meet the SoC-level performance, area and power consumption goals.

9.3.2. *Hardware debug and performance analysis*

Because a NoC "sees" most of the data communicated within a SoC, it is useful to be able to inspect the data traveling through it not only to debug SoC-level functionality but also to examine the on-chip dataflow to ensure the SoC is meeting performance requirements. To do this, a NoC must offer the designer different types of probes that can be configured to intercept data-in-transit and provide useful information back to the designer (during development) or system (during runtime). The types of information that can be determined based on this NoC data include errors, either internal to the NoC or created by attached IP blocks, and performance metrics, such as bandwidth and latency. The hardware probes can act like CPU performance counters, filtering on packet or transaction parameters and generating performance histograms and event statistics, or traces, which can be output to industry standard debug infrastructures, like AMBA CoreSight, or to internal memories for use at runtime.

The usefulness of these hardware mechanisms is enhanced when similar capabilities are implemented in the SystemC performance models generated by the NoC configuration software, not just the NoC RTL that becomes part of the SoC. In this way, the SoC architecture and design teams have better correlation between simulated and actual functionality and performance, which results in a more predictable chip behavior and design schedule.

9.3.3. *Functional safety and security*

9.3.3.1. *NoC functionality, failure modes and safety mechanisms*

Safety at the SoC level has become important in life-critical industries such as the automotive industry, where the ISO 26262 specification provides guidance related to functional safety implementation and processes for automotive electronics. Within a SoC, NoC functional safety mechanisms are used to detect, mitigate and correct faults and failures caused by random errors when a system is in operation.

Single-event effects (SEE) – electrical disturbances caused by cosmic rays and the ionization energy emitted when they interact with semiconductors – are the cause of random errors. These random errors can have transient (temporary) or permanent effects. Examples of random, transient effects – also known as "soft errors" – include single-bit upsets (SBU), such as "bit flips" in memory cells or logic flops, and single-event transients (SET), which are voltage glitches that may or may not cause an error. There are also instances where these can occur simultaneously, resulting in multiple bit upsets (MBU). "Hard errors" caused by SEEs resulting in permanent damage include single-event latch-up (SEL), single-event burnout (SEB) and single-event gate rupture (SEGR) (JESD86A 2006, pp. 3–4).

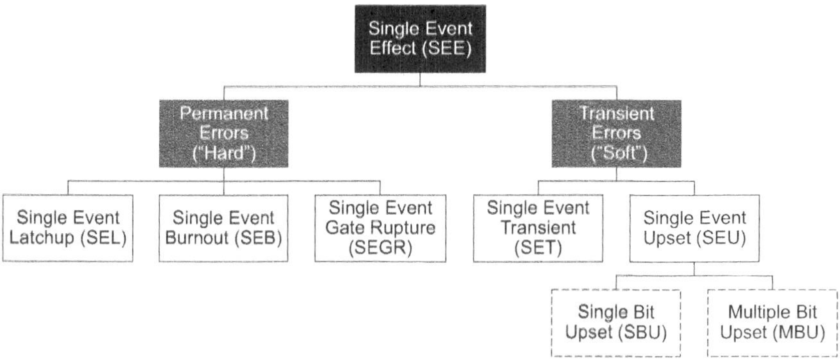

Figure 9.6. *Single-event effect (SEE) error hierarchy diagram (ISO26262-11 2018b; JESD86A 2006)*

Because the cause of these errors is natural physical phenomena and manufacturing defects, they will occur randomly, so it is important to detect and mitigate their effects to achieve and maintain system safety. This is especially important for the NoC interconnect, which sees most of the dataflow on the chip and can therefore detect and perhaps mitigate errors that occur within the NoC or even within the IP blocks attached to it. Today's state-of-the-art NoCs have safety-specific technical features within their products to do this. Below are examples of these features, also known as functional safety mechanisms:

- adding and checking parity or ECC bits added to on-chip communications traffic;
- duplicating logic and comparing results;
- triple-mode redundancy (TMR) or majority voting;
- communication time-outs;
- hardware checkers that verify correctness of operation;

– safety controller to gather error messages from throughout the system, make sense of them and communicate higher up the system;

– built-in self-test (BIST) for all functional safety mechanisms to detect latent faults.

The state-of-the-art NoC interconnect implements these features to provide data protection for automotive functional safety as well as reliability, enabling SoCs built with them to qualify for up to ISO 26262 ASIL D certification.

9.3.3.1.1. NoC functional safety analysis and FMEDA automation

Completing failure mode effects analysis (FMEA) and failure mode effects and diagnostic analysis (FMEDA) is particularly difficult for interconnect IP. First, because they are safety elements out of context (SEooC, as are all IP and generally chips), they therefore have to be validated against agreed assumptions of use (AoU). Also, interconnect IP is highly configurable, which makes reaching agreement on the assumptions of use, defining potential failure modes and effects and validating safety mechanisms even more challenging than it is for other IP, as safety assurance must be determined on the configuration built by the integrator/commercial interconnect IP user.

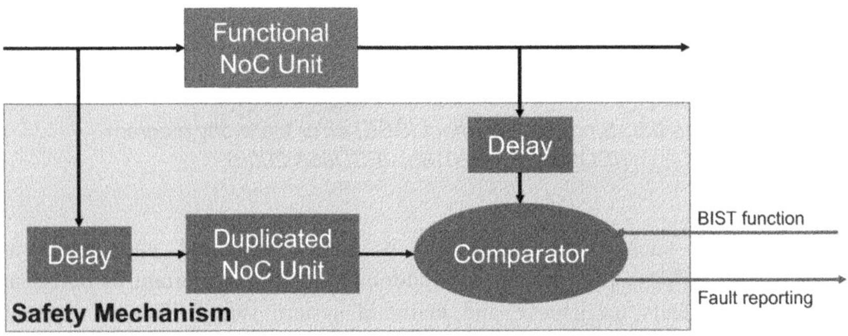

Figure 9.7. *Failure mode effects and diagnostic analysis (FMEDA) includes analysis of safety mechanisms, such as hardware unit duplication and a built-in self-test (BIST) (Arteris IP 2020)*

A state-of-the-art way to handle this complexity is to start with a qualitative, bottoms-up FMEA analysis, which can then guide the creation of a configuration-specific FMEDA (ISO26262-10 2018a, pp. 75–76). This requires modular NoC components that can be configured in specific hierarchies with safety mechanisms tied to those components, so that as you configure the IP, those safety mechanisms are automatically implemented in a manner that ensures safety for the overall function. For example, in the Arteris IP NoCs, the network interface unit (NIU) responsible for packetization can be duplicated (Figure 9.7). Then in operation, results from these units are continually compared; variances which would signal a failure are searched for. NoCs can also implement ECC or parity checks in the transport. Following the FMEA/FMEDA philosophy, these safety mechanisms are designed to counter the various potential failure modes identified by the IP vendor. And because of the modularity, functional safety diagnostic coverage of the entire NoC and the NoC-level failure mode distribution (FMD) can be calculated based on the concatenation of individual module coverage metrics.

The NoC IP should also provide BIST functions to check the logic for safety mechanisms at reset and during runtime. All of this safety status rolls up into a safety controller provided as part of the NoC, which can be monitored at the system level.

9.3.3.2. *NoC contributions to system-level security*

Security is often portrayed as a software- or system-level issue with most engineers' perception of "hardware security" focused on dedicated security IP blocks such as trusted boot controllers and security protocol or cryptographic accelerators. However, as in the enterprise computing world where a major component of "IT security" is network security, ensuring SoC-level security requires hardware security features in the network-on-chip interconnect. The most important NoC security capability is the firewall, similar to an IT equipment firewall, which is programmed by chip designers to allow certain data to pass when it is expected in a use case, and blocks or poisons data when it should not be in transit. Firewalls protect parts of the SoC from security intrusions that have found their way into the system through an attack surface, such as a memory (buffer overflow error), intentional power glitch or other man-made phenomena.

NoC firewalls filter traffic at runtime, only allowing traffic to transit through the firewall if it passes a chip designer-defined set of tests. These tests within firewalls can be programmed based on a set of use cases that are expected at runtime and are implemented as part of a system-level hardware and software security scheme. In addition to detecting the progression of data that is determined to be unsecure, there is flexibility in how the firewall reports the transgression, tags the data and other actions. One possible mechanism is that unallowed packets are "poisoned", indicating to logic blocks that receive the data that the data may be corrupted or otherwise suspect. This allows the SoC security architect to determine and limit what responses a hacker can glean from the system, making the hacker's job harder.

9.4. Hardware cache coherence

As the number and types of processing elements within modern SoCs have increased, it has become increasingly important for hardware to help manage the dataflow created by these IP blocks. Managing data access for multiple processors within a SoC is complicated by the sheer number of cores as well as the varying traffic patterns of dissimilar, heterogeneous cores. Implementing cache coherence at the SoC-level in hardware is the best way to create scalable coherency while enabling simpler software. However, because each coherent SoC architecture is different, the means to implement SoC-level cache coherence must be highly configurable. For these reasons, NoC technology has become the state of the art for multi-processor cache coherent systems.

NoC technology is ideally suited to implementing hardware cache coherence for many reasons. First, NoCs are highly configurable and can adapt to different transaction protocols, power domains and functional safety requirements. As explained in section 9.1.1, NoC architectures separate the transaction from the transport and physical mechanisms of the interconnect. This separation allows novel cache coherence features that would be extremely difficult to do in a technology other than NoC, for example, integrating IP using multiple types of protocols, such as AMBA ACE and CHI, in the same cache coherent system. Second, the hierarchical and modular nature of the NoC technology allows for the strategic addition of caches within a system to allow new capabilities. An example of this is a proxy cache, which allows multiple non-coherent IP blocks, such as artificial intelligence (AI) accelerators using the AXI protocol, to be connected into clusters that participate as first-class citizens within the cache coherent domain. Another benefit of NoCs in implementing cache coherence is that their modularity allows each NoC element to be placed within a chip floorplan to optimize for the desired mix of performance, power consumption and die area. This is in stark contrast to older bus- and crossbar-based "cache controller" IP, which had to be implemented as a monolithic IP block, often causing SoC routing congestion, timing closure and performance issues.

9.4.1. *NoC protocols, semantics and messaging*

NoC interconnects usually implement cache coherency with a directory protocol instead of a broadcast protocol. This allows for more point-to-point communications through the network, which lowers the bandwidth required as the number of coherent agents scales up. NoC interconnects can have multiple directories which track the location and state of shared data (cache lines) within the coherent domain (Figure 9.8).

Although a cache coherent NoC manages dataflow similar to any other NoC, the message types and sizes being communicated through the NoC differ in their nature from "traditional" NoCs. In addition to read-and-write messages and data common to protocols such as AXI and OCP, a cache coherent NoC carries messages implementing a protocol that describes the states of data that can be located within a set of memory locations within the system, including IP block caches, NoC caches and off-SoC memory. To do this, a cache coherent NoC has a native internal protocol that allows it to carry the semantics of the traffic of attached IP, which are using their own protocols. To accommodate multiple industry-standard protocols such as AMBA ACE and CHI, a NoC must have an underlying native protocol that is a "superset" protocol, implementing all of the required semantics for both protocols. The flexibility of the NoC technology allows for the communications of nearly any type of semantics, whether load/store, cache coherence, message passing or others not yet invented.

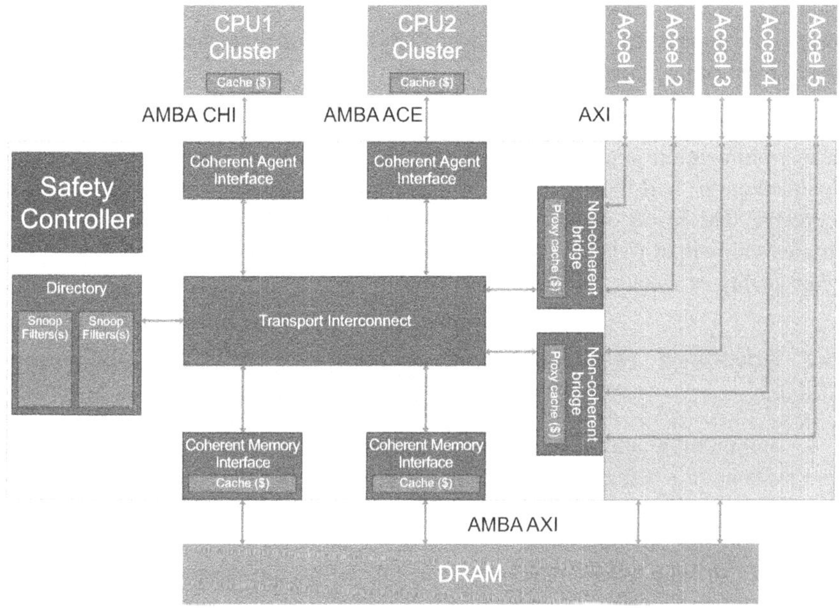

Figure 9.8. *A cache coherent NoC interconnect allows the integration of IP using multiple protocols (Arteris IP 2020)*

9.5. Future NoC technology developments

Network-on-chip has established itself as the foundation of nearly all SoCs having any semblance of complexity in size, semantics or power consumption requirements and will be the underlying in-silicon communications backbone for chips for years to come. However, time does not stand still and there are multiple avenues for the NoC technology innovation on the horizon. Here are a few.

9.5.1. *Topology synthesis and floorplan awareness*

As SoCs become larger and more complex with hundreds of initiators and targets, highly heterogeneous traffic patterns and chip floorplan sizes surpassing 600 square multimeters in today's sub-16 nm processes, it is becoming more difficult for chip architects to understand and analyze all the options required to configure the on-chip interconnect. In most use cases, restricting the alternatives to regular topologies such as meshes and rings to simplify the analysis results in suboptimal performance, power and area results. One solution is to provide more automation at earlier stages in the NoC configuration design flow by using more system-level information earlier in the flow to make decisions. In the future, NoC configuration tooling will import user-defined and production (LEF/DEF format) floorplans for the entire SoC; automatically generate a NoC topology that adapts to the floorplan while meeting performance, power and area metrics; and automatically configure pipelines in NoC logic and wire links to meet timing closure.

This level of automation requires innovations in the domains of graph theory, linear optimization and other mathematical constructs. Intelligent algorithmic implementation will be required to ensure configuration tooling runtimes are acceptable. The result of this innovation will be the capability to create even more complex chips than today's, in less development time while achieving an equal or higher quality of results.

9.5.2. *Advanced resilience and functional safety for autonomous vehicles*

As automobiles and other vehicles become more autonomous, the brains of these systems are being required to function in the presence of errors, faults and failures. In the current edition of ISO 26262, a system failure that affects safety can lead to a fallback fail-safe state where a driver assumes control of the vehicle. As autonomous driving technologies progress, this fallback process is being supplanted by the requirement for fail-operational, where the system remains available and functions despite faults. Implementing SoCs capable of fail-operational has at least two opportunities for technical innovation.

First, SoC architectures will add even more redundancy and checking to reduce common cause failures and increase fault tolerance. Many of the techniques used in aviation, space, military and medical applications will be adapted for on-chip implementation in these SoCs. Examples of techniques include triple-modular redundancy (TMR, also called 2-out-of-3 voting), increased diversity of processing and logic elements, and the addition of even more diagnostics and runtime testing. Second, the nature of diagnostics will change from *detecting* when something has failed to *predicting* when something might fail, before the failure occurs. An example of this is in-silicon sensors that detect the rate of transistor aging.

In all the cases above, the NoC interconnect will play a hugely important role because it sees all the data on the SoC and is physically implemented throughout the floorplan. This makes the NoC an ideal target for implementing these new functional safety mechanisms and diagnostics as well as integrating the information from them for on-chip analysis and action.

9.5.3. *Alternatives to von Neumann architectures for SoCs*

Most SoC architectures have evolved as instantiations of the von Neumann architecture where CPU functionality connects to a single memory hierarchy and a set of I/O devices. As Moore's law has slowed down and SoC frequencies have leveled off, SoC architectures have changed to adapt to this new reality and obtain better performance than was previously possible. Two innovation paths where the NoC technology plays a critical role are in parallel processing architectures and hierarchical cache coherence. Both allow more processing to occur simultaneously while keeping more of the data being acted upon within the SoC in local memories, rather than needing to be sent off-die and succumbing to the latency, power consumption and bandwidth penalties of off-die memory accesses.

9.5.3.1. *Massively parallel AI/machine learning architectures*

Artificial intelligence and machine learning applications have spearheaded a long-anticipated revolution in massively parallel computing architectures. Rather than being implemented solely in exotic supercomputers, these architectures are being created for mass-market chips that reside in data center servers, vehicles and consumer electronics. The current state of the art is related to hardware implementation of various flavors of neural networks, but this could change in the future. What will not change is the need to have multiple processing units (which today are usually the same but are becoming heterogeneous for greater efficiency) connected within a SoC with multiple memories.

Because of the ease of analysis and implementation, many of today's AI chips are organized using a NoC mesh or ring topology connecting tens or hundreds of a single type of processing element. Each processing element is a subsystem containing logic and memories, often connected together by its own internal NoC. As the number of processing elements increases, chip designers often choose to create hard macro tiles of synthesized parts of the mesh structure (Figure 9.9). These tiles are then connected at the top level to create an even larger NoC topology. As chips scale, being able to easily create and close timing of replicable tiles and hard macros using a NoC will become extremely important. Technical innovation opportunities will arise in how to implement internal NoC addressing methods to efficiently enable this

Another area of technical effort for these types of chips is in NoC transaction semantics, to be extended beyond the traditional reads and writes. For example, an efficient update of neural network weights to multiple processing elements requires the NoC to broadcast write data while meeting latency goals and not consuming egregious amounts of NoC bandwidth. Conversely, a NoC able to simultaneously broadcast read operations to multiple targets and aggregate the results using mathematical functions such as addition, min and max will offer great performance benefits in multiple applications (Jerger *et al.* 2017, pp. 148–149).

In addition to NoC tiling and semantics, AI chip architectures are evolving to use different types of processing elements (heterogeneous) which need to be connected in more complex and irregular topologies than simple meshes and rings. The topology synthesis and floorplan awareness technologies described earlier in section 9.5.1 will help accomplish this.

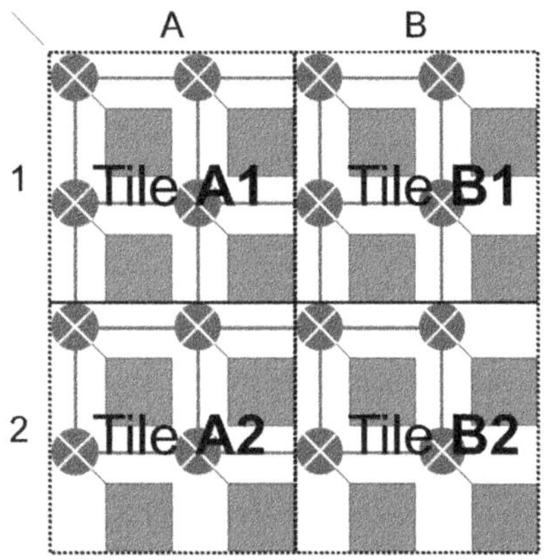

Figure 9.9. *NoC interconnects enable easier creation of hard macro tiles that can be connected at the top level for scalability and flexibility (Arteris IP 2020)*

9.5.3.2. *Hierarchical cache coherence*

Another way to work around the off-die memory bottlenecks of the typical von Neumann architecture is again to provide more in-SoC memory than traditional designs, as well as to do so in a way that is more software-friendly than the massively parallel architectures with explicit data exchange between units, described above. Hierarchical cache coherence will become mainstream over the coming decade to solve both these problems. In this scheme, cache coherent subsystems are created containing multiple processing elements and a subsystem-specific directory and cache, and many of these subsystems are then connected together in another cache coherent subsystem with its own directory and caches (Sorin *et al.* 2011, pp. 181–184). Although any of the caches within the overall system can be snooped by any coherent agent, architects will want to design the set of coherent subsystems so that data locality within subsystems is optimized.

Hierarchical coherency will benefit from the flexibility and configurability of the NoC technology, allowing support for heterogeneous processing elements and coherency clusters. Also, the hierarchical coherency subsystems can be synthesized into hard macros and tiles, as described in section 9.5.3.1, thereby enabling massive scalability of on-die cache coherent systems.

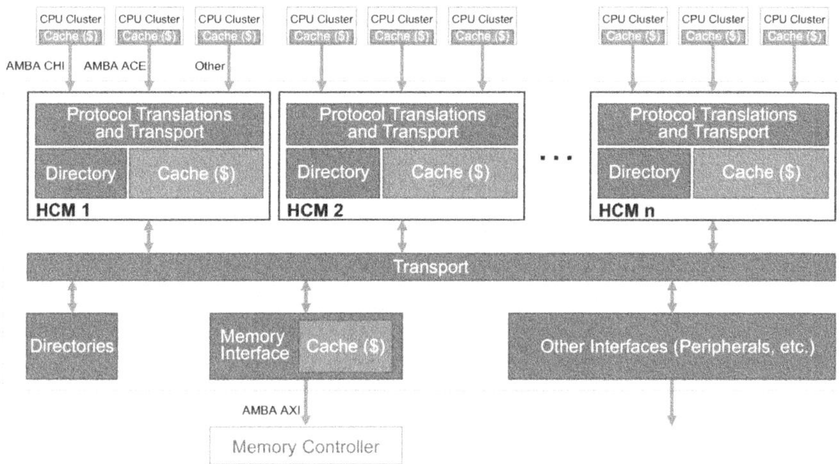

Figure 9.10. *Hierarchical coherency macros enable massive scalability of cache coherent system (Arteris IP 2020)*

Figure 9.10 shows an example of how this will be implemented within the cache coherent NoC. The hierarchical coherency subsystems are called hierarchical coherency macros (HCM) in the diagram. These can be the same or have heterogeneous contents, and each HCM contains its own directory with snoop filter(s) and cache. The multiple HCMs are connected within a separate cache coherent subsystem that ties them all together into one cache coherent NoC.

9.5.4. *Chiplets and multi-die NoC connectivity*

As massively parallel architectures using tiling and hierarchical cache coherence are developed, another need becomes obvious: how do we expand these systems beyond the confines of a single die? The NoC technology will play a critical role in making this possible.

Today's state-of-the-art multi-die and chiplet systems are being developed using somewhat *ad hoc* or bespoke connectivity solutions. This will soon change as inter-die connectivity standards evolve and stabilize. For the development of companion accelerator SoCs for server chips such as the Intel Xeon family, compute express link (CXL) is emerging as a first-generation chip-to-chip standard. The CXL standard allows one-way coherency with the x86 or other CPU chip acting as a master with CXL-attached accelerator chips being slaves. Another standard, called cache coherent interconnect for accelerators (CCIX), allows all the dies or chips connected using the interface to share a common view of memory and be able to snoop each other, therefore creating a multi-die cache coherent system. CCIX is expected to become a standard for the creation of chiplet SoCs, which contain multiple dies within the same chip package acting together as one larger chip.

The configurability and flexibility of the NoC interconnect technology makes it ideal for use in chips using CXL, CCIX or other die-to-die communication standards. One issue with these solutions, though, is the requirement to use a PHY as the underlying physical and transport mechanism. Both CXL and CCIX use the PCIe PHY as the physical layer connection mechanism with changes to parts of the upper PCIe layers (link, transaction and protocol) to implement the new standards. All-digital connectivity technology will emerge for chiplet/die-to-die connectivity within a chip package where long distances will not need to be traversed. This will be underpinned by a foundation of high-bandwidth NoCs within each chiplet, physically connected by through-silicon via (TSV), silicon-interposer and other 3D chip integration technologies. Using chiplet architectures while minimizing power consumption and maintaining on-die-level bandwidth and latency will require all-digital interconnects using the NoC technology.

9.5.5. *Runtime software automation*

One of the challenges of large-scale SoC development is keeping the associated runtime software in sync with the on-chip peripheral and processor hardware capabilities and versions. A NoC interconnect implementation has much of the information that is required to create the foundation of operating system device drivers and hardware abstraction layers (HAL), including items such as memory maps, register maps, peripheral access and command capabilities and mask and interrupt information. Information from the NoC configuration tooling, for instance, exported using the IP-XACT format, could be used to populate header file information for software device drivers and hardware abstraction layer components for industry-standard operating systems such as Linux (with udev) and AUTOSAR (with its Microcontroller Abstraction Layer, or MCAL, shown in Figure 9.11).

The benefits of using information from the NoC configuration to populate these parts of the software infrastructure are that it ensures that the software stays in sync with the implemented hardware capabilities (version control), reduces the chance of errors due to retyping copious amounts of information (systematic errors) and automates traceability for compliance to functional safety standards such as ISO 26262 for automotive and IEC 61508 for general electronics systems.

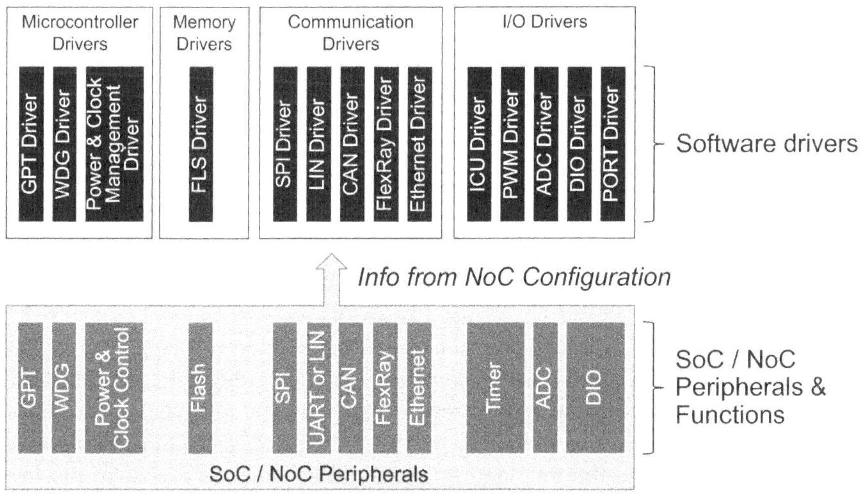

Figure 9.11. *An AUTOSAR MCAL showing how NoC configuration information will be used as a "golden source" for software driver header files. Adapted from Renesas (2020)*

9.5.6. *Instrumentation, diagnostics and analytics for performance, safety and security*

As we have seen in the previous sections, the NoC interconnect has a "God's eye" view of data flowing through a chip, and we have already implemented the technical means to see this data in real time using hardware probes and performance counters (see section 9.3.2), data checkers for functional safety (see section 9.3.3.1) and firewalls for security (see section 9.3.3.2). What if we could coordinate all this data together and analyze it for patterns that can detect or predict problems in performance and QoS, safety and security? What if we could use these inferences to then change settings within the NoC to alleviate issues, without requiring coordination with the system-level software? And what if we could give information to the system-level software about what is going on within the SoC dataflow so that the software can make more intelligent decisions for better use of chip resources?

This is one of next frontiers of NoC technology. These technologies, once combined in a useful format that is similar in concept to Waze or Google's live traffic maps, can be used to provide information to regulate the dataflow throughout the NoC, including links to the system level in order to control and synchronize the behavior of IP blocks transmitting data onto the NoC. As the number and types of initiators and targets attached to a NoC increase, we create a highly complex system that becomes impossible to understand without looking at it in its entirety. Integrating advanced NoC instrumentation and analytics at the system level will provide this necessary capability.

9.6. Summary and conclusion

The purpose of this chapter has been to explain how multi-processor SoCs have been enabled by NoC technology, first by showing the history and evolution of NoCs and the current state of the art, then by describing how the technology will progress as SoC architecture requirements change. Networks-on-chip play the critical role in on-chip dataflow, which affects the achievable bandwidth and latency of a chip. Managing this data traffic at the system level becomes the most important SoC engineering issue as chips scale in size and the numbers and types of processing elements increase. In the past, individual blocks of IP, such as processors and hardware accelerators, dominated the time and imaginations of engineers creating the latest generations of SoCs. Soon, the NoC interconnect will be widely accepted as most important IP required to create these large chips.

9.7. References

ARM (1999). AMBA specification (Rev 2.0) [Online]. Available: https://static.docs.arm.com/ihi0011/a/IHI0011A_AMBA_SPEC.pdf.

Arteris IP (2020). Arteris IP – The Nerwork-on-Chip (NOC) interconnect company [Online]. Available: http://www.arteris.com/.

Benini, L. and De Micheli, G. (2002). Networks on chips: A new SoC paradigm. *Computer*, 35(1), 70–78.

De Micheli, G. and Benini, L. (2017). Networks on chips: 15 years later. *Computer*, 50(5), 10–11.

Hennessy, J.L. and Patterson, D.A. (2019). Interconnection networks. *Book companion – Computer Architecture: A Quantitative Approach*, 6th edition. Morgan Kaufmann Publishers Inc., F1–F118.

ISO26262-10 (2018). Road vehicles – Functional safety – Part 10: Guidelines on ISO 26262, ISO standard 26262-10 ISO26262-10:2018, ISO – International Organization for Standardization [Online]. Available: https://www.iso.org/standard/68392.html.

ISO26262-11 (2018). Road vehicles – Functional safety – Part 11: Guidelines on application of ISO 26262 to semiconductors, ISO standard 26262-11 ISO26262-11:2018, ISO – International Organization for Standardization [Online]. Available: https://www.iso.org/standard/69604.html.

Jerger, M.E., Krishna, T., and Peh, L.-S. (2017). *On-Chip Networks: Second Edition. Synthesis Lectures on Computer Architecture*. Morgan & Claypool Publishers.

JESD86A (2006). Measurement and reporting of alpha particle and terrestrial cosmic ray-induced soft errors in semiconductor devices, Standard No. 89A JESD89A, JEDEC [Online]. Available: https://www.jedec.org/standards-documents/docs/jesd-89a.

Lecler, J.-J. and Baillieu, G. (2011). Application driven network-on-chip architecture exploration & refinement for a complex soc. *Des. Autom. Embedded Syst.*, 15(2), 133–158. Available: https://doi.org/10.1007/s10617-011-9075-5.

Renesas (2020). Microcontroller abstraction layer (MCAL) [Online]. Available: https://www.renesas.com/us/en/solutions/automotive/technology/autosar/autosar-mcal.html.

Sorin, D., Hill, M.D., and Wood, D.A. (2011). *A Primer on Memory Consistency and Cache Coherence. Synthesis Lectures on Computer Architecture*. Morgan & Claypool Publishers.

10

Minimum Energy Computing via Supply and Threshold Voltage Scaling

Jun SHIOMI[1] **and Tohru ISHIHARA**[2]

[1] *Department of Communications and Computer Engineering, Graduate School of Informatics, Kyoto University, Japan*
[2] *Department of Computing and Software Systems, Graduate School of Informatics, Nagoya University, Japan*

Dynamically scaling the supply voltage and threshold voltage of microprocessors is one of the most promising approaches for reducing the energy consumption of the processors without introducing fundamental performance degradation.

This chapter shows an algorithm for always keeping the processors running at the most energy-efficient operating point by appropriately tuning the supply voltage and threshold voltage under a specific performance constraint. This algorithm is applicable to a wide variety of microprocessor systems including high-end processors, embedded processors and tiny processors used in wireless sensor nodes.

This chapter also shows hardware and software properties required for effectively implementing the algorithm on practical microprocessor systems. For fully exploiting the energy efficiency achieved by the voltage scaling technique, we introduce a near-threshold circuit architecture for on-chip memory subsystems. This architecture is constructed based on standard cells, which makes it possible to run the entire processor including on-chip memory stably at near-threshold voltage. An energy management

For a color version of all figures in this book, see www.iste.co.uk/andrade/multi1.zip.

Multi-Processor System-on-Chip 1 – Architectures,
coordinated by Liliana ANDRADE and Frédéric ROUSSEAU.

policy which can be integrated on commercial real-time OSs for effectively implementing the run-time energy minimization algorithm is also summarized in this chapter.

Experimental results obtained using several prototype processor chips designed by us demonstrate the effectiveness of the algorithm and the near-threshold memory architecture.

10.1. Introduction

The Internet of Things (IoT), sometimes called machine-to-machine communication technology (M2M), has emerged as a new concept in which billions of computers and sensor devices are interconnected, enabling autonomous exchange of information. It requires multi-processor SoCs on the wireless sensor devices to handle much bigger and far more complex multimedia data than ever before. Both high computational power and high energy efficiency are thus extremely crucial to the digital processors embedded in these SoCs. Traditionally, wireless sensor devices use sub-threshold logic circuits that operate with a power supply voltage less than the transistors' threshold voltage. They will be suitable for specific applications which do not need high performance but require extremely low power consumption. For the IoT applications, however, the computational power of the sub-threshold logic circuits is no longer satisfiable.

Recently, as a solution to the above-mentioned issue, a concept of near-threshold computing has been proposed (Jain *et al.* 2012). It scales the supply voltage to the near-threshold region, which brings not only quadratic dynamic energy savings but also super-linearly reduced leakage energy, while maintaining the performance degradation of the processors to a minimum. In (Jain *et al.* 2012), a dual-V_{DD} design based on an IA-32 processor fabricated in 32 nm CMOS technology is presented, where on-chip memories stay at a V_{DD} of 0.55 V, while logic scales further down to 280 mV. Minimum energy operation for the logic parts is achieved in the near-threshold voltage region with the total energy reaching minima of 170pJ/cycle at 0.45 V. With the near-threshold computing, a 4.7x improvement in the energy efficiency can be achieved. Although caches use a 10-transistor bit cell, which is more robust to noises and variations than a 6- or 8-transistor bit cell, the minimum V_{DD} for the caches is 100 mV higher than the minimum energy operational point (i.e. 0.45 V). This means that the minimum V_{DD} for the on-chip memories is limiting the true minimum energy operation of the processor.

As a remedy to the issue of the relatively higher minimum V_{DD} in on-chip memories, standard-cell-based memories (SCMs) have been proposed (Teman *et al.* 2016; Shiomi *et al.* 2019). More detailed explanation on the SCM is provided in section 10.2. With SCM-based caches integrated on the processor, the minimum possible voltage of the processor can be lowered and the energy efficiency can be improved (Xu *et al.* 2019). Figure 10.1 shows the energy consumption results obtained for an RISC processor with an on-chip SCM cache implemented using the 65 nm SOI-CMOS process technology. As can be seen from the figure, a 6x improvement in the energy efficiency can be achieved in this case.

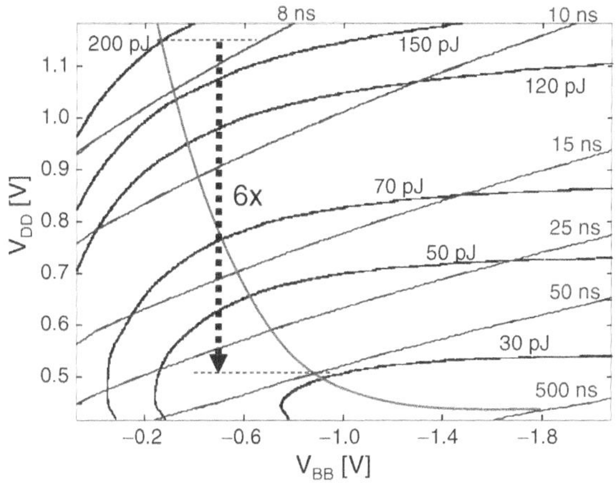

Figure 10.1. *Energy efficiency improvement by near-threshold computing*

Although the near-threshold computing is highly energy efficient, the performance of processors in the near-threshold region is not satisfiable for some classes of IoT applications, such as car navigation or image processing. To dynamically correspond to the variable workload, dynamic voltage and frequency scaling (DVFS) is widely used in modern microprocessors. DVFS quadratically reduces the dynamic energy by lowering the supply voltage in an off-peak situation of the application while it provides the peak performance of the processor with the highest available supply voltage when the application requires peak performance. Although DVFS reduces the dynamic energy consumption drastically and has been widely used in commercial microprocessors today, lowering the supply voltage down to the near-threshold region increases the delay of the circuit exponentially and thus increases leakage energy consumption. As a solution to this problem, variable threshold voltage scaling has been proposed (Kuroda *et al.* 1998). This exponentially reduces the leakage energy by dynamically scaling up the threshold voltage in a leakage-dominant situation. Simultaneous scaling of supply voltage (V_{DD}) and the transistor's threshold voltage (V_{TH}) is thus one of the most promising approaches of reducing the energy consumption of microprocessors without fundamental performance degradation. For the DVFS techniques, there is only one tuning knob (i.e. supply voltage) for a given workload. However, if we scale V_{DD} and V_{TH} simultaneously, there are two tuning knobs to handle at a time to minimize the energy consumption of the processor for a given workload. Therefore, we need a more complicated algorithm than the DVFS algorithm to find the optimal pair of V_{DD} and V_{TH} at runtime for a given workload. In this chapter, we refer to the method for finding the optimal pair of V_{DD} and V_{TH} at runtime, which minimizes the energy consumption of the processor for a given workload as minimum energy point tracking (MEPT). Section 10.3 first presents the basic theory behind MEPT. Algorithms for MEPT and their implementation examples are then presented. Finally, section 10.4 concludes this chapter.

10.2. Standard-cell-based memory for minimum energy computing

Even with the aid of the 10-transistor bit cell structure described in the previous section, on-chip SRAMs are the dominant source that limits the voltage scalability of processors. Thus, they may degrade the energy efficiency of the entire SoC. This section discuses voltage-scalable SCMs as an alternative to the conventional SRAM-based on-chip memories. Section 10.2.1 briefly presents the overview of low-voltage on-chip memories including low-voltage SRAMs and SCMs. Section 10.2.2 presets the design strategy for achieving area- and energy-efficient SCMs. Section 10.2.3 shows hybrid design, exploiting the advantages of both SRAMs and SCMs with minimum performance overhead. Section 10.2.4 presents a body biasing technique as an alternative to the power gating technique.

10.2.1. *Overview of low-voltage on-chip memories*

10.2.1.1. *Low-voltage SRAM*

One of the biggest challenges in the low-voltage operation is intra-die process variations such as random dopant fluctuation (RDF) (Asenov 1998), metal grain granularity (MGG) (Wang *et al.* 2011) and line edge roughness (LER) (Wang *et al.* 2011). These variabilities fluctuate transistors' threshold voltage (V_{TH}). The drive strength balance among access transistors, pull-down transistors and pull-up transistors in a bit cell is key to the stable readout/write operation of the conventional 6-transistor SRAM. However, since the V_{TH} fluctuation has an exponential impact on the drain current in the low-voltage region (Keller *et al.* 2014), it is hard to guarantee the stable operation of conventional 6T-SRAMs for the near-threshold region.

A number of papers thus propose specially designed SRAM structures for the low-voltage operation. For example, bit cell structures with redundant transistors have been widely studied, such as an 8-transistor cell structure (Chang *et al.* 2005) and a 10-transistor cell structure with the voltage boost technique (Chang *et al.* 2009). Redundant bit cells for an error-correcting code (ECC) are implemented into the SRAM structure in order to guarantee the stable operation, even though some bit cells suffer from operation failure (Dreslinski *et al.* 2010). A gate sizing strategy for bit cells is proposed in order to cancel out the process variation (Chen *et al.* 2010).

However, they still use the conventional *bit-line-based structure* where all the bit cells are interconnected with a single bit line, which results in a large parasitic capacitance on the bit line. As a result, even though the supply voltage can be aggressively downscaled with the aid of low-voltage SRAM structures, they inherently waste its dynamic energy consumption. This is because the parasitic capacitance on the bit line is charged/discharged in every readout/write operation. To solve this problem, a concept of SCM is presented in the next section as an alternative to the conventional low-voltage SRAM structures.

10.2.1.2. *Standard-cell-based memory (SCM)*

Another flexible approach is to design an on-chip memory via the hardware description language (HDL) and to synthesize it with conventional standard cells such as NAND cells and latch cells through commercial EDA (electronic design automation) tools for CMOS digital circuits. The flexible fully digital on-chip memory structure was referred to as a standard-cell-based memory (SCM) by Meinerzhagen *et al.* (2010) for the first time. A similar concept has been proposed by Wang and Chandrakasan (2005). A block diagram of the SCM with an $R \times C$-bit capacity is depicted in Figure 10.2. The SCM has a dual-port structure with one readout port and one write port in this diagram.

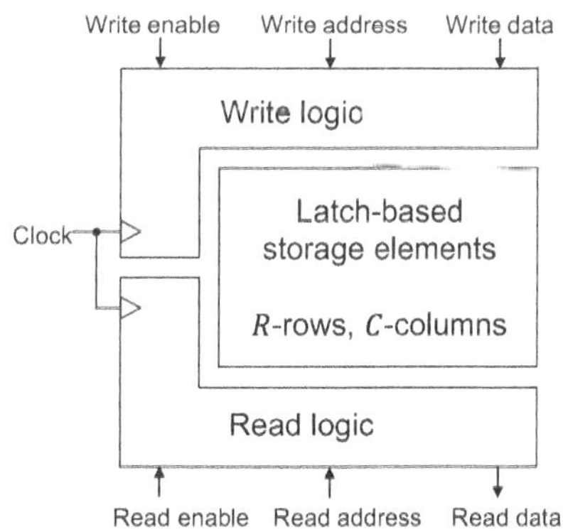

Figure 10.2. *Block diagram of the SCM with an $R \times C$-bit capacity*

The SCM consists of latch-based storage elements, write circuitry and readout circuitry. Latch-based logic gates such as latches and flip-flops are implemented into the storage elements array. The readout circuitry uses fully digital address decoders and readout circuits. R-to-1 readout multiplexer is typically implemented into the readout circuits. A clock gating array as well as write address decoders is typically implemented into the write circuitry. The clock gating array is used in order to activate the clock port of the storage elements, which are determined by the write address. An implementation example of SCM structures with $R = C = 4$ is shown in Figure 10.3 (Shiomi *et al.* 2019). The 4-to-1 readout multiplexers are implemented into the readout circuitry. Based on the one-hot signals generated by the read address decoder, the readout operation is performed. The clock port of the storage elements at each row is interconnected with a clock gating circuit. This clock gating circuit activates the storage elements based on the one-hot signal generated by the write address decoders. The latch cells labeled "Master latch" and "Slave latch (storage elements)" form the master-slave flip-flop. As a result, the write operation is performed digitally.

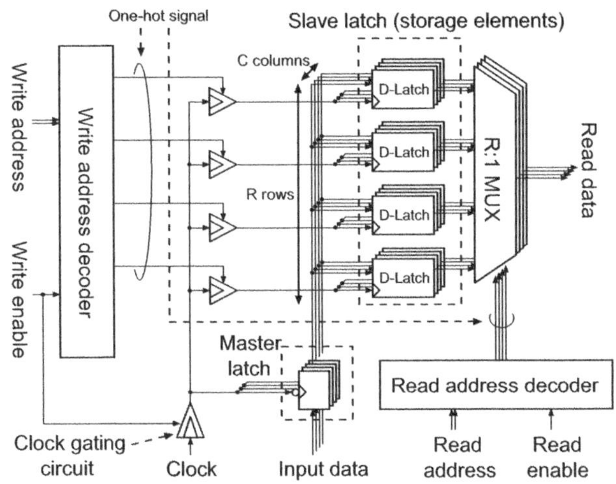

Figure 10.3. *An example of SCM structures ($R = 4$, $C = 4$)*

Since only digital structures are implemented into the SCM circuitry, the voltage scalability of the on-chip memories can be improved to the level of the conventional logic circuits without any design effort. Wang and Chandrakasan show that the latch-based fully digital structure enables on-chip memories to operate even in the sub-threshold region where the supply voltage (V_{DD}) is below the threshold voltage (V_{TH}) (Wang and Chandrakasan 2005), which enables the target processor to operate in the near-threshold voltage region without any functional failures.

As described in the previous section, bit-line structures used in SRAMs inherently waste the dynamic energy consumption at the bit line in each readout/write operation. Since SCM structures can easily be optimized by tuning the HDL code, with the aid of aggressive signal- and clock-gating techniques, the energy consumption of SCMs can be considerably smaller than that of low-voltage SRAMs with the bit-line structure (see section 10.2.2.1 for more details).

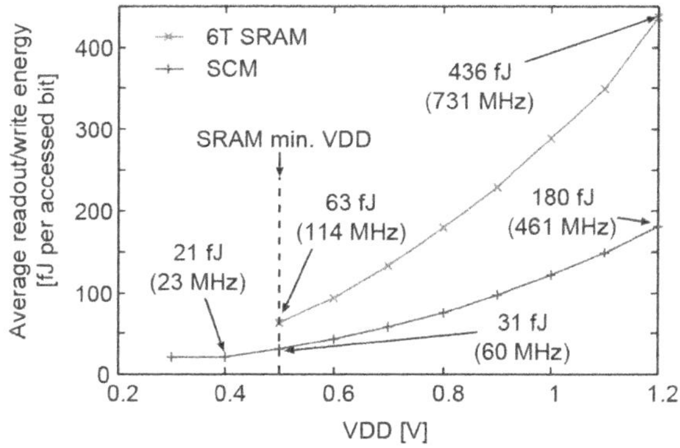

Figure 10.4. *Energy measurement results for two memories with a* 256×32 *capacity in a 65 nm low-Vt SOI-CMOS technology. 6T SRAM: 6-transistor SRAM. SCM: standard-cell-based memory*

For example, Figure 10.4 shows silicon measurement results of energy consumption per operation for a 6-transistor SRAM and an SCM (Shiomi *et al.* 2019), which are fabricated with a commercial 65 nm low-Vt SOI-CMOS process technology. The vertical axis is average energy consumption per accessed bit when the target memory readouts/writes pseudo-random numbers sequentially at the maximum operating speed. No body biasing techniques are applied to the memories. Note that we do not consider the sufficient yield for each on-chip memory since only single instances are measured in this experiment. The energy curve labeled "6T SRAM" implies that the conventional 6-transistor SRAM fails to operate below a 500 mV supply voltage before it reaches the optimum voltage point, which minimizes its energy consumption. On the other hand, the fully digital structures in SCMs enable the lowering of the minimum operating voltage up to at least 200 mV in comparison with the 6-transistor SRAM. As a result, the SCM can operate at the optimum voltage point, where its energy consumption is minimized. The SCM achieves three times better energy efficiency (21 fJ per bit) at a 400 mV voltage condition than the 6-transistor SRAM operating at a 500 mV voltage condition. The SCM also exhibits better energy consumption than the 6-transistor SRAM even though the supply voltage condition is the same. For example, the SCM exhibits about twice better energy consumption than the SRAM at the 500 mV supply voltage condition. This is because SRAM structures use bit-line-based structures with a large parasitic capacitance. Summarizing the above, the results imply that SCMs play an important role in energy-efficient near-threshold computing. Another advantage is their flexibility in its floorplan since they can be synthesized through commercial EDA tools for digital CMOS circuits. Shiomi *et al.* point out that, without any performance/area degradation, SCM can be implemented into various floorplans whose aspect ratio ranges from 1:4 to 4:1 (Shiomi *et al.* 2019). SCMs can be basically fit into any places in the chip, which maximizes the performance of the entire chip.

A potential drawback of the SCM is an area overhead. As for a small memory capacity, SCMs still exhibit better area efficiency than SRAMs. This is because peripheral circuits for SRAMs (e.g. sense amplifier) are the dominant source that determines the entire area of the target SRAM, while the SCM does not require these special peripheral circuits. Teman *et al.* point out that SCMs still exhibit better area efficiency than foundry-provided 6-transistor SRAMs in a 28 nm process technology if their capacity is not more than 1 kb (Teman *et al.* 2016). On the other hand, as for the SCM with more than 1 kb capacity, the situation becomes different. Shiomi *et al.* show that SCMs exhibit at least about twice larger area than the conventional 6-transistor/low-voltage SRAMs (Shiomi *et al.* 2019) under a condition where the memory capacity is not less than 16 kb. The area penalty of SCMs directly results in the loss of the capacity for the on-chip memories. This results in increasing the access frequency of the energy-consuming off-chip memory (Xu *et al.* 2019). As a result, the energy consumption of the entire SoC may drastically increase if we simply implement SCMs into the entire on-chip memory system. As a remedy to this problem, the next section briefly presents a design strategy to improve the area efficiency while exploiting the good energy efficiency of SCMs.

10.2.2. *Design strategy for area- and energy-efficient SCMs*

Improving area efficiency is a must in taking advantage of SCMs. This section briefly presents design strategies for achieving area-efficient SCMs in three different design stages.

10.2.2.1. *Architectural-level design strategy*

Since SCMs can be synthesized using the HDL code, fine-grained tuning of their circuit structure is available with the minimum design effort. In order to maximize the area and energy efficiencies of the SCMs, architectural-level design strategies have thus been widely studied. Meinerzhagen *et al.* examined various SCM structures including the readout/wire logic and storage elements (Meinerzhagen *et al.* 2010). As for the readout logic, they show that a multiplexer-based logic exhibits smaller area and lower power consumption than a tristate-buffer-based logic. The tristate-buffer-based logic uses one tristate buffer per storage cell where the input of each tristate buffer is interconnected with the output of the corresponding storage cell. All tristate buffer outputs in the same column are interconnected with a bit line such as SRAMs. The readout operation is performed by activating one tristate buffer based on the one-hot signal generated by the read address decoder. If the column length (i.e. the number of words) becomes large, stronger drivers consuming large area are required in a tristate-buffer-based logic to improve readout delay. On the other hand, a multiplexer-based logic uses 2-to-1 multiplexers interconnected to operate as a binary selection tree. Since the load capacitance of each multiplexer is much smaller than that of a bit line in the tristate-buffer-based logic, a high drive strength in each multiplexer is not required. The multiplexer-based readout logic thus achieves lower energy and area than those in the tristate-buffer-based readout logic. The multiplexer-based readout logic also plays an important role in terms of the timing yield (Shiomi *et al.* 2015). In the near-threshold voltage region, the magnitude of the intra-die random delay fluctuation in each logic gate is considerably larger than that in the super-threshold region. On the other hand, the critical path of the multiplexer-based readout circuit is serially connected 2-to-1 multiplexers. As a result, if we consider the critical path delay, the delay can be obtained by summing up the random delay fluctuation, which averages out the total delay fluctuation. The multiplexer-based readout circuit thus also exhibits better timing yields than tristate-buffer-based logic.

As for the write logic, a clock gating array is implemented as described in the previous section. In order to reduce the energy consumption at a clock tree, fine-grained clock gating circuits are implemented in order to activate the minimum necessary clock buffers (Teman *et al.* 2016). Latches or flip-flops are typically implemented into storage arrays. Meinerzhagen *et al.* point out that latch cells achieve better area and energy efficiencies than flip-flops since latches have fewer transistors (Meinerzhagen *et al.* 2010). A number of SCMs thus use the multiplexer-based readout logic with latch-based storage elements (Wang and Chandrakasan 2005; Andersson *et al.* 2014; Teman *et al.* 2016; Shiomi *et al.* 2019).

10.2.2.2. *Physical placement optimization*

Since the SCM has a systematic array structure, designers can effectively reduce its area (and thus energy consumption) by a *controlled placement* technique. Teman *et al.* propose a controlled placement technique where all the bit cells and readout multiplexers are manually aligned (Teman *et al.* 2016) in the place-and-route design phase. They show that the controlled placement technique can reduce the routing congestion. As a result, the SCM area can be reduced by 25% on average, in comparison with the non-controlled conventional implementations. A similar concept is proposed by Yoshida and Usami (2017). They manually isolate latch cells and other circuits, such as readout multiplexers, into different substrate islands in the place-and-route phase, which enables not only reducing the routing congestion but also applying aggressive reverse body biasing to latch cells only. Yoshida and Usami point out that readout paths typically become critical paths of SCMs. Therefore, they can reduce a large amount of leakage energy without degrading their operating speed.

10.2.2.3. *Cell-level optimization*

Area optimization in the standard cells leads to improvement in the entire SCM area. Several papers thus propose the cell-level area optimization technique

(Andersson *et al.* 2013; Shiomi *et al.* 2019). Latch cells and readout multiplexers are the dominant source that determines the entire area of the SCM. Andersson *et al.* (2013) thus designed and added a special standard cell into a conventional standard-cell library. It consists of: 1) two latch gates for 2-bit storage elements, and 2) two NAND2 gates used in the readout multiplexer of the SCM. The four logic gates are interconnected with one another in a cell layout design phase. Since the latch gates and NAND2 gates dominate the entire area of the SCM, the special cell enables us to improve the area efficiency (and thus energy consumption). Shiomi *et al.* propose a concept of a minimum height cell (MHC) (Shiomi *et al.* 2019). Based on the fact that SCMs do not require complex logic functions but require simple ones only (e.g. readout multiplexers and address decoders), they design a simple standard-cell library with almost minimum necessary logics. This simplification enables us to optimize the physical structure of the standard cells. They designed an MHC library where all the logic gates have the minimum height to construct CMOS logic gates in order to reduce the area and thus the energy consumption.

10.2.3. *Hybrid memory design towards energy- and area-efficient memory systems*

The SCM exhibits better energy efficiency with several times larger area than the SRAM. As a remedy to this, it is natural to design the hybrid memory structure using both SCMs and SRAMs in order to optimize energy efficiency with minimum area overhead. Gautschi *et al.* propose the parallel ultra-low power (PULP) architecture for a multi-core system with a special data memory referred to as a tightly coupled data memory (TCDM) (Gautschi *et al.* 2017). Each bank in a TCDM consists of energy-efficient SCM blocks with a small capacity and area-efficient SRAM blocks. If the target processor does not require high performance, the SRAM blocks on the TCDM and most of the cores are disabled, which enables the processor to operate as an energy-efficient single-core processor. By downscaling the supply voltage to the near-threshold region, energy-efficient operation can be achieved. If a high performance is required, the disabled memories and cores wake up while increasing the supply voltage. In order to take advantage of the energy-efficient SCM, Conti *et al.* present a software-level technique that assigns data with high access frequency into the SCM in the TCDM (Conti *et al.* 2016).

The hybrid memory structure also plays an important role in reducing the energy consumption of caching systems. One of the early studies related to the hybrid caching system is by Kin *et al.* (1997). They propose a concept of a filter cache with a relatively small memory capacity of about 256 bytes. The key point is that the energy consumption of the filter cache is much smaller than a typical cache with a large capacity. They show that by inserting the filter cache between the CPU core and the level-1 (L1) cache, the L1 access is filtered by the filter cache. As a result, the total power dissipation of the caching system can be effectively reduced at the cost of degrading execution time. Dreslinski *et al.* (2008) propose a reconfigurable hybrid cache consisting of a near-threshold SRAM block and 6-transistor SRAM blocks. If the processor does not require high performance, the near-threshold SRAM block and 6-transistor SRAM blocks operate as an energy-efficient filter cache and an L1 cache, respectively. In the high-speed mode, all the memory blocks are simultaneously accessed as a fully associative cache. In the same way, Xu *et al.* (2019) propose a two-level instruction cache memory structure consisting of a level-0 (L0) filter SCM and an L1 SRAM. An SCM operating in the near-threshold region is implemented as the lower-level cache rather than the conventional 6-transistor-SRAM-based L1 cache operating in the super-threshold region. They point out that a hit ratio of the cache has an approximately logarithmic relationship with a cache capacity. Therefore, even though the L0 cache capacity is not sufficiently large to store the entire instruction code of the target program, its hit ratio is not considerably low. Combining the L0 SCM cache and the L1 SRAM cache, they point out that it is possible to effectively reduce the access frequency of energy-consuming higher-level memories (e.g. off-chip memory access) while fully exploiting the energy-efficient L0 cache.

10.2.4. *Body biasing as an alternative to power gating*

Reducing the leakage power introduced by billions of transistors on a chip is one of the most important challenges for achieving energy-efficient near-threshold computing. Power gating is one of the most effective approaches to eliminate the leakage current in idle modules. However, if we simply apply power gating to volatile on-chip memories, including SCMs, critical data stored in them are lost. For example, if data stored in cache lines are lost by power gating, they may have to access main memories in order to get the lost data again immediately after they are powered up, which degrades the energy efficiency and performance of the entire SoC. Many papers thus focus on developing a method to find dead cache lines which are not used again during execution (Kaxiras *et al.* 2001). Another approach is to use non-volatile memories. For example, in Fujita *et al.* (2014) and Nakada *et al.* (2015), processor architectures integrating non-volatile memories are proposed. Before applying power gating, critical data stored in the volatile memories are transferred to non-volatile memories. However, as well as increasing the design cost, the major drawback in this approach is the latency overhead. This is because target volatile on-chip memories have to transfer the critical data from/to non-volatile memories every time they are powered up/down.

As an alternative method to power gating, aggressive body biasing can exponentially reduce leakage current without losing data stored in on-chip memories. In Yamamoto *et al.* (2013) and Fukuda *et al.* (2014), they show that, with the aid of aggressive reverse body biasing, the standby power can be up to 1,000 times smaller than the normal mode without losing data stored in SRAMs. Therefore, the latency overhead is minimum in comparison with the power gating technique. Combining the minimum energy point tracking technique presented in the next section, body biasing plays an attractive role in achieving energy-efficient computing.

10.3. Minimum energy point tracking

One of the most effective approaches for reducing the energy consumption of microprocessors is dynamic supply and threshold voltage scaling. Techniques for dynamically scaling the supply voltage (V_{DD}) and/or threshold voltage (V_{TH}) under dynamic workloads of microprocessors have thus been widely investigated over the past 20 years (Flautner *et al.* 2002; Basu *et al.* 2004; Yan *et al.* 2005). We refer to the best pair of V_{DD} and V_{TH}, which minimizes the energy consumption of the processor under a given operating condition, as a minimum energy point (MEP). V_{TH} of transistors can be dynamically changed by tuning the back-gate bias voltage (V_{BB}) of the gates. Although the dynamic V_{DD} and V_{TH} scaling is a promising technique, it is not trivial to find the MEP when the processor is running. This is because the MEP is highly dependent on the operating conditions such as chip temperature, activity factor, process variation and performance required for the processor, as shown in Figure 10.5. For example, if the performance requirement for the processor is low, the supply voltage of the processor can be lowered. In the nanometer process, the manufacturing process condition widely varies, which causes a die-to-die leakage power variation. In addition, the energy consumption of the processor is highly dependent on chip temperature and activity factor, where the activity factor is the probability that the circuit node transitions from 0 to 1 in a clock cycle (Weste and Harris 2010). The MEP curve widely shifts depending on the above-described operating conditions, as shown in Figure 10.5. The important observation from Figure 10.5 is that the shape of the MEP curve is almost unchanged over the different operating conditions although the coordinate of the curve is widely shifted depending on the conditions (Takeshita *et al.* 2016). This section presents a simple but effective algorithm for closely tracking the MEP of the processor under a wide range of operating conditions such as chip temperature, activity factor, process variation and performance required for the processor.

10.3.1. *Basic theory*

10.3.1.1. *Necessary conditions for minimum energy computing*

Designers generally face the problem of achieving minimum energy under a delay constraint. Equivalently, the power consumption of the system is limited by battery or cooling considerations and the designer seeks to achieve minimum delay under an energy constraint. To achieve those goals, the designer seeks the best pair of V_{DD} and V_{TH}. Figure 10.6 shows contours of delay and energy. The best pair of V_{DD} and V_{TH}, which we refer to as a minimum energy point (MEP) at a given delay, is where the delay and energy contours are tangent (Weste and Harris 2010).

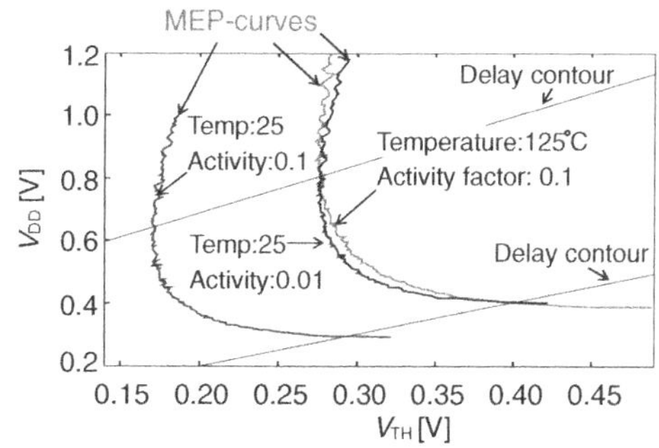

Figure 10.5. *Minimum energy point curves*

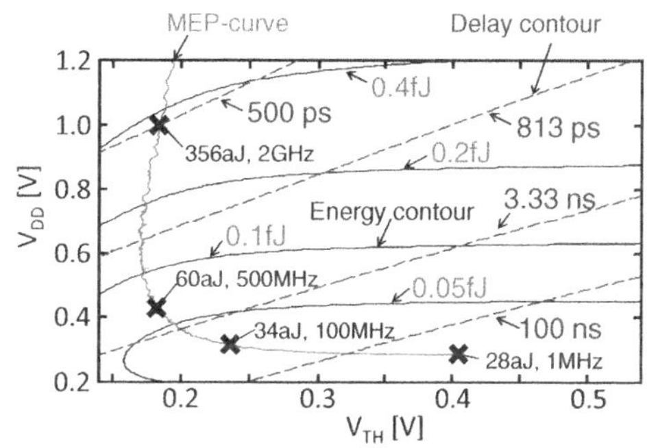

Figure 10.6. *Energy and delay contours*

In Figure 10.6, a 50-stage fanout-4 inverter chain is used to reflect the behavior of a microprocessor pipeline. The energy consumption of the circuit consists of dynamic energy E_d and static energy E_s, as shown in [10.1]. E_d is a quadratic function of V_{DD}, as shown in [10.2]. k_1 and k_2 are fitting coefficients. E_s is exponential to V_{TH} and linear to the delay D and V_{DD}, as shown in [10.3]. N_S is $n_i \cdot \phi_t$, where n_i is an ideal factor of MOSFET, which is typically between 1 and 2, and ϕ_t is the thermal voltage, which is 26 mV at room temperature. Since the value of n_i in a target 28 nm technology is about 1.6, the value of N_S in [10.3] is approximately 42 mV at room temperature.

$$E = E_d + E_s. \quad [10.1]$$

$$E_d = k_1 {V_{DD}}^2. \quad [10.2]$$

$$E_s = k_2 D V_{DD} e^{-\frac{V_{TH}}{N_S}}. \quad [10.3]$$

When $V_{\rm DD} \gg V_{\rm TH}$, the delay can be accurately modeled using alpha power law MOSFET model (Sakurai and Newton 1990), as shown in [10.4], where $V_{\rm DT} = V_{\rm DD} - V_{\rm TH}$. The value of α is around 1.3 in nanometer technologies. When $V_{\rm DD}$ is near-threshold ($V_{\rm DD} \simeq V_{\rm TH}$) and sub-threshold ($V_{\rm DD} < V_{\rm TH}$), the delay can be approximated as exponential functions of $V_{\rm DT}$ (Keller *et al.* 2014) as shown in [10.5]. The parameters k_3 and k_4 are fitting coefficients.

$$D = \frac{k_3 V_{\rm DD}}{{V_{\rm DT}}^{\alpha}}. \qquad (V_{\rm DD} \gg V_{\rm TH}) \qquad [10.4]$$

$$D = k_4 V_{\rm DD} e^{-\frac{V_{\rm DT}}{N_{\rm S}}}. \qquad (V_{\rm DD} \leq V_{\rm TH}) \qquad [10.5]$$

The minimum energy point is determined by a pair of $V_{\rm DD,opt}$ and $V_{\rm TH,opt}$, where the energy consumption of the circuit is minimized under a specific performance constraint. It can be found at a point where the contours of delay and energy are tangent. Thus, at the minimum energy point, the following equation holds as a necessary condition for the minimum energy computing:

$$\frac{-\frac{\partial E_{\rm d}}{\partial V_{\rm TH}} - \frac{\partial E_{\rm s}}{\partial V_{\rm TH}}}{\frac{\partial E_{\rm d}}{\partial V_{\rm DD}} + \frac{\partial E_{\rm s}}{\partial V_{\rm DD}}} = \frac{\frac{\partial D}{\partial V_{\rm TH}}}{-\frac{\partial D}{\partial V_{\rm DD}}}. \qquad [10.6]$$

The left and right of [10.6] represent gradients of the energy contour and the delay contour at the minimum energy point, respectively.

10.3.1.2. *Sub- and near-threshold regions*

If we look at a sub- or near-threshold region where $V_{\rm DD}$ is less than or very close to $V_{\rm TH}$, the ON current increases exponentially as $V_{\rm TH}$ decreases. Thus, the delay D decreases exponentially, as shown in [10.5]. The leakage current exponentially increases as $V_{\rm TH}$ decreases, as shown in [10.3], and offsets the decrease in the delay D. Therefore, $E_{\rm s}$ does not change in the sub-threshold region (Calhoun *et al.* 2005). This characteristic is also seen in [10.7], which can be obtained by substituting [10.5] into [10.3].

$$E_{\rm s} = k_2 k_4 V_{\rm DD}^2 \exp\left(-\frac{V_{\rm DD}}{N_{\rm S}}\right). \qquad [10.7]$$

Multiplying [10.6] by $\left(-\frac{\partial E_{\rm d}}{\partial V_{\rm TH}} - \frac{\partial E_{\rm s}}{\partial V_{\rm TH}}\right) \frac{\partial D}{\partial V_{\rm TH}}$ gives us the following relation:

$$\left(\frac{\partial E_{\rm d}}{\partial V_{\rm DD}} + \frac{\partial E_{\rm s}}{\partial V_{\rm DD}}\right) \frac{\partial D}{\partial V_{\rm TH}} = \left(\frac{\partial E_{\rm d}}{\partial V_{\rm TH}} + \frac{\partial E_{\rm s}}{\partial V_{\rm TH}}\right) \frac{\partial D}{\partial V_{\rm DD}}. \qquad [10.8]$$

Since $E_{\rm d}$ and $E_{\rm s}$ are independent from $V_{\rm TH}$ in the sub-threshold region (Calhoun *et al.* 2005), $\frac{\partial E_{\rm d}}{\partial V_{\rm TH}}$ and $\frac{\partial E_{\rm s}}{\partial V_{\rm TH}}$ are zero in the sub-threshold region. On the other hand, the delay D depends on the threshold voltage (V_{TH}). Therefore, $\frac{\partial D}{\partial V_{\mathrm{TH}}} \neq 0$ holds for all V_{TH}s in the sub-threshold region. Thus, [10.8] can be converted into [10.9]:

$$\left(\frac{\partial E_{\mathrm{d}}}{\partial V_{\mathrm{DD}}} + \frac{\partial E_{\mathrm{s}}}{\partial V_{\mathrm{DD}}}\right) \frac{\partial D}{\partial V_{\mathrm{TH}}} = 0 \Leftrightarrow \frac{\partial E_{\mathrm{d}}}{\partial V_{\mathrm{DD}}} + \frac{\partial E_{\mathrm{s}}}{\partial V_{\mathrm{DD}}} = 0. \quad [10.9]$$

By partially differentiating [10.2] and [10.7] with respect to V_{DD}, the following relations are obtained:

$$\frac{\partial E_{\mathrm{d}}}{\partial V_{\mathrm{DD}}} = 2k_1 V_{\mathrm{DD}} = \frac{2E_{\mathrm{d}}}{V_{\mathrm{DD}}}, \quad [10.10]$$

$$\frac{\partial E_{\mathrm{s}}}{\partial V_{\mathrm{DD}}} = 2k_2 k_4 V_{\mathrm{DD}} \exp\left(-\frac{V_{\mathrm{DD}}}{N_{\mathrm{S}}}\right) - \frac{k_2 k_4 V_{\mathrm{DD}}^2}{N_{\mathrm{S}}} \exp\left(-\frac{V_{\mathrm{DD}}}{N_{\mathrm{S}}}\right)$$
$$= \frac{2E_{\mathrm{s}}}{V_{\mathrm{DD}}} - \frac{E_{\mathrm{s}}}{N_{\mathrm{S}}}. \quad [10.11]$$

Therefore, [10.9], [10.10] and [10.11] give us the following equation, which holds as a necessary condition for the minimum energy point operation in the sub-threshold region:

$$\frac{E_{\mathrm{d}}}{E_{\mathrm{s}}} = \frac{V_{\mathrm{DD}}}{2N_{\mathrm{S}}} - 1. \quad [10.12]$$

On the other hand, the $E_{\mathrm{d}}/E_{\mathrm{s}}$ ratio is also simply calculated as $k_1/k_2k_4 \cdot \exp\left(V_{\mathrm{DD}}/N_{\mathrm{S}}\right)$ from [10.2] and [10.7]. Therefore, the following equation holds in the sub-threshold region:

$$\frac{E_{\mathrm{d}}}{E_{\mathrm{s}}} = \frac{k_1}{k_2 k_4} \exp\left(\frac{V_{\mathrm{DD}}}{N_{\mathrm{S}}}\right) = \frac{V_{\mathrm{DD}}}{2N_{\mathrm{S}}} - 1. \quad [10.13]$$

Since the middle and the right of [10.13] are exponential and linear to V_{DD}, respectively, there are at most two V_{DD} values which satisfy [10.13], as shown in Figure 10.7. However, the lower one is about $2N_{\mathrm{S}}$, which is less than 102 mV in the target process technology at room temperature. This is too small for a supply voltage to ensure a sufficient noise margin. If we do not adopt this low V_{DD}, there is only one practical V_{DD} which satisfies [10.13]. Therefore, the $V_{\mathrm{DD,opt}}$ in the sub-threshold region is stacked at a specific value if N_{S} is fixed. This is the reason why the minimum energy curve in the sub-threshold region is horizontal (see Figure 10.6). Thus, if we find a pair of V_{DD} and V_{TH}, which satisfies [10.13], that is only the pair which minimizes the energy consumption in the practical voltage region. Therefore, [10.13] is a necessary and sufficient condition for the minimum energy point computing in the sub-threshold region.

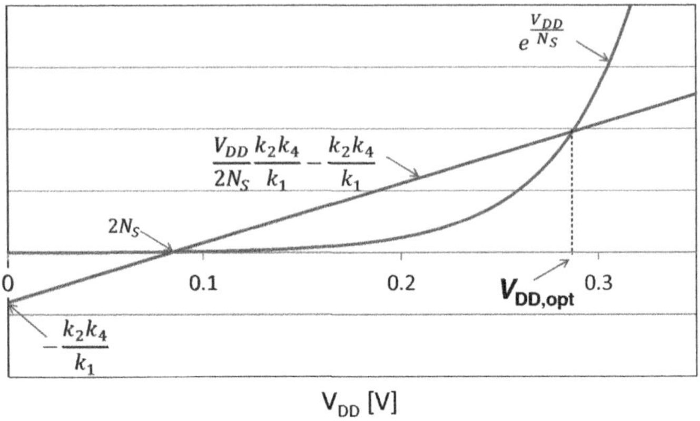

Figure 10.7. *Minimum energy point in near- or sub-threshold region*

10.3.1.3. *Super-threshold region*

By partially differentiating [10.4] with respect to $V_{\rm DD}$ and $V_{\rm TH}$, respectively, the following relations can be obtained.

$$\frac{\partial D}{\partial V_{\rm DD}} = \frac{-\alpha V_{\rm DD} + (V_{\rm DD} - V_{\rm TH})}{V_{\rm DD}\,(V_{\rm DD} - V_{\rm TH})} D, \qquad [10.14]$$

$$\frac{\partial D}{\partial V_{\rm TH}} = \frac{\alpha}{V_{\rm DD} - V_{\rm TH}} D. \qquad [10.15]$$

From [10.15], $D = \frac{V_{\rm DD}-V_{\rm TH}}{\alpha} \cdot \frac{\partial D}{\partial V_{\rm TH}}$ can be obtained. Therefore, substituting $D = \frac{V_{\rm DD}-V_{\rm TH}}{\alpha} \cdot \frac{\partial D}{\partial V_{\rm TH}}$ into [10.14] gives us the following relation:

$$-\frac{\partial D}{\partial V_{\rm DD}} = \frac{\alpha V_{\rm DD} - (V_{\rm DD} - V_{\rm TH})}{\alpha V_{\rm DD}} \frac{\partial D}{\partial V_{\rm TH}}. \qquad [10.16]$$

As with the discussion in section 10.3.1.2, $\frac{\partial E_{\rm d}}{\partial V_{\rm DD}}$, $\frac{\partial E_{\rm d}}{\partial V_{\rm TH}}$, $\frac{\partial E_{\rm s}}{\partial V_{\rm DD}}$ and $\frac{\partial E_{\rm s}}{\partial V_{\rm TH}}$ can be obtained using [10.2] and [10.3]:

$$\frac{\partial E_{\rm d}}{\partial V_{\rm DD}} = 2k_1 V_{\rm DD} = \frac{2E_{\rm d}}{V_{\rm DD}}, \qquad [10.17]$$

$$\frac{\partial E_{\rm d}}{\partial V_{\rm TH}} = 0, \qquad [10.18]$$

$$\frac{\partial E_{\rm s}}{\partial V_{\rm DD}} = \frac{\partial D}{\partial V_{\rm DD}} \frac{E_{\rm s}}{D} + \frac{E_{\rm s}}{V_{\rm DD}}, \qquad [10.19]$$

$$\frac{\partial E_{\rm s}}{\partial V_{\rm TH}} = \frac{\partial D}{\partial V_{\rm TH}} \frac{E_{\rm s}}{D} - \frac{E_{\rm s}}{N_{\rm S}}. \qquad [10.20]$$

By substituting [10.16] into the right-hand side of [10.6], it can be expressed using α, V_{DD} and V_{TH} only. In the same way, from [10.14], [10.15], [10.17], [10.18], [10.19] and [10.20], the left-hand side of [10.6] can also be expressed using α, V_{DD}, V_{TH}, N_{S}, E_{d} and E_{s} only. By the substitutions, [10.6] can be converted to [10.21], which holds as a necessary condition for the minimum energy point computing in the super-threshold region.

$$\frac{E_{\mathrm{d}}}{E_{\mathrm{s}}} = \frac{\alpha V_{\mathrm{DD}} - (V_{\mathrm{DD}} - V_{\mathrm{TH}})}{2N_{\mathrm{S}}\alpha} - \frac{1}{2}. \quad [10.21]$$

As shown in Figure 10.6, the MEP in the super-threshold region moves vertically significantly if we change the performance constraint of the target circuit. As described in section 10.3, MEPs also depend on chip temperature and activity factor even though the performance constraint is fixed. Based on [10.21], for typical parameters of CMOS circuits, the $E_{\mathrm{d}}/E_{\mathrm{s}}$ ratio changes in the range between 2 and 7. For example, if we set the performance constraint of the target circuit as 500 ps, $V_{\mathrm{DD,opt}}$ and $V_{\mathrm{TH,opt}}$ are 1.0 V and 0.18 V, respectively (see Figure 10.6). In this condition, the ratio of the static energy consumption to the total energy consumption (E_{s}/E) is about 20%. On the other hand, if we increase the constraint from 500 ps to 10 ns, $V_{\mathrm{DD,opt}}$ and $V_{\mathrm{TH,opt}}$ shift to 0.35 V and 0.25 V, respectively. As a result, the value of E_{s}/E increases to about 27%.

Multiplying [10.21] by $E_{\mathrm{s}}/V_{\mathrm{DD}}$ gives the following relation.

$$\frac{E_{\mathrm{d}}}{V_{\mathrm{DD}}} = \left(\frac{\alpha - (1 - V_{\mathrm{TH}}/V_{\mathrm{DD}})}{2N_{\mathrm{S}}\alpha} - \frac{1}{2V_{\mathrm{DD}}}\right) E_{\mathrm{s}}. \quad [10.22]$$

According to [10.2] and [10.3], the left-hand side of [10.22] is a linear function of V_{DD}, while the right hand side is an exponential function of V_{TH}. For satisfying [10.22] for all the MEPs in super-threshold region, the change of V_{DD} should be much smaller than that of V_{TH}. This is the reason why the minimum energy curve in the super-threshold region is vertical.

If we set the performance constraint as D_0, [10.21] can be converted to [10.23].

$$\frac{k_1 V_{\mathrm{DD}}}{k_2 D_0 \exp\left(-\frac{V_{\mathrm{TH},D_0}}{N_{\mathrm{S}}}\right)} = \frac{(\alpha - 1)V_{\mathrm{DD}} + V_{\mathrm{TH},D_0}}{2N_{\mathrm{S}}\alpha} - \frac{1}{2}, \quad [10.23]$$

where V_{TH,D_0} is the threshold voltage when the circuit delay D is D_0. Note that V_{TH,D_0} can be expressed as [10.24], and it can be approximated as a linear function of V_{DD}.

$$V_{\mathrm{TH},D_0} = V_{\mathrm{DD}} - \left(\frac{k_3}{D_0} V_{\mathrm{DD}}\right)^{\frac{1}{\alpha}}. \quad [10.24]$$

Similar to the discussion in section 10.3.1.2, the left-hand side of [10.23] shows an exponential trend, while the right-hand side can be approximated as a linear function. Therefore, there are at most two V_{DD} values that satisfy [10.23], as shown in Figure 10.8. However, the lower one is not a practical value since it is typically in the range between $2N_S$ and $4N_S$. If we do not adopt this low V_{DD}, there is only one practical V_{DD} which satisfies [10.23]. Therefore, [10.21] is a necessary and sufficient condition for the minimum energy point operation in the super-threshold region.

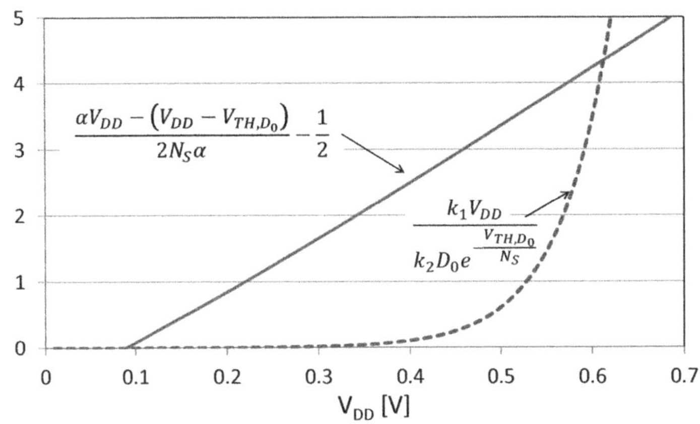

Figure 10.8. *Minimum energy point in super-threshold region*

10.3.2. *Algorithms and implementation*

As described in section 10.3.1, now we know that [10.6] is the necessary and sufficient condition for minimum energy computing. Based on this condition, we can come up with a very simple algorithm for finding the minimum energy point of the processor at runtime.

If we denote the slope of constant delay contours by s_d, which corresponds to the right-hand side of [10.6], it can be converted into [10.25] by partially differentiating the circuit delay represented by [10.4] with respect to V_{DD} and V_{TH}, respectively. Note that $s_d \approx 1$ when $V_{DD} \leq V_{TH}$ (Takeshita *et al.* 2016). If we refer to the slope of constant energy contours at any MEP as s_e, it can be expressed as [10.26] by partially differentiating [10.2] and [10.3] with respect to V_{DD} and V_{TH}, respectively.

$$s_d = \frac{\frac{\partial D}{\partial V_{TH}}}{-\frac{\partial D}{\partial V_{DD}}} = \frac{\alpha V_{DD}}{\alpha V_{DD} - (V_{DD} - V_{TH})} \qquad [10.25]$$

$$s_e = \frac{-\dfrac{\partial E}{\partial V_{TH}}}{\dfrac{\partial E}{\partial V_{DD}}} = \frac{-\dfrac{\partial E_d}{\partial V_{TH}} - \dfrac{\partial E_s}{\partial V_{TH}}}{\dfrac{\partial E_d}{\partial V_{DD}} + \dfrac{\partial E_s}{\partial V_{DD}}} = \frac{E_s V_{DD}}{(2E_d + E_s)N_s} \qquad [10.26]$$

The value of s_d can be easily obtained if we know V_{DD} and V_{TH} of the processor. The values of α in [10.25] and N_s in [10.26] can be obtained through model fitting for the target process technology at a specific temperature. Note that the value of N_s is proportional to the absolute temperature. The value of s_e can be estimated at runtime by measuring the temperature, dynamic and static energy values. Once E_d, E_s and temperature are measured at runtime, the algorithm identifies the direction of stepping voltage. Figure 10.9 shows the basic concept of the algorithm. Suppose a pair of V_{DD} and V_{TH} is on the delay contour which satisfies a performance requirement. At the MEP, $s_d = s_e$ since [10.6] holds. If $s_d > s_e$, both V_{DD} and V_{TH} should be stepped down to get close to the MEP. Contrarily, if $s_d < s_e$, both V_{DD} and V_{TH} should be stepped up toward the MEP.

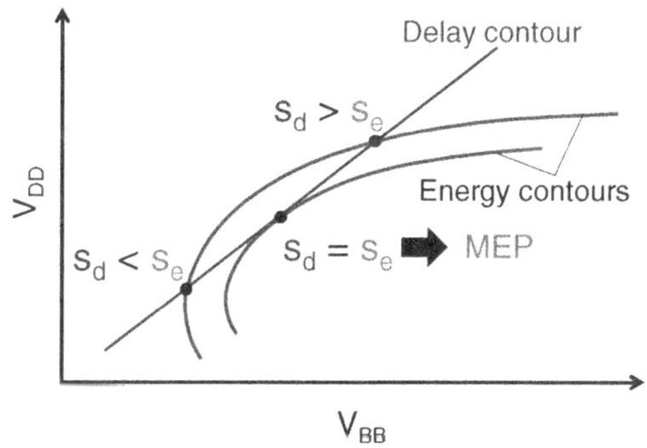

Figure 10.9. *The concept of minimum energy point tracking algorithm*

The algorithm presented above can be easily implemented with a critical-path monitor, a dynamic power sensor, a static power sensor and a temperature sensor. An MEP tracking processor chip integrating all digital power sensors, a temperature sensor and a critical-path monitor is presented in Hokimoto *et al.* (2018). It uses a set of performance counters to estimate the dynamic power consumption at runtime. Krishnaswamy *et al.* (2015) reported that the dynamic power consumption can be accurately estimated using performance counters, which represent the activity factor of the processor. The maximum error of their approach is less than 5%. A leakage sensor which estimates runtime static power consumption is proposed in Islam *et al.* (2015). The leakage sensor is based on a ring oscillator, which is driven by the sub-threshold leakage current. Since the oscillation frequency is proportional to the leakage current, the latter can be estimated by counting the oscillation frequency. V_{DD} and V_{BB} supplied to the processor are also given to the leakage monitor so that the leakage current of the processor can be accurately represented by the leakage current of the monitor. If we supply fixed constant V_{DD} and V_{BB} to the leakage monitor, it can be worked as a temperature sensor because the leakage current in this case depends only on the temperature (Islam *et al.* 2015). The critical-path monitor is used to replicate the critical-path delay of the processor. With the delay tracking techniques such as those presented in Kuroda *et al.* (1998), Nomura *et al.* (2006) and Park and Abraham (2011), the pair of V_{DD} and V_{BB} can be dynamically tuned so that the critical-path delay of the processor is closely tracking the target clock cycle time.

More simplified MEP tracking approaches are proposed in von Kaenel *et al.* (1994), Nomura *et al.* (2006) and Jeongsup Lee *et al.* (2019). Those approaches are based on empirical observation that the energy consumption under a delay constraint is minimized when leakage is about half dynamic power (von Kaenel *et al.* 1994; Nose and Sakurai 2000; Markovic *et al.* 2004). Note that this observation is based on experimental investigation for specific process technologies. The algorithm can be quite similar to that presented in Figure 10.9. When the leakage is half of the dynamic power, the current operating point is the MEP. If the leakage is less than half of the dynamic power, both V_{DD} and V_{TH} should be stepped down to get close to the MEP. On the other hand, if the leakage is more than half of the dynamic power, both V_{DD} and V_{TH} should be stepped up toward the MEP. In practice, the algorithm has to work not only to achieve the minimum energy consumption but also to satisfy the performance demands. Therefore, V_{DD} and V_{TH} should be controlled by simultaneously taking both the dynamic-to-static power ratio and the performance constraint into consideration.

10.3.3. *OS-based approach to minimum energy point tracking*

If we consider developing a holistic approach to improving the energy efficiency of the system without sacrificing the quality of services, an OS-based approach is indispensable. For example, aggressive voltage scaling by the MEPT algorithm sometimes increases sensitivity to noises and degrades the reliability of the system, which can be a fatal disadvantage to a specific type of application such as a mission critical system. If the algorithm is implemented by a software program, it is possible to flexibly coordinate the requirements of energy efficiency and system reliability appropriately. Moreover, software implementation makes it very simple for porting the algorithm into not only power managers for IoT devices but also operating systems running on general-purpose processors used in data centers and cloud servers. This section presents an OS-based approach towards MEPT, which keeps underlying microprocessors always running at MEP with coordinately considering other requirements such as system reliability and real-time responsiveness.

Before discussing the OS-based MEPT, we summarize some of key technologies for OS-based dynamic voltage and frequency scaling (DVFS) approaches that have already been applied to commercial products. Based on the observation that the system is rarely fully used, energy management systems can successfully trade performance for energy without causing the software to miss its deadlines. DVFS algorithms can be categorized into interval-based algorithms and task-based algorithms. Task-based algorithms analyze the deadline and workload of each task and adjust processor performance for each task based on the analysis results. Interval-based algorithms measure the processor utilization of past time intervals and use this information to set the current performance level. For predictable workloads, task-based algorithms save more energy and provide more suitable performance than interval-based algorithms because they use more precise information of the tasks. Based on the predefined deadline of each task, they provide near-optimal performance to each task separately. However, the workload is rather unpredictable in a variety of systems, from simple cell phones to more complex personal computers and servers. Moreover, the implementation of a task-based algorithm is more complex and their overhead is higher compared to that of interval-based algorithms. As a result, it is hard to adopt task-based algorithms into general-purpose operating systems. If we look at embedded operating systems, the first-order priority is real-time responsiveness, and reducing the energy consumption is a second-order priority in many classes of them. If it involves time overheads for voltage transition and/or additional computational time for finding the best voltage setting, which prevent the OSs from real-time responsiveness, DVFS algorithms are rather difficult to be incorporated into the OSs. As a result, DVFS algorithms are rarely incorporated into commercial embedded real-time OSs.

Interval-based algorithms, on the other hand, are easy to implement, have little overhead and can be implemented transparently (Weiser *et al.* 1994; Pering *et al.* 1998). In other words, they require no modification of the existing application and work well even without providing detailed information on each task. Thus, commercial operating systems, such as Linux and Microsoft Windows, use interval-based algorithms for their performance management.

For example, the enhanced Intel SpeedStep running on Pentium III Mobile processors decide whether the performance should be raised or lowered based on the CPU utilization in past time intervals. In the case of performance raising, the voltage will first increase and then the frequency will increase. In the case of performance lowering, the frequency change will occur first and then the voltage is adjusted to this frequency later (Gochman *et al.* 2003). With the algorithm, the processor can keep running even when the voltage is transitioning. Note that the clock frequency is generated using a critical-path replica which can closely and quickly replicate the critical-path delay of the processor. For example, a critical-path monitor (CPM) presented in (Park and Abraham 2011) is composed of multiple critical-path replicas (CPRs) and a timing checker. Whenever the input signal enters, it goes through all of the CPRs in parallel, and the worst delay appears at the output. Then the output of the CPRs reaches the timing checker, and it compares the CPR delay with the clock cycle time. Because the CPM samples the delay of the CPRs at every clock cycle, it can instantaneously and continuously check the speed of a chip. If the CPR delay is faster than that of the clock cycle time, the system will increase its clock frequency or decrease the supply voltage of the processor in order to exploit the timing margins. If slower than clock cycle, the system will decrease its clock frequency or increase the supply voltage of the processor to prevent real critical paths from failing. MEPT algorithms most suitably fit into the interval-based algorithms used in general-purpose systems. Based on the above-mentioned DVFS algorithm, a simple MEPT algorithm running on general-purpose systems can be designed, as shown in Figure 10.10. In the case of performance lowering, the frequency is first decreased and then a pair of V_{DD} and V_{TH} is found using the delay tracking method, such as that presented in Kuroda *et al.* (1998), Nomura *et al.* (2006) and Park and Abraham (2011), which adjust the speed of the processor to the predetermined frequency using the critical-path replica. In the case of performance raising, the first step is increasing V_{DD} and decreasing V_{TH} to specific amounts. Then, the corresponding frequency is determined, using the delay tracking method. If the pair of V_{DD} and V_{TH} found by the delay tracking method is not the MEP, the MEPT algorithm finds the best pair of V_{DD} and V_{TH}, which minimizes the energy consumption for the frequency. According to the discussion in sections 10.3.1.2 and 10.3.1.3, the MEP curve in near- and sub-threshold regions is horizontal, and that in the super-threshold region is nearly vertical with a specific slope, as shown in Figure 10.5. With this knowledge on the shape of the MEP curve, we can estimate the stepping direction for V_{DD} and V_{TH}. However, the estimation may not be accurate. If there is an error in the estimation, the operating point can be moved to the MEP using the MEPT algorithm, as shown in Figure 10.9. With this simple algorithm, the processor can be always running at the MEP even in the case that the chip temperature, activity factor, process variation, transistor's aging status and workload of the applications running on the processor widely vary.

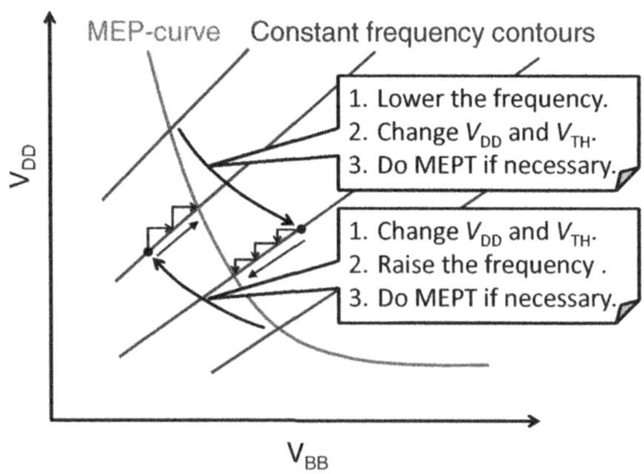

Figure 10.10. *An OS-based algorithm for MEPT*

10.4. Conclusion

Near-threshold computing is one of the most promising approaches for achieving high-performance and energy-efficient computation of digital processors. However, on-chip memories such as conventional SRAMs may limit the voltage scalability of the digital processor on the SoC. As a result, it is hard to fully exploit the high-performance and energy-efficient computation introduced by NTC.

In this chapter, first, standard-cell-based memories (SCMs) are presented as an alternative to the conventional SRAMs. The fully digital structure in the SCMs improves the voltage scalability even down to the sub-threshold region, at the cost of area overhead. As a remedy to the area overhead, various design strategies in different design stages are presented. Then, the reverse body biasing technique is presented as an alternative to the power gating technique. This technique does not require transferring data stored in the memory into non-volatile memories while sufficiently reducing the leakage power of the target digital processor.

Although the near-threshold computing with the SCMs is highly energy efficient, some classes of applications, such as car navigation, gaming applications and cloud services, cannot be satisfied with the performance of near-threshold computing. Instead of always running at near-threshold voltage, a more adaptive mechanism for dynamically tracking the required performance of application programs is preferable for many IoT applications. Thus, approaches for dynamically tracking the minimum energy point of the processors has been widely investigated over the past several decades. This chapter introduced a concept of minimum energy point tracking (MEPT), which responds to the dynamic workload of applications running on microprocessors. Based on the necessary and sufficient conditions for the minimum energy point operation, the MEPT algorithm that finds the optimal supply and threshold voltages at run-time has been presented. This algorithm is well suited to existing OS-driven interval-based dynamic voltage and frequency scaling (DVFS) techniques for general-purpose systems. The interval-based MEPT algorithm which can be integrated into existing OSs may keep SoCs always running at the MEP even in the case that the operating conditions such as chip aging status, chip temperature, activity factor, chip-to-chip delay and workload widely vary. However, there is still significant room to investigate more sophisticated MEPT algorithms for complicated SoCs such as multi- or many-core system-on-chips, which one of our future works should be to tackle.

10.5. Acknowledgments

This work was supported by JST CREST Grant Number JPMJCR18K1 and KAKENHI grant-in-aid for scientific research 19K21531 from JSPS. The authors acknowledge the support of the VLSI Design and Education Center (VDEC) at the University of Tokyo.

10.6. References

Andersson, O., Mohammadi, B., Meinerzhagen, P., Burg, A., and Rodrigues, J. (2013). Dual-VT 4kb sub-VT memories with <1 pW/bit leakage in 65 nm CMOS. *European Solid State Circuits Conference*, 197–200.

Andersson, O., Mohammadi, B., Meinerzhagen, P., and Rodrigues, J. (2014). A 35 fJ/bit-access sub-VT memory using a dual-bit area-optimized standard-cell in 65 nm CMOS. *European Solid State Circuits Conference*, 243–246.

Asenov, A. (1998). Random dopant induced threshold voltage lowering and fluctuations in sub-0.1 μm MOSFET's: A 3-D "Atomistic" simulation study. *IEEE Transactions on Electron Devices*, 45(12), 2505–2513.

Basu, A., Lin, S.-C., Wason, V., and Mehrotrat, A. (2004). Simultaneous optimization of supply and threshold voltages for low-power and high-performance circuits in the leakage dominant era. *Proceedings of the 41st Design Automation Conference*, 884–887.

Calhoun, B.H., Wang, A., and Chandrakasan, A. (2005). Modeling and sizing for minimum energy operation in subthreshold circuits. *IEEE Journal of Solid-State Circuits*, 40(9), 1778–1786.

Chang, I.-J., Kim, J.-J., Park, S.P., and Roy, K. (2009). A 32 kb 10T sub-threshold SRAM array with bit-interleaving and differential read scheme in 90 nm CMOS. *IEEE Journal of Solid-State Circuits*, 44(2), 650–658.

Chang, L., Fried, D., Hergenrother, J., Sleight, J., Dennard, R., Montoye, R., Sekaric, L., McNab, S., Topol, A., Adams, C., Guarini, K., and Haensch, W. (2005). Stable SRAM cell design for the 32 nm node and beyond. *Symposium on VLSI Technology*, 128–129.

Chen, G., Sylvester, D., Blaauw, D., and Mudge, T. (2010). Yield-driven near-threshold SRAM design. *IEEE Transactions on Very Large Scale Integration Systems (VLSI)*, 18(11), 1590–1598.

Conti, F., Rossi, D., Pullini, A., Loi, I., and Benini, L. (2016). PULP: A ultra-low power parallel accelerator for energy-efficient and flexible embedded vision. *Journal of Signal Processing Systems*, 84(3), 339–354.

Dreslinski, R.G., Chen, G.K., Mudge, T., Blaauw, D., Sylvester, D., and Flautner, K. (2008). Reconfigurable energy efficient near threshold cache architectures. *International Symposium on Microarchitecture*, 459–470.

Dreslinski, R.G., Wieckowski, M., Blaauw, D., Sylvester, D., and Mudge, T. (2010). Near-threshold computing: Reclaiming Moore's law through energy efficient integrated circuits. *Proceedings of the IEEE*, 98(2), 253–266.

Flautner, K., Mudge, T., and Blaauw, D. (2002). Combined dynamic voltage scaling and adaptive body biasing for lower power microprocessors under dynamic workloads. *Proceedings of International Conference on Computer Aided Design*, 721–725.

Fujita, S., Nomura, K., Noguchi, H., Takeda, S., and Abe, K. (2014). Novel nonvolatile memory hierarchies to realize "normally-off mobile processors". *Proceedings of Asia and South Pacific Design Automation Conference*, 6–11.

Fukuda, T., Kohara, K., Dozaka, T., Takeyama, Y., Midorikawa, T., Hashimoto, K., Wakiyama, I., Miyano, S., and Hojo, T. (2014). A 7ns-Access-Time 25 μW/MHz 128kb SRAM for low-power fast wake-up MCU in 65 nm CMOS with 27fA/b retention current. *IEEE International Solid-State Circuits Conference*, 236–237.

Gautschi, M., Schiavone, P.D., Traber, A., Loi, I., Pullini, A., Rossi, D., Flamand, E., Gürkaynak, F.K., and Benini, L. (2017). Near-threshold RISC-V core with DSP extensions for scalable IoT endpoint devices. *IEEE Transactions on Very Large Scale Integration (VLSI) Systems*, 25(10), 2700–2713.

Gochman, S., Ronen, R., Anati, I., Berkovits, A., Kurts, T., Naveh, A., Saeed, A., Sperber, Z., and Valentine, R. (2003). The Intel Pentium M processor: Microarchitecture and performance. *Intel Technology Journal*, 7(2), 21–36.

Hokimoto, S., Shiomi, J., Ishihara, T., and Onodera, H. (2018). All-digital on-chip heterogeneous sensors for tracking the minimum energy point of processors. *Proceedings of IEEE International Conference on Microelectronic Test Structures*, 128–133.

Islam, M., Shiomi, J., Ishihara, T., and Onodera, H. (2015). Wide-supply-range all-digital leakage variation sensor for on-chip process and temperature monitoring. *IEEE Journal of Solid-State Circuits*, 50(11), 2475–2490.

Jain, S., Khare, S., Yada, S., Ambili, V., Salihundam, P., Ramani, S., Muthukumar, S., Srinivasan, M., Kumar, A., Gb, S., Ramanarayanan, R., Erraguntla, V., Howard, J., Vangal, S., Dighe, S., Ruhl, G., Aseron, P., Wilson, H., Borkar, N., De, V., and Borkar, S. (2012). A 280 mV-to-1.2V wide-operating-range IA-32 processor in 32nm CMOS. *IEEE International Solid-State Circuits Conference*, 66–68.

Jeongsup Lee, J., Zhang, Y., Dong, Q., Lim, W., Saligane, M., Kim, Y., Jeong, S., Lim, J., Yasuda, M., Miyoshi, S., Kawaminami, M., Blaauw, D., and Sylvester, D. (2019). A 6.4pJ/cycle self-tuning cortex-M0 IoT processor based on leakage-ratio measurement for energy-optimal operation across wide-range PVT variation. *IEEE International Solid-State Circuits Conference*, 314–316.

von Kaenel, V., Pardoen, M., Dijkstra, E., and Vittoz, E. (1994). Automatic adjustment of threshold and supply voltages for minimum power consumption in CMOS digital circuits. *Proceedings of IEEE Symposium on Low Power Electronics*, 78–79.

Kaxiras, S., Zhigang Hu, and Martonosi, M. (2001). Cache decay: Exploiting generational behavior to reduce cache leakage power. *Proceedings of 28th Annual International Symposium on Computer Architecture*, 240–251.

Keller, S., Harris, D., and Martin, A. (2014). A compact transregional model for digital CMOS circuits operating near threshold. *IEEE Transactions on Very Large Scale Integration (VLSI) Systems*, 22(10), 2041–2053.

Kin, J., Gupta, M., and Mangione-Smith, W.H. (1997). The filter cache: An energy efficient memory structure. *International Symposium on Microarchitecture*, 184–193.

Krishnaswamy, V., Brooks, J., Konstadinidis, G., McAllister, C., Pham, H., Turullols, S., Shin, J., Yifan, Y., and Haowei, Z. (2015). Fine-grained adaptive power management of the SPARC M7 processor. *IEEE International Solid-State Circuits Conference*, 1–3.

Kuroda, T., Suzuki, K., Mita, S., Fujita, T., Yamane, F., Sano, F., Chiba, A., Watanabe, Y., Matsuda, K., Maeda, T., Sakurai, T., and Furuyama, T. (1998). Variable supply-voltage scheme for low-power high-speed CMOS digital design. *IEEE Journal of Solid-State Circuits*, 33(3), 454–462.

Markovic, D., Horowitz, M., and Brodersen, R. (2004). Methods for true energy-performance optimization. *IEEE Journal of Solid-State Circuits*, 39(8), 1282–1293.

Meinerzhagen, P., Roth, C., and Burg, A. (2010). Towards generic low-power area-efficient standard cell based memory architectures. *IEEE International Midwest Symposium on Circuits and Systems*, 129–132.

Nakada, T., Shimizu, T., and Nakamura, H. (2015). Normally-off computing for IoT systems. *International SoC Design Conference*, 147–148.

Nomura, M., Ikenaga, Y., Takeda, K., Nakazawa, Y., Aimoto, Y., and Hagihara, Y. (2006). Delay and power monitoring schemes for minimizing power consumption by means of supply and threshold voltage control in active and standby modes. *IEEE Journal of Solid-State Circuits*, 41(4), 805–814.

Nose, K. and Sakurai, T. (2000). Optimization of V_{DD} and V_{TH} for low-power and high speed applications. *Proceedings of Asia and South Pacific Design Automation Conference*, 469–474.

Park, J. and Abraham, J. (2011). A fast, accurate and simple critical path monitor for improving energy-delay product in DVS systems. *Proceedings of IEEE/ACM International Symposium on Low-Power Electronics and Design*, 391–396.

Pering, T., Burd, T., and Brodersen, R. (1998). The simulation and evaluation of dynamic voltage scaling algorithms. *Proceedings of IEEE/ACM International Symposium on Low-Power Electronics and Design*, 76–81.

Sakurai, T. and Newton, R. (1990). Alpha-power law MOSFET model and its applications to CMOS inverter delay and other formulas. *IEEE Journal of Solid-State Circuits*, 25(2), 584–594.

Shiomi, J., Ishihara, T., and Onodera, H. (2015). An energy-efficient on-chip memory structure for variability-aware near-threshold operation. *International Symposium on Quality Electronic Design*, 23–28.

Shiomi, J., Ishihara, T., and Onodera, H. (2019). Area-efficient fully digital memory using minimum height standard cells for near-threshold voltage computing. *Integration, the VLSI Journal*, 65, 201–210.

Takeshita, T., Ishihara, T., and Onodera, H. (2016). Guidelines for effective and simplified dynamic supply and threshold voltage scaling. *Proceedings of International Symposium on VLSI Design, Automation and Test*, 1–4.

Teman, A., Rossi, D., Meinerzhagen, P., Benini, L., and Burg, A. (2016). Power, area, and performance optimization of standard cell memory arrays through controlled placement. *ACM Transactions on Design Automation of Electronic Systems*, 21(4), 1–25.

Wang, A. and Chandrakasan, A. (2005). A 180-mV subthreshold FFT processor using a minimum energy design methodology. *IEEE Journal of Solid-State Circuits*, 40(1), 310–319.

Wang, X., Brown, A., Cheng, B., and Asenov, A. (2011). Statistical variability and reliability in nanoscale finFETs. *IEEE International Electron Devices Meeting*, 5.4.1–5.4.4.

Weiser, M., Welch, B., Demers, A., and Shenker, S. (1994). Scheduling for reduced CPU energy. *Proceedings of Symposium on Operating Systems Design and Implementation*, 13–23.

Weste, N. and Harris, D. (2010). *CMOS VLSI Design: A Circuits and Systems Perspective*, 4th edition. Addison-Wesley, Boston.

Xu, H., Shiomi, J., Ishihara, T., and Onodera, H. (2019). On-chip cache architecture exploiting hybrid memory structures for near-threshold computing. *IEICE Transactions on Fundamentals of Electronics, Communications and Computer Sciences*, E102.A(12), 1741–1750.

Yamamoto, Y., Makiyama, H., Shinohara, H., Iwamatsu, T., Oda, H., Kamohara, S., Sugii, N., Yamaguchi, Y., Mizutani, T., and Hiramoto, T. (2013). Ultralow-voltage operation of silicon-on-thin-BOX (SOTB) 2Mbit SRAM down to 0.37 V utilizing adaptive back bias. *Symposium on VLSI Technology*, T212–T213.

Yan, L., Luo, J., and Jha, N. (2005). Joint dynamic voltage scaling and adaptive body biasing for heterogeneous distributed real-time embedded systems. *IEEE Transactions on Computer Aided Design*, 24(7), 1030–1041.

Yoshida, Y. and Usami, K. (2017). Energy-efficient standard cell memory with optimized body-bias separation in silicon-on-thin-BOX (SOTB). *IEICE Transactions* on Fundamentals of Electronics, Communications and Computer Sciences, E100.A(12), 2785–2796.

11

Maintaining Communication Consistency During Task Migrations in Heterogeneous Reconfigurable Devices

Arief WICAKSANA[1], Olivier MULLER[1], Frédéric ROUSSEAU[1] and Arif SASONGKO[2]

[1] *Université Grenoble Alpes, CNRS, Grenoble INP, TIMA, 38000 Grenoble, France*
[2] *Bandung Institute of Technology (ITB), Indonesia*

The capacity to deploy, halt and migrate applications between virtualized processing nodes has proven to be consequential in future cloud provisioning. FPGAs as accelerators have been gaining traction in the cloud due to the high performance, energy efficiency and on-the-fly programmability offered. But the FPGA virtualization in cloud is currently restricted to certain families and vendors of FPGAs, and naturally, migrating applications between processing nodes that have different FPGAs is troublesome. Task migration between heterogeneous FPGAs presents a risk in the communication during the extraction and restoration of the memory contents related to hardware tasks. To maintain a consistent execution after task migration, the communication between tasks in the processing nodes must be efficiently managed.

In this chapter, we present a communication methodology in hardware context switching suitable for cloud infrastructure integrating heterogeneous FPGAs. The communication mechanism follows the Kahn process network (KPN) model to

For a color version of all figures in this book, see www.iste.co.uk/andrade/multi1.zip.

Multi-Processor System-on-Chip 1 – Architectures,
coordinated by Liliana ANDRADE and Frédéric ROUSSEAU.

provide a deterministic behavior in the system and ensure the resumption of execution without any error. We introduce a communication protocol that manages the communication data and FPGA task context in hardware task migration. The proposed communication solution for hardware context switching between heterogeneous FPGAs is implemented in the Zynq and Intel FPGA platforms. The result shows that communication protocol ensures a consistent execution to migrated applications. A small overhead in resource utilization and data footprint is detected in exchange for up to 98% reduction in the context switch latency (best-case scenario), which facilitates scheduling and provisioning in cloud services. In future works, an integration of our solution into large-scale systems is feasible.

11.1. Introduction

11.1.1. *Reconfigurable architectures*

In recent years, the trend in computing architecture has been moving towards parallelism and multi-processing systems. We cannot rely solely upon Moore's law (Moore 2006) anymore due to the slowdown in silicon scaling technology. This phenomenon results in significant developments of multi-processor system-on-chip (MPSoC) technology in many applications, including mainstream devices, automotive, household appliances and Internet of Things (IoT) devices. To obtain higher efficiency and performance, today's MPSoC architecture contains multi-core CPUs and heterogeneous computing resources, for example graphics processing units (GPUs), field-programmable gate arrays (FPGAs) and application-specific integrated circuits (ASICs). These resources, likely to run heavy and repetitive tasks in order to accelerate the overall execution within the system, are known as accelerators.

FPGAs as hardware accelerators have been gaining popularity thanks to their high performance and low energy consumption. FPGAs are often coupled with CPUs to form reconfigurable architectures (Estrin 1960). By moving performance-critical and computational-heavy applications from CPUs to FPGAs, a 10×–100× typical acceleration can be obtained. This significant acceleration is thanks to the hardware implementation of specific algorithms on FPGA logic units. Compared to most GPUs, FPGAs also work with much lower energy consumption, especially in bit-level executions.

Due to the growth of cloud computing, both in size and service diversity, there has been growing interest in using FPGAs in data center infrastructures. This has been shown in many cloud-FPGA implementations (Byma *et al.* 2014; Fahmy *et al.* 2015; Putnam *et al.* 2015). In such systems, FPGAs must be treated as a virtualized resource in the computing nodes to fully exploit their capabilities. However, dynamic task allocation on FPGAs, especially for load balancing and power management purposes requires significant efforts due to their hardware implementations. One way to avoid such an issue is to provide excellent scheduling methodology in the cloud-FPGA infrastructures. Otherwise, abundant use of FPGAs in the systems is needed, which is costly and non-optimized. For these reasons, we think that a feature that allows interruption and migration of a running task should be supported on FPGAs.

11.1.2. *Contribution*

Task migration between FPGAs can be supported in the system using hardware context switch methodology. Besides task configuration on FPGAs, hardware context switching enables saving and restoring FPGA snapshots or contexts so that a task can be interrupted and resumed from the same point afterwards. A context from an FPGA includes the states of the interrupted task (circuit) stored in the FPGA memory elements, for example registers or block RAMs (BRAMs). Furthermore, hardware context switching on FPGAs requires managing the communication between tasks. This communication management is crucial in order to provide consistent execution and avoid any error in the results, due to moving FPGA tasks in the system. Previous works have been providing solutions of hardware context switching with little discussion on the communication issues. Hence, many existing solutions inflict serious performance limitations on the system.

In this chapter, we present the challenges related particularly to the communication between tasks in the FPGA virtualization. We propose a communication methodology in hardware context switching to enable task migration between processing nodes, such as in the cloud-FPGA environment or any reconfigurable architecture. The proposed solution not only ensures execution consistency of the migrated tasks but also satisfies the state-of-the-art hardware context switch method for heterogeneous FPGAs. Compared to the existing solutions, our methodology can work with dynamic task scheduling approaches. Performance-wise, the proposed method offers much less latency compared to existing solutions in the literature, even though a rather simple scheduler is used in the experiments. It also guarantees that a hardware context switch can be done within a deadline. The experiments to evaluate the performance of our method are done using cryptography and image processing applications.

Section 11.2 of this chapter provides some definitions and describes the challenges in this work. Section 11.3 presents our review of the related works found in the literature. Section 11.4 details our proposed solution; the implementation of which is presented in section 11.5. Section 11.6 will show the experiments and results, while section 11.7 concludes the chapter with possible future works.

11.2. Background

This section presents the background of the work presented in this chapter where we mainly address communication during hardware context switching on heterogeneous FPGAs. To avoid ambiguity in the rest of the chapter, we will describe the relevant terminology used to explain our work. This section will also present the problem and its associated challenges.

11.2.1. *Definitions*

For years, much work has been done to enable hardware context switching on FPGAs. Most of the time, the same terminology is used to explain an almost similar or even a different concept. Some of the definitions used in this chapter are issued in (Hauck and DeHon 2010), a reference publication in reconfigurable computing. For the rest of the text, the following definitions will be applied.

11.2.1.1. *Hardware context switch*

The term hardware context switch often refers to replacing tasks or circuits on FPGAs (Scalera and Vázquez 1998). For stateless tasks such as mathematical operators, this can be done by modifying the configuration on FPGAs with traditional or dynamic partial reconfiguration approaches. In more general terms, hardware context switching also includes saving and restoring the snapshot or context of the FPGA for a task with states at the interruption point. This is triggered by the system scheduler when it must interrupt an ongoing task for any reason and replace it with another, i.e. a preemption request. Here, we consider the tasks on FPGAs run with state machines. Hence, context saving and restoring are necessary in the hardware context switch.

11.2.1.2. *Hardware task context*

Hardware task context can be described as the snapshot of a task on an FPGA. It allows an interrupted task to resume execution from the interruption point in a hardware context switch. We identify a task that runs on a processor as a software task. The context of a software task is the content of the processor registers when it runs. In contrast, a hardware task context consists of the state information stored in the FPGA registers and memories. This is due to the initial hypothesis where a task on an FPGA works with a state machine and not just a stateless operator. The size of a context depends on the live variables of the task.

11.2.1.3. *Communication channel*

In this chapter, we consider that a heterogeneous reconfigurable system can implement one or more tasks which are connected to one other via communication channels. A communication channel is defined as a physical link between hardware tasks and memory elements in the system. It includes a buffer, often implemented as FIFOs, to store data when they are being transferred between two tasks. Depending on the interface and the communication protocol used by each task, the communication channel may also perform a protocol conversion.

11.2.2. *Problem scenario and technical challenges*

In a heterogeneous computing architecture, FPGAs are often used to execute streaming-based applications, for example multimedia, signal processing and deep neural networks. Instead of implementing these applications as a monolithic system, they are often put in pipelines with the software parts in CPU to increase the overall performance with parallelism (Goldstein *et al.* 1999). We consider that the software and hardware tasks run in parallel on available resources and perform a communication via FIFO channels in the system. This hypothesis is depicted in Figure 11.1. Each task can be allocated in the available resource in the system. A task can have one or multiple inputs and outputs with limited-size FIFO buffers that temporarily store the data in the communication channels.

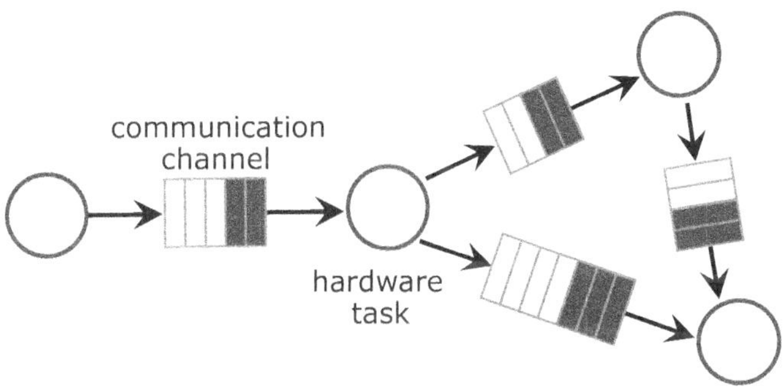

Figure 11.1. *Tasks in a heterogeneous computing architecture communicate with each other via FIFO-based channels*

For load balancing and power management purposes, the capability to migrate allocated tasks in a heterogeneous computing architecture is necessary. This is particularly so when a static scheduling policy cannot be applied or there is a lack of computation resources for the tasks allocated in the system. As such, a task can be interrupted and replaced by another task before finishing its execution. Here, we focus on hardware context switching as a method of migrating tasks from an FPGA to another. We consider that the solutions of software task migration have already been covered since the technology in this area is quite established.

When a request for a hardware context switch is given to an FPGA, the task will be interrupted and its communication flow will be suspended. We recognize two scheduling approaches in a task interruption (Guan *et al.* 2008): preemptive and non-preemptive (cooperative). For non-preemptive scheduling, a task voluntarily yields control of computing resources to another task, without being initiated by another process. In this approach, a task can ensure communication consistency by emptying the FIFO channel before being replaced by another task on the FPGA. For preemptive scheduling, a running task can be interrupted without requiring its cooperation in order to exploit the occupied resources for another task. This may happen immediately (Wheeler *et al.* 2001; Kalte and Porrmann 2005) or at checkpoints (Koch *et al.* 2007; Bourge *et al.* 2016). In this approach, ensuring communication consistency is non-trivial. It is hard to know whether a task has ongoing transfers in the FIFO, both for the input and output sides.

To illustrate the complexity of the challenges in hardware context switching, we present a scenario where a task configured on an FPGA must be replaced with another task. This is described in Figure 11.2. The hardware task V communicates with one predecessor task and one successor task. On the input side of V, there are some data (tokens) from *predecessor* which are not yet consumed by V. On the output side of V, there are tokens to be consumed by *successor* as well. The hardware context switch of V must guarantee continuity of the communication flow between these tasks.

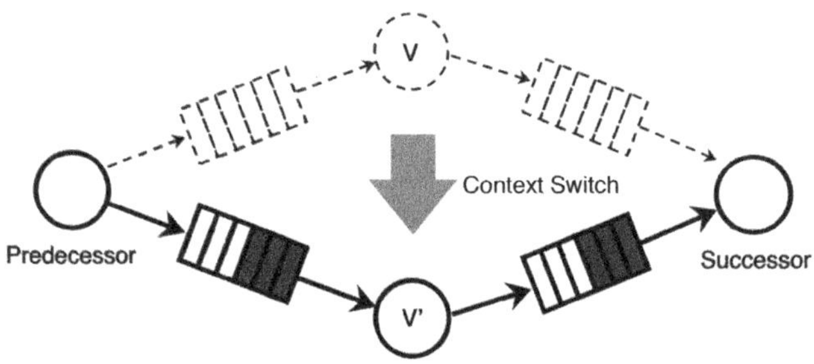

Figure 11.2. *Hardware context switch on a task with FIFO-based communication channels*

When V is interrupted, it stops consuming the tokens from *predecessor* and producing the tokens to *successor*. However, *predecessor* will keep sending new tokens to the communication channel, until the buffer is full, while V is being replaced. The same thing happens for *successor* where its execution flow will continue only when there are still tokens in the buffer. A mechanism to manage the communication in a hardware context switch, without stopping the other tasks in the system, is necessary. The unconsumed tokens in the communication FIFOs must be managed to avoid any data loss. We must ensure that the new task replacing V will not consume the tokens for V.

Finally, when V is restored to V', the communication channels with *predecessor* and *successor* must be re-established. The token transferred from *predecessor* must be redirected to a new input channel of V' since the old channel does not exist anymore. In the same way, *successor* should now receive tokens from the new output channel of V'. For *predecessor* and *successor*, the interruption of V should be as transparent as possible and the communication between the tasks should continue as before when the context switch is performed.

Managing the communication is as important as extracting and restoring the task context. To guarantee consistency and maintain continuity in the communication, the previously explained challenges should be overcome. In the next section, we will present some works which are related to our study on communication in the hardware context switch.

11.3. Related works

11.3.1. *Hardware context switch*

For years, some research works have been trying to support hardware context switching on FPGAs. The literature presents two families of hardware context switch methods that can be implemented on commercial FPGAs. The first family is known as the configuration-based method (Jozwik *et al.* 2010; Morales-Villanueva and Gordon-Ross 2013). This method reads back the configuration bitstream, including the hardware task context, from the FPGA configuration port. The configuration-based method is considered the most popular solution for enabling hardware context switching on FPGAs.

The configuration-based method offers low performance overhead to the hardware tasks programmed on FPGAs since it uses an existing support (Simmler *et al.* 2000). However, it is only supported by certain FPGA families from certain vendors, for example Virtex FPGA family from Xilinx. Furthermore, knowledge of the bitstream format of the FPGA is indispensable to separate the task context from the task configuration, which takes more than 90% of the bitstream in FPGAs (Kalte and Porrmann 2005). Otherwise, the entire configuration and context must be extracted together, which takes significant time and they can only be restored to an identical FPGA. Before extracting a task context from the FPGA configuration port, communication with external memory must also be stopped, for instance, by freezing related clocks, in order to avoid read/write data inconsistency.

One of our initial hypothesises is the use of heterogeneous FPGAs in the system which is not supported by the configuration-based method. For this reason, we require the second family of hardware context switch methods, which is known as the design-based method (Koch *et al.* 2007; Bourge *et al.* 2016). The design-based method supports hardware context switching by instrumenting hardware tasks to access their register and memory contents. In contrast to the configuration-based method, the design-based method does not depend on the FPGA architecture, thereby supporting the use of different FPGAs in the system. The design-based method adds structures which enable access to task states and live variables stored in the FPGA

(Wheeler *et al.* 2001). The drawback, however, is that the added structures introduce an overhead to the task on FPGAs.

Bourge *et al.* (2016) present a method to add context extraction capability using a high-level synthesis (HLS) tool. Their solution offers an autonomous instrumentation of the hardware task that targets FPGA implementation. In addition, the notion of a checkpoint is used to select the states where a hardware context switching can occur, which is also done autonomously by the HLS tool. As a result, the method in (Bourge *et al.* 2016) provides much better performance and less resource overhead compared to previous works. This solution, however, only instruments the hardware task without the communication channels. It handles only the context extraction and restoration, although managing the flow of communication when a hardware context switch occurs is also very important (Shih *et al.* 2010).

The work in Vu *et al.* (2016) presents a design-based context switch method with communication management of sorts. It proposes waiting until the communication channels are idle (empty) before extracting the task context from FPGAs. In order to boost the idle condition of the channels, they throttle the communication data transfer, for instance, by blocking new data in the input FIFOs. But this approach introduces some performance limitations. First, the scheduler has to wait until the FIFOs of the hardware task are empty, which adds latency in a hardware context switch. To reduce this latency, FIFO size has to be as small as possible. Second, the solution requires hardware tasks to send and receive communication data with a regular rate to predict the waiting time to context switch. The hardware context switch will be blocked if one of the successor tasks does not consume the communication data in the output FIFO.

11.3.2. *Communication management*

Besides the ability to extract the task context from FPGAs, the method for managing the communication is also very important in a hardware context switch. The communication management between tasks is normally embedded in a reconfigurable computing framework which integrates CPUs and FPGAs. Although some frameworks are known to be adapted to FPGA use in reconfigurable systems, only a few support hardware context switching, for example ReconOS (Lübbers and Platzner 2008) and Rainbow (Jozwik *et al.* 2013). ReconOS (Lübbers and Platzner 2008) controls the communication with the hardware tasks using modules in the operating system (OS). It employs an infrastructure using FIFO buffers. Although ReconOS is conceived for reconfigurable systems, the initial version lacks the support of preemptive scheduling policies in hardware context switching. Subsequently, an extension of ReconOS to support hardware context switching using a configuration-based method, was presented (Happe *et al.* 2015). Rainbow (Jozwik *et al.* 2013) also provides communication management in a reconfigurable system with the configuration-based context switch. Nevertheless, both ReconOS and Rainbow frameworks do not offer the possibility of relocating tasks to a different FPGA. Thus, they are not compatible with heterogeneous reconfigurable systems.

The task migration study in the multi-processor system-on-chip (MPSoC) context also discusses the communication issue between tasks. El-Antably *et al.* (2015) presents a mechanism to avoid token loss when a task is being migrated from one processor to another. Their system uses FIFO-based channels to handle communication between tasks. To ensure any task can be migrated without causing inconsistency, two copies of communication data (tokens) are prepared on the sender and receiver side. When a migration occurs, the unprocessed tokens will be flushed as since the receiver task sends acknowledgment when consuming every token. However, the implementation of such a method may substantially increase the communication traffic since acknowledgment is required at every token consumption. Another approach in MPSoC is to perform token recollection from the communication channels in a task migration (Scherer 2011). It offers an interesting concept and may be suitable for hardware context switch application in reconfigurable systems.

In summary, some efforts are still required to support hardware context switching in heterogeneous FPGAs. The methodology presented in Bourge *et al.* (2016) shows an efficient method of hardware task context extraction, which is not limited to a certain FPGA family. However, the lack of communication management in the proposed method forbids context switching when there is ongoing communication. The solution in Vu *et al.* (2016) is not optimized, especially when the system scheduler does not have control on the communication channels. Furthermore, the existing frameworks (Lübbers and Platzner 2008; Jozwik *et al.* 2013; Happe *et al.* 2015) support only homogeneous FPGAs. In the next section, we present our communication solution to support hardware context switching on heterogeneous FPGAs.

11.4. Proposed communication methodology in hardware context switching

In a static scheduling approach, the tasks in heterogeneous reconfigurable architectures are allocated based on resource availability. As explained in section 11.2.2, these tasks can be run in parallel to increase overall performance and to communicate with each other via FIFO channels, which offers some challenges in hardware context switching. The scheduling methodology in reconfigurable architectures is not within the scope of this chapter. However, we focus on how to provide the ability to safely interrupt a running task which is a part of the communication network in the system, and continue the execution later in the same or in a different FPGA resource. This objective presents some challenges which depend on the hardware task context and the communication management between tasks. As we know, it takes huge effort to extract a hardware task context from an FPGA and restore it to another FPGA. By doing so, there is a high risk of inconsistency in communication, which can lead to faulty execution.

First of all, we use the Kahn process network (KPN) specification to model the communication between tasks (Kahn 1974). KPN is considered to be the most general form of process networks. In KPN, a task is considered to be an autonomous process that works concurrently with other processes in the network, communicating with one another through channels. The KPN specification suggests that communication channels can be described as infinite FIFOs with non-blocking write and blocking read. In practice, both write and read operations are blocking since a finite amount of memory is used (Geilen and Basten 2003). The tasks in KPN are described as nodes and the communication channels which connect them as arcs with FIFOs.

In the KPN specification, a deterministic behavior can be offered when the communication of a task is suspended. The input and output (I/O) communication data are processed based on their availability in the FIFOs. This specification allows a flow control of the communication channel between two tasks, which is necessary when a context switch occurs. Particularly when a hardware task is migrated from one FPGA to another, the I/O communication should be suspended and picked up correctly from where it left off. The mechanism to perform this action should therefore be part of a hardware context switch operation.

An issue may arise when one of the tasks in KPN is removed, whereas the rest of the network continues. There are two possibilities: whether the I/O FIFOs are moved to the new resource together with the migrating task, as shown in Figure 11.2, or whether they stay with the predecessor and successor tasks. The latter allows the execution to continue until the input FIFO of the successor task is empty and the output FIFO of the predecessor task is full. However, I/O FIFOs of a hardware task are often located inside the FPGA in a reconfigurable architecture (Jozwik *et al.* 2013; Happe *et al.* 2015) to reduce communication latency.

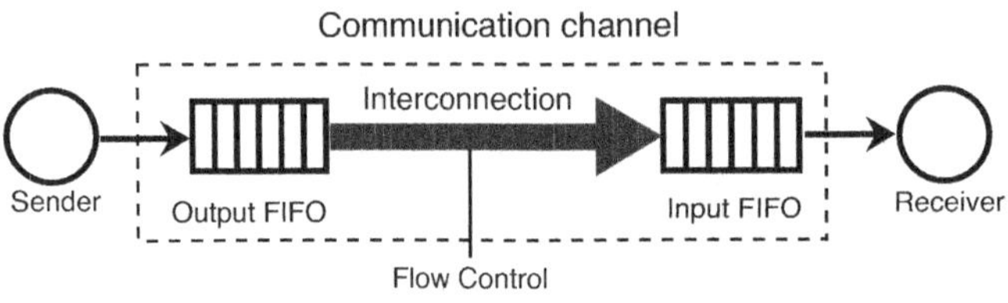

Figure 11.3. *Modified communication channel to support hardware context switching*

For this reason, we propose a different representation of the communication channels in the system. Figure 11.3 describes a modified communication channel between two tasks: sender and receiver. We consider a FIFO on each side of the tasks with an interconnection between them. As a result, we are able to control the task execution flow by manipulating the interconnection between input and output FIFOs. When the receiver task is being migrated, the sender task may still continue the execution until the output FIFO is eventually full. Similarly, the receiver task may consume the data stored in its input FIFO until it is empty when the sender task is migrated. Although this approach guarantees consistent communication to tasks in a hardware context switch, it has a performance limitation due to the size of the FIFOs. To precisely define the depths of the FIFOs used on each side of a task, its I/O rate should be known. Otherwise, the FIFOs will be filled (output FIFO) or empty (input FIFO) too early. In future works, this policy should be optimized in order to offer better performance in the data transfer between FIFOs.

In our approach, the task context extraction can begin immediately when a preemptive request is given to the task. An interrupt command will be sent to the task and to the interconnection between FIFOs. If an ongoing transfer exists, the link between FIFOs will be disconnected after the transfer finishes, to avoid new tokens arriving while handling the switched task FIFOs. This is our first step to guarantee consistency in communication during a context switch. After the FIFOs are disconnected, the task context will be extracted using context extraction techniques. The proposed communication mechanism can be adapted to any existing context switch technique, such as scan-chain extraction (Koch *et al.* 2007; Bourge *et al.* 2016), and it supports heterogeneous architectures.

After the task context is extracted, we can be sure that the switched task will not process any tokens on both input and output FIFOs. The communication FIFOs can be flushed by extracting all the tokens that are not processed by the moving task. The infrastructure to support the FIFO extraction will be detailed in section 11.5. The extracted tokens will then be transmitted to the FIFOs of the switched task when it is restored. Using this token extract/restore mechanism, copies of communication tokens on both sides of FIFOs are not required. Furthermore, the process is transparent to the predecessor and successor task. After the FIFOs of the switched task are flushed, the resource occupied by the moving task and the FIFOs are then released and can be used for other tasks. The rest of the system may continue the execution with their FIFOs located in the communication channel.

Finally, when a task needs to continue, its task context and the FIFOs are restored to the allocated resource. Then, the link between FIFOs in the communication channel can be re-established. With this protocol, we can guarantee that the I/O data are always transmitted in the correct order and there will be no data loss in the process. To summarize, our methodology offers the support of hardware context switching at any time during the execution, even when there are ongoing communication transfers. Furthermore, we ensure the communication consistency of a task being migrated from one FPGA to another.

The described solution can be presented as a hardware context switch protocol which manages the communication data. Figure 11.4 depicts the steps in the proposed protocol, when applied to a hardware task with one successor and one predecessor task. This protocol is generalized for $\sum_{n=1}^{N} m_n$ FIFOs, where N is the number of tasks communicating with the task T (successor and predecessor sides) and m_n is the number of FIFOs of T which relates to task n. The steps in the proposed protocol are described as follows:

1) Disconnect the link between FIFOs on each side of the tasks connected to the migrated task;

2) Interrupt the migrated task and store its context and communication data in the intermediate storage;

3) Restore the context and communication data of the migrated task to an available resource;

4) Re-establish the communication with FIFOs between the migrated task and its predecessor and successor tasks.

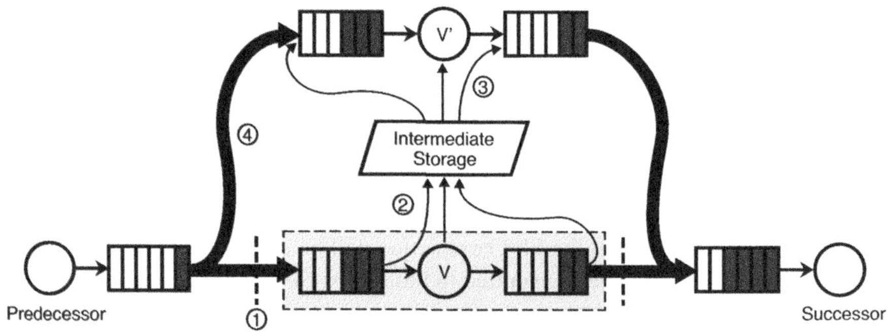

Figure 11.4. *The proposed communication protocol in hardware context switching*

11.5. Implementation of the communication management on reconfigurable computing architectures

In this section, we highlight the implementation of the communication infrastructure on FPGAs to support hardware context switching on heterogeneous FPGAs. The communication infrastructure follows the protocol explained in section 11.4 to maintain the consistency in communication. This implementation will use commercial FPGAs to take advantage of the vendor tools already available. The vendor tools, such as for place-and-route, will optimize the performance of the tasks implemented on hardware. Without overlay techniques, hardware context switching on heterogeneous FPGAs can be provided using design-based solutions (Koch *et al.* 2007; Bourge *et al.* 2016).

The development of the communication infrastructure is presented in two aspects. First, the structure that allows extraction and restoration of the communication data or tokens located in FIFO channels during the hardware context switch must be implemented. We take this approach to ensure no data loss in the process. As such, we introduce FIFOs with reconfigurable channels. Second, we develop a mechanism to control the hardware structure, including communication and context data management and flow control. This will be implemented using finite-state machines in the interface between hardware tasks. We assume that the hardware task context is accessible from a specialized port in the task since we use design-based solutions.

11.5.1. *Reconfigurable channels in FIFO*

To support hardware context switching, extracting and restoring the context and communication data is necessary on heterogeneous FPGAs. There are two types of FIFO used in the system: I/O FIFO and context FIFO. The context switch support in these FIFOs can be made possible by using one of the two approaches. In the first approach, the data stored in FPGA can be accessed by direct reading from the FPGA memory resource used by the FIFOs. However, applying such a method requires pinpointing the exact resources used by the design, and extracting from the FPGA configuration port. In the second approach, the communication data is accessed through modification of FIFOs in register-transfer level (RTL) implementation. We prefer this approach since it gives an abstraction of the physical implementation which enables the use of many different FPGAs in heterogeneous systems.

We introduce multiple channels in the FIFOs that can be configured according to the required purpose at a given time. Figure 11.5 describes the FIFOs used for the task context and the communication data in the system. The I/O FIFO is used to temporarily store and transfer communication data from one task to another. The communication flow goes from the input port to the output port in normal execution. However, when a hardware context switch occurs, the communication data can be extracted from *context switch output* (cs output), and similarly, they can be restored to *context switch input* (cs input). We make the communication channels in the FIFOs reconfigurable in order to support the protocol presented in section 11.4. Similarly, the channels in the context FIFO (see Figure 11.5) are configurable to support the proposed management of the task context. The context FIFO consists of save and restore FIFOs that share an interface to the hardware task.

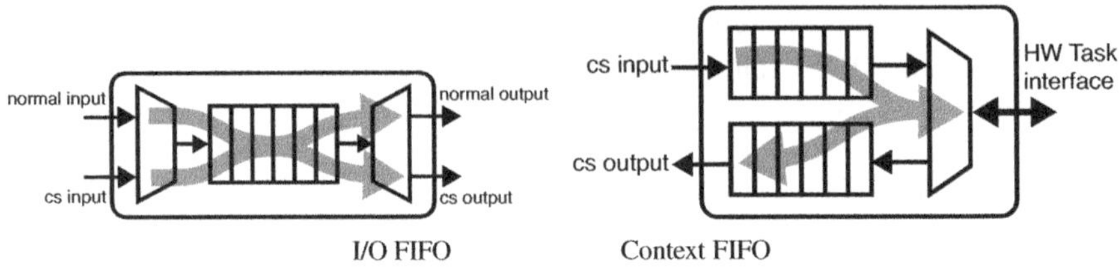

Figure 11.5. *FIFOs in the communication channel*

11.5.2. *Communication infrastructure*

This section highlights the implementation of the proposed methodology as a communication infrastructure on FPGAs. We develop a structure which complements existing communication topology, in order to support hardware context switching on the FPGAs while maintaining the communication consistency. We consider our system to be a heterogeneous computing system with CPU(s) and FPGA(s) running tasks on each resource in parallel. We use the main memory to share data between the tasks on the CPU and the tasks on the FPGA. Such an architecture can be implemented using standard communication topologies, for example bus and network-on-chip (NoC).

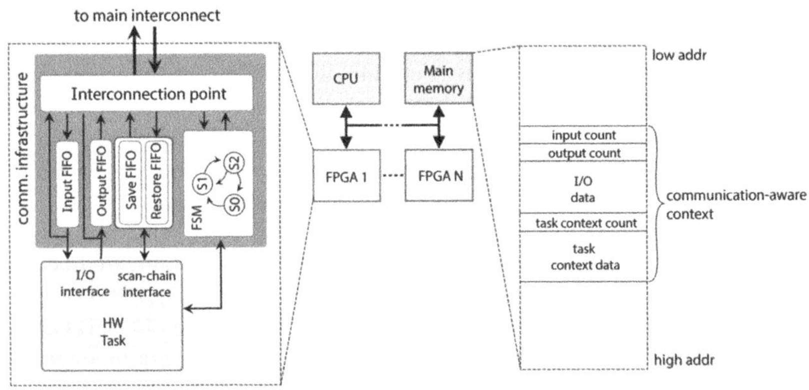

Figure 11.6. *Reconfigurable architecture with the proposed communication structure*

Figure 11.6 illustrates the proposed communication infrastructure on FPGAs. We consider each hardware task to be connected to the main interconnect through a communication infrastructure. A system scheduler manages task allocation and hardware context switching in the system. In every communication infrastructure, I/O FIFOs and context FIFOs (presented as save and restore FIFOs) with reconfigurable channels are used. It also integrates a finite-state machine (FSM) that controls the communication flow of the tasks on FPGAs.

When a context switch occurs in a hardware task, the FSM in the communication infrastructure adapts the FIFO channels and interrupts the flow to extract the non-processed data. The FSM then forwards the command to perform a hardware context switch to the hardware task. The interconnection point will block the data transfer to the outside of the FPGA. According to KPN specification, the other tasks in the system may continue their execution and a deterministic behavior can be maintained even though a task is removed from the network. The channels in I/O FIFOs will be configured to extract the data from cs output (see Figure 11.5). The communication data will be saved at a special region in the main memory. Meanwhile, context FIFO stores the context of the hardware task following the context extraction request. After the I/O FIFOs finish flushing all the data, the context FIFO sends the task context to the main memory as well. The communication data and the task context are stored together as a *communication-aware context* in the main memory in a format shown in Figure 11.6. The format includes a header of each data to indicate the type and the size of transmitted data.

The task migration and hardware context switch requests come from the system scheduler to the CPU in the architecture. The CPU then forwards these requests to the FPGA and controls the communication infrastructure to follow the aforementioned protocol. We develop a Linux kernel module on the CPU that can adjust the communication link and trigger the extraction and restoration of the task context and communication data. This module holds the information of the connections between tasks in a table and updates them according to the ongoing and interrupted tasks. It also controls the FSM in the communication infrastructure to start the execution and perform a hardware context switch operation. It accepts interrupt signals (IRQ) from the communication infrastructure due to certain conditions, for example end of execution and end of extraction.

11.6. Experimental results

11.6.1. *Setup*

To evaluate the performance and overhead of our communication method supporting hardware context switching, we propose a series of experiments. We select the benchmark applications from the CHStone suite (Hara *et al.* 2008) and the inverse discrete cosine transform (IDCT) application. We consider applications that perform streaming communication, such as cryptography and image compression, to be are sufficient to test the proposed methodology. Since we want to support heterogeneous FPGAs in the architecture, design-based technique is preferred to enable hardware context switching on FPGAs. The HLS-based method (Bourge *et al.* 2016) is used to automatically instrument the tasks with the infrastructure that allows extraction and restoration of the hardware task contexts from a dedicated interface, in addition to their original I/O interfaces. This method is implemented as a CP3 plugin in the HLS tool AUGH (Prost-Boucle *et al.* 2014).

A Xilinx ZC706 Evaluation Board and an Intel Arria V SoC FPGA are used as the platform in the experiments. The ZC706 board features a Xilinx xc7z045 Programmable SoC from the Zynq-7000 family: a dual-core ARM Cortex-A9 and a Kintex FPGA. The Arria V SoC FPGA board also provides a dual-core ARM Cortex-A9 and an FPGA fabric. Both platforms are chosen due to their similarity in the product line and price range. They are both SoC-based platforms that include a CPU, an FPGA, an external memory and various peripheral devices and interfaces.

In both platforms, the FPGA, the CPU, the memory and all the peripheral devices are connected via AXI interconnection. The ARM processor is responsible for FPGA reconfiguration through the configuration port. It is also used to run another task that communicates with the task on the FPGA and to schedule a context switch scenario in the experiments. The processor runs a Linux Linaro distribution at its default frequency 667 MHz. Most of the hardware tasks in the experiments can be synthesized at 100 MHz, but we set the FPGA working frequency to 50 MHz in the experiments for simplicity.

11.6.2. *Experiment scenario*

Figure 11.7 presents the scenario followed in the experiments for both platforms. We consider that the system uses preemptive scheduling to provide dynamic task allocations. This scenario is used for every benchmark application and repeated 1,000 times for statistical reports. It begins by running the hardware task execution on the FPGA. In the middle of the execution, a preemption request is sent at a random time to the FPGA. We define the time from the arrival of preemption request to the beginning of context extraction as the latency in hardware context switching (preemption latency). In the proposed method, the task context and the data in communication channels are extracted together. At a later time, the context and communication data are loaded to another FPGA and the execution continues until the end of the application. The time between the context extraction and restoration is the storing time of the data outside of the FPGA and therefore is not considered in the experiments.

The execution scenario is implemented in a bash script. In the first part, we run the explained scenario to extract and restore the task on the same FPGA to measure the performance and the overhead of the proposed method. We mainly use the ZC706 platform in the evaluation to measure the context switch time and the latency. The FPGA configuration time is not included in the measurement since we consider that it is done by the beginning of the execution. The average time to configure the FPGA on a ZC706 board from the ARM processor is 0.546 s. For a fair evaluation, the scheduler is not optimized to find the timing when the communication channels contain minimum data. In the second part, we run the scenario to migrate a hardware task from ZC706 to Arria V SoC, and vice versa.

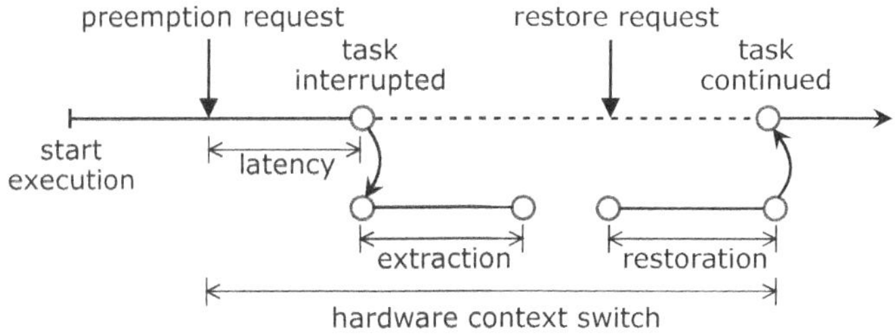

Figure 11.7. *Hardware context switch scenario in the experiments*

For the experiments, we implement our version of conventional communication management without the ability of hardware context switching (Basic). The Basic version handles normal I/O flow between the tasks. Another version that follows the solution in Vu *et al.* (2016) for the hardware context switch is also implemented (CS). In CS, the preemption request can be acknowledged at any time but the context extraction can only begin when the communication channels are idle. As such, the communication between the task being moved is cut from its predecessors and waits until the successor tasks consume all the tokens in the FIFO channels. These two versions, together with our proposed solution (which will be referred as CSComm, for Context Switch with Communication management), are implemented on the ZC706 and on the Arria V SoC. We assume that no partial reconfiguration is involved in the experiments. The next section presents this comparison in terms of resource overhead.

11.6.3. *Resource overhead*

Table 11.1 shows a comparison of the resource utilization by the three versions of communication wrapper on the ZC706 FPGA. These versions of communication wrapper are synthesized without the hardware tasks on Vivado Design Suite 2015.3. Each version implements an FSM, an interconnection point and FIFOs, as presented in Figure 11.6, except for the Basic version, which does not need context FIFO. In this work, the interconnection point is implemented using an AXI4 IP interface due to the platforms used. Other types of interface can be used, for instance, a PCIe endpoint IP when the FPGA is physically connected to the CPU via PCIe. For comparison, we present, in Table 11.2, the synthesis result of the three wrappers for the Arria V SoC FPGA on Quartus II 15.0.

	Basic	CS	CScomm
LUT	683	1095	1309
FF	707	1133	1216
BRAM	1	7	7

Table 11.1. *Resource utilization of the communication wrapper in Basic, CS and CSComm versions on the ZC706 FPGA*

	Basic	CS	CScomm
ALM	476	910	1053
REG	760	1382	1478
M10K	2	26	26

Table 11.2. *Resource utilization of the communication wrapper in Basic, CS and CSComm versions on the Arria V SoC FPGA*

The comparison between the three versions of communication wrapper is performed between the logic units (LUT) and the memory elements inside the FPGA (FF and BRAM). The wrappers do not use any DSP resource on the FPGA. The difference in the resource utilization for the three wrappers is due to the FSM, FIFOs and multiplexers, since they use the same AXI4 IP interface. From Table 11.1, we can see that the CS version consumes almost three times as many logic resources as the Basic version, excluding the AXI4 IP, which uses 454 LUTs and 590 FFs. We want to show here that the complexity of the communication infrastructure increases in order to support hardware context switching, although it may be influenced by the designers' coding ability. Likewise, the CSComm version adds LUT and FF utilizations on the FPGA due to the communication data management in the FIFO channels. However, the complexity added to the infrastructure was not high, since we only observe an increase of approximately 33% (214) for LUTs and 15% (83) for FFs in the ZC706.

The other important components in the communication infrastructure are the I/O and context FIFOs. They consume the BRAMs on the ZC706 FPGA (or M10K on the Arria V SoC). In the experiments, we fix the bus data width to 32-bit. As such, 1 BRAM 36K (Xilinx) or 4 M10Ks (Intel) is consumed for every 4 KB of data in a 1024×32-bit FIFO. As we use KPN to model the communication, it does not need a timestamp in the communication data. Although it is not optimized to certain IPs, the size of the communication FIFOs is fixed in the experiments. We fix the I/O FIFO size to 256×32-bit, which is sufficient for all the benchmark applications used in the experiments. The context FIFO size, however, depends on the context size of each hardware task. To give an idea of how many memory elements (BRAM or M10K) are needed, we show the context size of the benchmark applications in Table 11.3. Therefore, for simplicity reasons, we fix both save and restore FIFOs to 3072×32-bit in CS and CSComm versions.

Application	Context size in bytes
adpcm	1996
aes	2668
blowfish	4408
gsm	844
idct	628
motion	8400
sha	496

Table 11.3. *Size of hardware task context for each benchmark application*

To summarize, both hardware context switching and communication management supports introduce a resource overhead to the wrapper in FPGAs. This overhead is due to more complex FSM and multiplexers to handle the protocol revised in section 11.4. We consider the overhead here is modest since one-third of the utilization is already dedicated to the AXI interface. Regarding the memory elements, the addition communication management has no impact on resource utilization. Compared to the total resources available on the ZC706 platform, the proposed communication infrastructure consumes less than 1%. For comparison, we synthesize benchmark applications for the ZC706 platform and find that they use between 2,495 and 11,776 LUTs. Compared to these values, our communication wrapper costs less in terms of logic consumption.

11.6.4. *Impact on the total execution time*

The proposed communication method introduces more overhead of the data footprint in hardware context switching. This can be observed in the total execution time. In this work, we measure the total execution time, since a hardware task runs until it terminates. Based on our evaluation scenario, a context switch occurs at a random time in the middle of the execution. Thus, the total execution time includes the time required to perform a hardware context switch.

Table 11.4 presents a comparison of the average total execution time between the CS and CSComm versions. The data in the table is presented in FPGA cycles. Since the execution is repeated 1000 times, we are also able to obtain its standard deviation and coefficient of variation (CV), which represents how large the data spread is. As expected, we can see that CSComm produces a higher total execution time on average, which is due to the extract and restore process of the communication data. Depending on the I/O FIFO size in the communication infrastructure, the overhead may become significant compared to the execution time without hardware context switching. In the experiments, *idct*, which has a short execution time, suffers a higher overhead in CSComm.

	CS			CScomm			
	$\bar{x}$	σ	CV	$\bar{x}$	σ	CV	Overhead
adpcm	20879	19	0.09	20933	24	0.12	**0.26%**
aes	14184	12	0.09	14237	13	0.09	**0.37%**
blowfish	146255	30	0.02	146583	180	0.12	**0.22%**
gsm	12548	18	0.15	12619	49	0.39	**0.57%**
idct	1302	11	0.87	1421	97	6.86	**9.19%**
mpeg2	13009	37	0.29	13268	112	0.85	**1.99%**
sha	331767	13	0.004	332435	216	0.07	**0.2%**

Note: This table presents the average value ($\bar{x}$), the standard deviation (σ) and the coefficient of variation (CV) of total execution time when a single context switch operation is performed. The measurements were done on the FPGA with the CS and CSComm versions of the communication infrastructure. The measurements were repeated 1000 times.

Table 11.4. *Comparison of total execution time (FPGA cycles) and its variation on the CS and CSComm versions of communication infrastructure*

The CSComm version of the communication infrastructure also increases the standard deviation and CV of the total execution time compared to the CS version. The total execution time of CSComm obtained in the experiments varies according to the amount of communication data in the FIFOs when a hardware context switch occurs. If I/O FIFOs are empty when the hardware task is interrupted for context switch, we will obtain the minimum total execution time of CSComm. On the other hand, if I/O FIFOs are full when the task is interrupted, we will obtain the maximum total execution time of CSComm. We suppose that our scheduling technique is not optimized and the scheduler cannot know the state of the FIFOs before triggering a hardware context switch operation.

In contrast to CSComm, the total execution time of CS has less dependency on the communication data. If the I/O FIFOs in CS are not empty when a context switch request is received, the extraction cannot be started and must wait. However, the task execution continues during the waiting period, which makes the execution after the restore process shorter. This will be explained in detail in section 11.6.5. In conclusion, the overhead due to the proposed communication methodology is negligible in most of the benchmark applications, except for *idct*, which has a very short execution time. On average, the proposed communication methodology only adds less than 2% of the average total execution time for all benchmark applications.

11.6.5. *Impact on the context extract and restore time*

In this section, we analyze the impact of the proposed communication solution (in the CSComm version) on the extraction and restoration time in hardware context switching. The context extraction and restoration processes in CSComm include the extraction and restoration of task context and its communication data that we refer to as communication-aware context (see section 11.5.2). On the other hand, the context in CS does not include the communication data from I/O FIFOs. As expected, we observe an overhead in the proposed solution, as shown in Table 11.5 for the ZC706 platform. However, the extraction time in the table shows only small increases in CSComm compared to CS for most of the applications. Even though our solution involves the communication data extraction, the overhead is not significant. We take advantage of the behavior of scan-chain extraction provided by a CP3 plugin, where there is no data transfer between adjacent checkpoint states. As such, the task context extraction can be started from I/O FIFOs when a preemptive request arrives. By the time the task arrives at a checkpoint and the extraction of task context can be started, a certain amount of I/O data has already been extracted. For several applications, this leads to an extraction overhead less than 0.1%.

App	Extract time			Restore time			Total
	CS	CSComm	Overhead	CS	CSComm	Overhead	Overhead
adpcm	1151	1152	0.07%	1174	1229	4.7%	2.41%
aes	1485	1486	0.06%	1510	1562	3.45%	1.77%
blowfish	2400	2401	0.04%	2429	2720	12%	6.05%
gsm	527	547	3.74%	551	608	10.39%	7.14%
idct	417	501	20.18%	442	539	21.94%	21.08%
motion	4494	4495	0.02%	4519	4776	5.69%	2.86%
sha	396	714	80.33%	422	745	76.28%	78.24%

Table 11.5. *Comparison of average extraction and restoration times (FPGA cycles) in hardware context switching between CS and CSComm in ZC706*

From the experiments, we find that there exist conditions where extracting I/O data before the task context is less beneficial. First, the overhead due to the communication data extraction becomes significant when a task has a small context size. Among the benchmark applications, *sha* has the smallest context size. Extracting the communication data from such a task when hardware context switching occurs adds a significant overhead in time. The overhead also increases if the I/O rate of the task is low enough that it causes the I/O data to stay in the FIFOs longer. This is shown by *idct* in the experiments. On a task with a small context size and a high I/O traffic, for example, *blowfish*, *adpcm* and *motion*, the overhead of the proposed method is negligible.

In the task restoration process, the I/O data and context are restored sequentially to the empty FIFOs in the communication infrastructure. This explains the higher overhead in the restore time of CSComm for all applications in the experiments since more data are involved in this process. Furthermore, the FSM in the communication infrastructure has to retrieve the header of the communication-aware context from the main memory before restoring the context to the hardware task. This procedure adds approximately 25 clock cycles for all the tested applications. As a result, Table 11.5 shows that the restore time is longer than the extract time.

In conclusion, the overhead in the extraction and restoration processes in hardware context switching is almost inevitable as the proposed communication management increases the amount of data to be transferred. On top of that, the overhead will be more noticeable on a task that has a small context size. Nevertheless, the overhead is limited to the size of I/O FIFOs used in the architecture. The worst-case additional extract and restore times can be predicted from the size of the FIFOs, since we ensure that hardware context switching can be done at any time. Reducing the I/O FIFO size will thus decrease the time overhead in the extraction and restoration of the communication-aware context.

11.6.6. *System responsiveness to context switch requests*

The proposed communication methodology increases the system responsiveness to context switch requests. This is related to the preemption latency shown in Figure 11.7. In the figure, the preemption latency is defined as the time between the arrival of the preemption request and the beginning of context (and communication data) extraction. There are two components in the preemption latency in our system. The first component is caused by the checkpoint architecture. In order to start the context extraction from a hardware task, the execution flow has to reach a checkpoint state. This latency depends on the implementation of scan-chain structures in the task. The second latency component is due to the ongoing communication flows which prevent a hardware task from being context switched. In CS, this latency is shown by the delay in the context switch until there is no communication data temporarily stored in the channels. In contrast, CSComm removes this delay, since it can handle the situation where the communication channels are still occupied when a hardware context switch request is given.

Table 11.6 describes the comparison of average preemption latency provided by the CS and CSComm versions, presented in FPGA cycles. The table only presents the measurement done in the ZC706 board since the results in the Arria V SoC are similar. We also present the total of hardware context switch time, which includes the latency and the extraction and restoration process explained in the previous section (see Figure 11.7). Finally, the ratio shows the percentage of the latency compared to the hardware context switch.

App	CS			CSComm		
	Latency	Context Switch	Ratio	Latency	Context Switch	Ratio
adpcm	147	2472	5.93%	6	2387	0.23%
aes	148	3143	4.71%	17	3066	0.56%
blowfish	20017	24847	80.56%	56	5,177	1.08%
gsm	178	1256	14.16%	14	1168	1.16%
idct	180	1039	17.3%	10	1050	0.93%
motion	876	9889	8.86%	7	9278	0.07%
sha	10153	10971	92.54%	7	1467	0.51%

Table 11.6. *Comparison of average context switch latency (FPGA cycles) between CS and CSComm in ZC706*

In general, we can see that the proposed communication method has successfully reduced the latency in context switch. This is shown by the lower value of latency in the CSComm version compared to the CS version. For instance, the experiment with *blowfish* shows a 98% reduction in the latency ratio between CSComm and CS. In CS, the latency is due to the I/O data stored in communication channels when a hardware context switch occurs. Although the input stream is blocked in order to prevent new data to arrive at input FIFO, the scheduler still has to wait until the task finishes the input read process. It is worth noting that we use external memory to implement a very large FIFO of the successor task in the experiments so the output data can be transferred immediately to empty the output FIFO of the task. In a real application, the output FIFO may not be empty while the successor task does not consume the data. As a result, the latency will become much higher than the results in our experiments, Despite this, the latency of benchmark applications shown in the table is still significant, particularly for applications with high communication traffics, for example *blowfish* and *sha*. For applications with less traffic in communication, the latency is 147 cycles or more. This shows the cost of the penalty due to the communication.

In CSComm, the latency in preemption is significantly reduced, to 56 cycles or less on all the benchmark applications. This value reflects the initial latency caused by the checkpoint architecture since the context switch requests do not have to be stalled anymore when the task has ongoing communication flows. This result shows the evident benefit of our solution. While maintaining the communication data consistency in a migrated task, it also significantly reduces the latency in a hardware context switch operation.

If we compare the ratio of latency to total preemption time between CS and CSComm, we observe substantial differences between CS and CSComm. This ratio correlates to the system responsiveness to preemption requests, which is frequently important in a system with a strict time constraint. When the latency ratio is high in hardware context switching, the system has low responsiveness to preemption requests. CS spends more time waiting until the communication channels do not contain data rather than performing the context extraction and restoration. In Table 11.6, the latency of CS takes up to 92.54% of the context switch time, whereas CSComm reduces the latency to 1.16% and below for the benchmark applications. This result means that our communication method provides high system responsiveness to the hardware context switch.

With the reduction of latency in CSComm, we obtain a shorter context switch time for most of our benchmark applications. However, more obvious differences between CS and CSComm regarding the total preemption time are shown in *blowfish* and *sha*. The overhead caused by extracting and restoring the I/O communication data became negligible thanks to the reduction in the preemption latency. Nonetheless, the overhead to the data footprint is still higher for *idct* due to its small context size and less traffic in data exchange.

Finally, the proposed communication method also improves the predictability in the hardware context switch. The comparison of the context switch time between CS and CSComm is described using boxplots of the entire measurement of context switch time (preemption time) in Figure 11.8. The box describes 50% of the data, and the whiskers represent the maximum and minimum values in our measurements. The line in the middle of the box is the median data. A wide range of values in the plots shows a high variation in the context switch time.

Overall, the results show that the CS version causes a higher variation in the context switch time than CSComm. In the CS version, the extraction of task context can only be started after the communication channels are idle and the FIFOs are empty. This requires the system scheduler to know the states of the I/O FIFOs before issuing a hardware context switch request. The moment when there is no communication depends on the application run in the system, and this is hard to predict. Thanks to our communication method, we reduce the dependency of hardware context switching on the communication state and bound the context switch time. It now depends on the size of the I/O FIFOs.

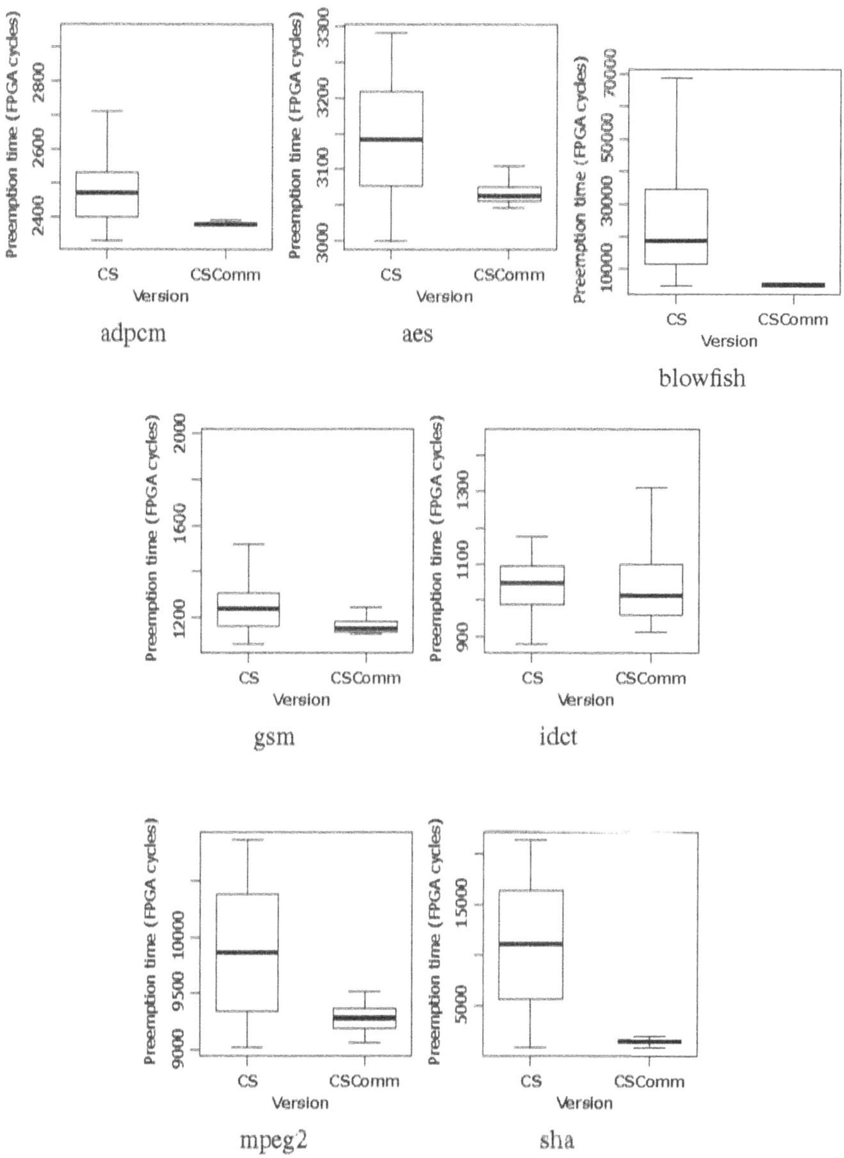

Figure 11.8. *Comparison of hardware context switch (preemption) time between CS and CSComm*

To summarize, the proposed solution offers the necessary high responsiveness in a hardware context switch in addition to preserving the consistency in the execution. It provides a predictability to the hardware context switch time. There is less variation in the context switch time, and the scheduler can always send a context switch request to hardware tasks at any time.

11.6.7. *Hardware task migration between heterogeneous FPGAs*

In this work, we present the method to manage the communication in hardware context switching to support hardware task migration between FPGAs. We emphasize that the heterogeneity aspect exists between FPGAs used in the system. Due to upgrade, cost and maintenance processes, the FPGAs used can be heterogeneous. For this reason, we perform hardware task migration between heterogeneous FPGAs, using the proposed methodology. Figure 11.9 presents the architecture of the task migration experiments. We use two reconfigurable SoC platforms: a ZC706 board from Xilinx and an Arria V SoC board from Intel. A host workstation is added as a system scheduler, the one that makes a decision to migrate tasks from ZC706 to Arria V SoC, and vice versa. They are connected to an Ethernet network with a disk mounted on the NFS protocol to store the application bitstreams. For each application, there are two bitstreams which correspond to the two SoC boards. However, the designs in the SoCs are identical.

We experiment with two types of connection between the two SoCs. These two connections are used to transfer the task context and the communication data from one FPGA to another. Figure 11.9 shows the difference when the NFS and the SSH protocols are used. For the NFS protocol, a disk was mounted in the network as a sharing medium between the platforms. For the SSH protocol, the context was transferred from an SoC platform to another using the SSH protocol. In both experiments, the host starts the flow by launching the application on one of the platforms. Then a migration request is sent to both platforms; the interruption command is given to the running platform, whereas the command to configure the FPGA is sent to the destination. Figure 11.10 depicts the timeline of the task migration experiment. A script on the host runs the entire migration flow and the transfers of task context and communication data. For each step in the task migration, the host triggers the CPU on each SoC platform to use the module to control the communication infrastructure. Note that this implementation is far from optimal since each step is blocking. A more optimized method can be provided in future works, for instance, using MPI packets to trigger the migration and send the context in parallel.

Table 11.7 describes the task migration time between the ZC706 and the Arria V SoC. The t_{extract} and t_{restore} values are different from the ones presented in section 11.6.5. Instead of measuring the cycles inside the FPGA, we take the overall extraction and restoration processes time from the SoC. We perform the experiments in two directions and present the average in the table.

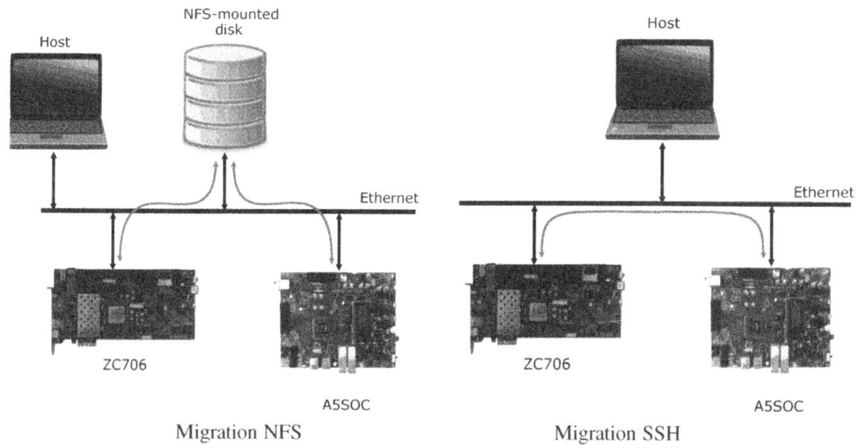

Figure 11.9. *Hardware task migration between heterogeneous reconfigurable SoCs*

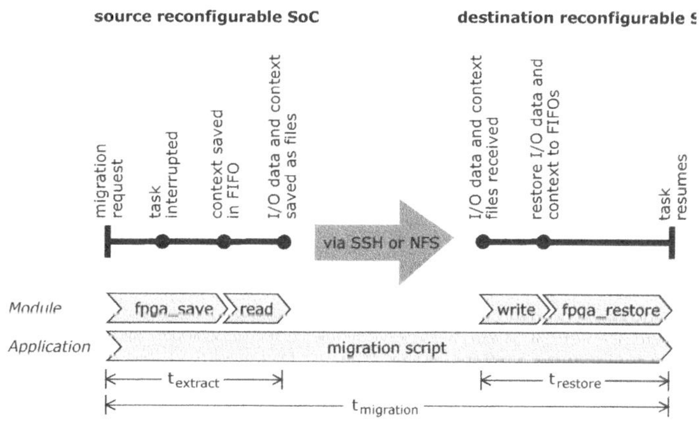

Figure 11.10. *Task migration timeline*

The t_{extract} value is 0.03 seconds or less, on average, to perform the process of reading the task context and the communication data as well as saving them as files. The t_{restore} requires roughly the same period to restore the tasks and resume the execution. The $t_{\text{migration}}$ shows the total required time as well as the cost of transferring the files from one platform to another. We can see, in the table, that the NFS transfer took a shorter amount of time, thanks to asynchronous transfer through an NFS-mounted disk in both platforms. The SSH transfer, on the other hand, performed handshakes before a synchronous transfer from a sender to a receiver. The values shown in Table 11.7 are the average values for both migration directions. As a result, they do not include the FPGA reconfiguration time due to the difference in each platform. Note that the reconfiguration of the ZC706 takes 0.546 seconds while the Arria V SoC takes 1.23 seconds.

	$t_{extract}$ (s)	$t_{restore}$ (s)	$t_{migration}$ (s) NFS	SSH
adpcm	0.026	0.021	0.577	0.759
aes	0.025	0.021	0.577	0.760
blowfish	0.030	0.027	0.606	0.969
gsm	0.025	0.020	0.572	0.757
idct	0.024	0.020	0.574	0.762
motion	0.028	0.027	0.606	0.972

Table 11.7. *Task migration time between A5SOC and ZC706*

Throughout this section, we show that task migration on heterogeneous reconfigurable systems can be supported using the proposed solution. Although further development is still necessary to improve the performance of the method, we present two advantages of our work. First, we show that the proposed mechanism is functional in heterogeneous reconfigurable systems and is capable of preserving the communication data integrity during the migration. Second, we show that our solution is generic by implementing the same mechanism in different heterogeneous platforms. Although some differences due to memory mapping must be considered when implementing the software, the hardware architectures are identical.

11.7. Conclusion

This chapter introduces a communication method in order to support hardware context switching on heterogeneous reconfigurable systems. The proposed solution maintains the consistency of hardware task execution by managing the task context and its I/O communication data. This work is based on a state-of-the-art task context extraction solution using the design-based method. This design-based method is required in order to use heterogeneous FPGAs in the system. In recent years, the FPGA overlay method is gaining interest for this objective, but the performance of such a method still needs to be improved in order to be applied for a real application. The design-based method in Bourge *et al.* (2016), on the other hand, offers a very low overhead in performance, but to integrate it into a reconfigurable architecture, a communication infrastructure is necessary.

This work deals with the communication data integrity that prevents a hardware context switch on FPGAs if the task still has ongoing communication flows. Hence, the existing solutions so far proposed a hardware context switch that works under very strict constraints, i.e. absence of communication data in the channels or at the end of communication flows. These conditions cause a performance drop and unpredictability in the process or the inability to context switch hardware tasks at the intended time. The communication method proposed in this work handles the task context and I/O communication data together when a hardware context switch occurs. A protocol is introduced and implemented as a communication infrastructure and a kernel module to control the flow on FPGAs.

We proposed a series of experiments to evaluate the protocol and the implementation of the proposed methodology. Reconfigurable SoC platforms with different FPGAs were used. We implemented a system that followed our communication solution and a system which used the existing solution for comparison purposes. As expected from the additional communication data management in a context switch operation, our solution modestly increases the time required to extract and restore the context, thus the total execution time. Despite the overhead in the context extraction and restoration time, we had successfully shown that our communication significantly reduced the latency in the preemption. This means that the responsiveness of the system towards preemption requests increases. Consequently, the total preemption time decreases and is more predictable in the system with our solution.

Regardless of the advantage and the low performance offered by the proposed solution, there is still room for improvement. In future work, the focus should be on the optimization of hardware context switching on several tasks at the same time. The method could also be implemented on a distributed system to improve the scalability. A network-based communication topology, such as network-on-chip (NoC), can be used in the reconfigurable architecture to improve the efficiency of communication. Finally, a task migration algorithm is also necessary to improve the performance of the system.

11.8. References

Bourge, A., Muller, O., and Rousseau, F. (2016). Generating efficient context-switch capable circuits through autonomous design flow. *ACM Transactions on Reconfigurable Technology and Systems (TRETS)*, 10(1), 9.

Byma, S., Steffan, J.G., Bannazadeh, H., Garcia, A.L., and Chow, P. (2014). FPGAs in the cloud: Booting virtualized hardware accelerators with openstack. *2014 IEEE 22nd Annual International Symposium on Field-Programmable Custom Computing Machines (FCCM)*, IEEE, 109–116.

El-Antably, A., Gruber, O., Rousseau, F., and Fournel, N. (2015). Transparent and portable agent based task migration for data-flow applications on multi-tiled architectures. *2015 International Conference on Hardware/Software Codesign and System Synthesis (CODES + ISSS)*, IEEE, 183–192.

Estrin, G. (1960). Organization of computer systems: The fixed plus variable structure computer. *IRE-AIEE-ACM '60 (Western): Papers presented at the May 3-5, 1960, western joint IRE-AIEE-ACM computer conference*, ACM, New York, 33–40.

Fahmy, S.A., Vipin, K., and Shreejith, S. (2015). Virtualized FPGA accelerators for efficient cloud computing. *2015 IEEE 7th International Conference on Cloud Computing Technology and Science (CloudCom)*, IEEE, 430–435.

Geilen, M. and Basten, T. (2003). Requirements on the execution of kahn process networks. *Programming Languages and Systems*, 319–334.

Goldstein, S.C., Schmit, H., Moe, M., Budiu, M., Cadambi, S., Taylor, R.R., and Laufer, R. (1999). PipeRench: A coprocessor for streaming multimedia acceleration. *ACM SIGARCH Computer Architecture News*, 27(2), 28–39.

Guan, N., Deng, Q., Gu, Z., Xu, W., and Yu, G. (2008). Schedulability analysis of preemptive and nonpreemptive EDF on partial runtime-reconfigurable FPGAs. *ACM Transactions on Design Automation of Electronic Systems (TODAES)*, 13(4), 56.

Happe, M., Traber, A., and Keller, A. (2015). Preemptive hardware multitasking in ReconOS. *International Symposium on Applied Reconfigurable Computing*, Springer, 79–90.

Hara, Y., Tomiyama, H., Honda, S., Takada, H., and Ishii, K. (2008). CHStone: A benchmark program suite for practical C-based high-level synthesis. *2008 IEEE International Symposium on Circuits and Systems – Engineering the Environmental Revolution (ISCAS 2008)*, IEEE, 1192–1195.

Hauck, S. and DeHon, A. (2010). *Reconfigurable Computing: The Theory and Practice of FPGA-based Computation: Volume 1*. Morgan Kaufmann, Burlington.

Jozwik, K., Tomiyama, H., Honda, S., and Takada, H. (2010). A novel mechanism for effective hardware task preemption in dynamically reconfigurable systems. *2010 International Conference on Field Programmable Logic and Applications (FPL)*, IEEE, 352–355.

Jozwik, K., Honda, S., Edahiro, M., Tomiyama, H., and Takada, H. (2013). Rainbow: An operating system for software-hardware multitasking on dynamically partially reconfigurable FPGAs. *International Journal of Reconfigurable Computing*, 2013, 5.

Kahn, G. (1974). The semantics of a simple language for parallel programming. *Information Processing*, 74, 471–475.

Kalte, H. and Porrmann, M. (2005). Context saving and restoring for multitasking in reconfigurable systems. *2005 International Conference on Field Programmable Logic and Applications (FPL)*, IEEE, 223–228.

Koch, D., Haubelt, C., and Teich, J. (2007). Efficient hardware checkpointing: Concepts, overhead analysis, and implementation. *Proceedings of the 2007 ACM/SIGDA 15th International Symposium on Field Programmable Gate Arrays (FPGA)*, ACM, 188–196.

Lübbers, E. and Platzner, M. (2008). Communication and synchronization in multithreaded reconfigurable computing systems. *Proceedings of the 2008 International Conference on Engineering of Reconfigurable Systems & Algorithms*, ERSA 2008, 83–89.

Moore, G.E. (2006). Cramming more components onto integrated circuits, Reprinted from Electronics, volume 38, number 8, April 19, 1965, pp. 114 ff. *IEEE Solid-State Circuits Newsletter*, 3(20), 33–35.

Morales-Villanueva, A. and Gordon-Ross, A. (2013). On-chip context save and restore of hardware tasks on partially reconfigurable FPGAs. *2013 IEEE 21st Annual International Symposium on Field-Programmable Custom Computing Machines (FCCM)*, IEEE, 61–64.

Prost-Boucle, A., Muller, O., and Rousseau, F. (2014). Fast and standalone design space exploration for high-level synthesis under resource constraints. *Journal of Systems Architecture*, 60(1), 79–93.

Putnam, A., Caulfield, A.M., Chung, E.S., Chiou, D., Constantinides, K., Demme, J., Esmaeilzadeh, H., Fowers, J., Gopal, G.P., and Gray, J. (2015). A reconfigurable fabric for accelerating large-scale datacenter services. *IEEE Micro*, 35(3), 10–22.

Scalera, S.M. and Vázquez, J.R. (1998). The design and implementation of a context switching FPGA. *Proceedings of the IEEE Symposium on FPGAs for Custom Computing Machines (FCCM '98)*, IEEE, 78–85.

Scherer, T. (2011). Task Migration for Multi-Processor Systems, Semester Thesis, Swiss Federal Institute of Technology Zurich.

Shih, K.-J., Sun, H.-Y., and Hsiung, P.-A. (2010). Dynamic hardware-software task switching and relocation mechanisms for reconfigurable systems. *2010 IET International Conference on Frontier Computing. Theory, Technologies and Applications*, IET, 157–162.

Simmler, H., Levinson, L., and Männer, R. (2000). Multitasking on FPGA coprocessors. *International Workshop on Field Programmable Logic and Applications (FPL)*, Springer, 121–130.

Vu, H.G., Kajkamhaeng, S., Takamaeda-Yamazaki, S., and Nakashima, Y. (2016). CPRtree: A tree-based checkpointing architecture for heterogeneous FPGA computing. *2016 Fourth International Symposium on Computing and Networking (CANDAR), IEEE, 57–66.*

Wheeler, T., Graham, P., Nelson, B., and Hutchings, B. (2001). Using design-level scan to improve FPGA design observability and controllability for functional verification. *International Conference on Field Programmable Logic and Applications (FPL)*, Springer, 483–492.

List of Authors

René AHLSDORF
Institute for Communication Technologies and Embedded Systems
RWTH Aachen University
Germany

Liliana ANDRADE
TIMA Lab
Université Grenoble Alpes
France

Gerd ASCHEID
Institute for Communication Technologies and Embedded Systems
RWTH Aachen University
Germany

Rotem BEN HUR
Electrical Engineering Department
Technion – Israel Institute of Technology
Haifa
Israel

Andreas BYTYN
Institute for Communication Technologies and Embedded Systems
RWTH Aachen University
Germany

Benoît DUPONT DE DINECHIN
Kalray S.A.
Grenoble
France

Adi ELIAHU
Electrical Engineering Department
Technion – Israel Institute of Technology
Haifa
Israel

Antoine FARAVELON
TIMA Lab
Université Grenoble Alpes
France

Ran GINOSAR
Electrical Engineering Department
Technion – Israel Institute of Technology
Haifa
Israel

Olivier GRUBER
LIG Lab
Université Grenoble Alpes
France

Ameer HAJ ALI
Electrical Engineering and Computer Science Department
University of California
Berkeley
USA

Andreas HERKERSDORF
Chair of Integrated Systems
Department of Electrical and Computer Engineering
Technical University of Munich
Germany

Tohru ISHIHARA
Department of Computing and Software Systems
Graduate School of Informatics
Nagoya University
Japan

K. Charles JANAC
Arteris IP
Campbell
USA

Ahmed JERRAYA
Cyber Physical Systems Programs
CEATech
Grenoble
France

Shahar KVATINSKY
Electrical Engineering Department
Technion – Israel Institute of Technology
Haifa
Israel

Oliver LENKE
Chair of Integrated Systems
Department of Electrical and Computer Engineering
Technical University of Munich
Germany

Olivier MULLER
TIMA Lab
Université Grenoble Alpes
France

Lars NOLTE
Chair of Integrated Systems
Department of Electrical and Computer Engineering
Technical University of Munich
Germany

Frédéric PÉTROT
TIMA Lab
Université Grenoble Alpes
France

Sven RHEINDT
Chair of Integrated Systems
Department of Electrical and Computer Engineering
Technical University of Munich
Germany

Frédéric ROUSSEAU
TIMA Lab
Université Grenoble Alpes
France

Arif SASONGKO
Bandung Institute of Technology (ITB)
Indonesia

Jun SHIOMI
Department of Communications and
Computer Engineering
Graduate School of Informatics
Kyoto University
Japan

Akshay SRIVATSA
Chair of Integrated Systems
Department of Electrical and
Computer Engineering
Technical University of Munich
Germany

Yankin TANURHAN
Solutions Group
Synopsys, Inc.
Mountain View
USA

Pieter VAN DER WOLF
Solutions Group
Synopsys, Inc.
Eindhoven
The Netherlands

Arthur VIANES
Kalray S.A.
Grenoble
France

Arief WICAKSANA
TIMA Lab
Université Grenoble Alpes
France

Thomas WILD
Chair of Integrated Systems
Department of Electrical and
Computer Engineering
Technical University of Munich
Germany

Author Biographies

René Ahlsdorf received his BSc degree from RWTH Aachen University in 2017. He joined the Institute for Communication Technologies and Embedded Systems (ICE) of RWTH Aachen University in 2016 as a student researcher, while pursuing an MSc degree. His current research interests lie in the optimization of deep learning algorithms for embedded systems with network-on-chip-based interconnects and the exploration of artificial intelligence in healthcare, particularly in the field of therapy guidance.

Gerd Ascheid received his diploma and PhD Doctor of Engineering in Electrical Engineering (Communication Eng.) from RWTH Aachen University. In 1988, he was co-founder and managing director of a start-up, which was acquired by Synopsys, Inc., a California-based EDA market leader. During his nine years as Director/Senior Director at Synopsys, he was responsible for IC design projects for advanced digital communication systems. In 2003, he joined RWTH Aachen University as a Distinguished Professor at the Institute for Communication Technologies and Embedded Systems. His research interest is in wireless communication and sensor signal processing algorithms, in particular, using machine learning algorithms and methods, and in application-specific integrated platforms. He has co-authored three books, published numerous papers and co-founded several successful start-ups.

Rotem Ben Hur received her BSc in Electrical Engineering from the Technion – Israel Institute of Technology in 2014. In 2012, she joined Elbit Systems as an FPGA designer. Since 2015, she has been working towards a PhD degree (direct track) at the Andrew and Erna Viterbi Faculty of Electrical Engineering, Technion – Israel Institute of Technology. Her current research is focused on novel architectures for logic with emerging memory technologies.

Andreas Bytyn received his BSc and MSc degrees from RWTH Aachen University in 2011 and 2014, respectively. In 2014, he joined the Institute for Communication Technologies and Embedded Systems (ICE) of RWTH Aachen University as full-time Research Assistant, pursuing his PhD degree. His current research interests are the exploration of programmable hardware architectures for deep learning, with respect to their performance and energy efficiency. This includes both

VLSI design aspects and higher-level considerations related to dataflow optimization and the exploration of memory subsystems. Futhermore, the interaction between algorithmic optimizations, such as network quantization and pruning, with the underlying hardware architecture, is also part of his research.

Benoît DUPONT DE DINECHIN is the Chief Technology Officer of Kalray. He is the Kalray VLIW core main architect and the co-architect of the Multi-Purpose Processing Array (MPPA) processor. Benoît also defined the Kalray software roadmap and contributes to its implementation. Before joining Kalray, Benoît was in charge of Research and Development of the STMicroelectronics Software, Tools, Services division, and was promoted to STMicroelectronics Fellow in 2008. Prior to STMicroelectronics, Benoît worked at the Cray Research Park (Minnesota, USA), where he developed the software pipeliner of the Cray T3E production compilers. Benoît earned an engineering degree in Radar and Telecommunications from the Ecole Nationale Supérieure de l'Aéronautique et de l'Espace (Toulouse, France) and a doctoral degree in Computer Systems from the Université Pierre and Marie Curie (Paris) under the direction of Professor P. Feautrier. He completed his post-doctoral studies at McGill University (Montreal, Canada) at the ACAPS Laboratory led by Professor G.R. Gao. Benoît has authored 12 patents in the area of computer architecture and published over 55 conference papers, journal articles and book chapters in the areas of parallel computing, compiler design and operations research.

Adi ELIAHU received her BSc in Electrical Engineering, summa cum laude, from the Technion – Israel Institute of Technology in 2018. Since then, she has been working towards an MSc degree at the Andrew and Erna Viterbi Faculty of Electrical Engineering at the Technion. Her current research focuses on logic with memristors, and designing architectures for low-power systems using emerging non-volatile memory technologies.

Antoine FARAVELON received his PhD degree in Computer Science from Université Grenoble Alpes, France, in 2018. His interests during his PhD focused mostly on accelerating Dynamic Binary Translation, in the context of full system emulation. This led to a focus on accelerating emulation of virtual memory accesses by leveraging existing host hardware accelerators. Since the latter part of 2018, he has been working on embedded OSes and tools at GreenWaves Technologies.

Ran GINOSAR is a professor of Electrical Engineering at the Technion – Israel Institute of Technology. He received his BSc in Electrical Engineering and Computer Science at the Technion in 1978 and his PhD in Electrical Engineering and Computer Science at Princeton University in 1982. He conducted research at Bell Laboratories from 1982–1983 and has served on the Technion Faculty since 1983. He has spent research periods at the University of Utah and at Intel Research in Oregon, USA. He has co-founded several semiconductor and computer architecture companies. His research interests include VLSI computer architecture, parallel processing and asynchronous logic.

Olivier Gruber received his PhD degree in Computer Science from Pierre and Marie Curie University, France, in 1992. He spent 2 years in INRIA as the lead on the continuation of the research effort of his thesis subject: exploiting micro-kernels for designing efficient parallel and persistent object-oriented systems. From 1994 to 2007, he was part of the IBM research staff. First, at the Almaden Research Center, California, from 1994 to 1996, where he worked on DB2 and the Garlic query middleware. From 1997 to 2007, he worked at the Watson Research Center, New York, where he worked on the very first Java-based webserver by IBM, a research effort that sparked the WebSphere product family. He also worked on the OSGi standard as one of the core framework architects and initiated the Eclipse adoption of OSGi as its plugin platform, via the creation of the Equinox incubator where he coded the early prototypes. In 2007, he accepted a full-time professor position at Grenoble Alpes University, France, where he specialized in actor-based and component-based systems, combined with advanced virtualization. Since 2016, as "millenials" begin entering their university years, he has been leading an effort to improve their learning and enjoyment of software development, experimenting with new teaching skills and methods.

Ameer Haj Ali is currently a PhD student at the Department of Electrical Engineering and Computer Science at the University of California, Berkeley. He completed his MSc studies at the Andrew and Erna Viterbi Faculty of Electrical Engineering at the Technion – Israel Institute of Technology in 2018. He received his BSc degree in Computer Engineering, summa cum laude, in 2017 from the Technion – Israel Institute of Technology. In 2015/2016, he worked as a chip designer at Mellanox Technologies. His current research is focused on auto-tuning, reinforcement learning, machine learning for systems and systems for machine learning, ASIC design and high-performance computing.

Andreas Herkersdorf is a professor within the Department of Electrical and Computer Engineering and is also affiliated to the Department of Informatics at the Technical University of Munich (TUM). He received a Doctorate degree from ETH Zurich, Switzerland in 1991. Between 1988 and 2003, he held technical and management positions with the IBM Research Laboratory in Rueschlikon, Switzerland. Since 2003, Dr. Herkersdorf has been the head of the Chair of Integrated Systems at TUM. He is a senior member of the IEEE, member of the DFG (German Research Foundation) Review Board and serves as an editor for Springer and De Gruyter journals for design automation and information technology. His research interests include application-specific multi-processor architectures, IP network processing, Network-on-Chip and self-adaptive fault-tolerant computing.

Tohru Ishihara received his engineering doctorate in Computer Science in 2000, from Kyushu University. For the following three years, he was a research associate at the University of Tokyo. From 2003 to 2005, he was with Fujitsu Laboratories of America as research staff for an Advanced CAD Technology Group.

From 2005 to 2011, he was with Kyushu University, and with Kyoto University as an associate professor, for the following seven. In October 2018, he joined Nagoya University where he is currently a professor in the Department of Computing and Software Systems. His research interests include low-power design methodologies and power management techniques for embedded systems. One of his publications was recognized in the 2015 IEEE/ACM International Symposium on Low Power Electronics and Design (ISLPED) as "the 1st Most Cited Paper in 20 Years of ISLPED". Dr. Ishihara is a member of the IEEE, ACM, IPSJ and IEICE.

K. Charles JANAC is President and CEO of Arteris IP where he is responsible for growing and establishing a strong global presence for the company which is pioneering the concept of the NoC technology. His career spans 20 years and multiple industries including electronic design automation, semiconductor capital equipment, nanotechnology, industrial polymers and venture capital. In the first decade of his career, he held various marketing and sales positions at Cadence Design Systems (NYSE: CDN) where he helped build it into one of the 10 largest software companies in the world. He joined HLD Systems as president, shifting the company's focus from consulting services to IC floor planning software and building the management, distribution and customer support organizations. He then formed Smart Machines, manufacturer of semiconductor automation equipment, and sold it to Brooks Automation (NASDAQ: BRKS). After a year as Entrepreneur-in-Residence at Infinity Capital, a leading early-stage venture capital firm, where he consulted on information technology investment opportunities, he joined Nanomix as President and CEO, helping build this nanotechnology start-up. He holds BSc and MSc degrees in Organic Chemistry from Tufts University and an MBA from Stanford Graduate School of Business.

Shahar KVATINSKY is an Associate Professor at the Andrew and Erna Viterbi Faculty of Electrical Engineering, Technion – Israel Institute of Technology. Shahar received his BSc degree in Computer Engineering and Applied Physics and his MBA degree in 2009 and 2010, respectively – both from the Hebrew University of Jerusalem – and his PhD degree in Electrical Engineering from the Technion – Israel Institute of Technology, in 2014. From 2006 to 2009, he worked as a circuit designer at Intel. From 2014 to 2015, he was a post-doctoral research fellow at Stanford University. He is an editor of *Microelectronics Journal* and has been the recipient of numerous awards, including the 2019 Krill Prize for Excellence in Scientific Research, the 2015 IEEE Guillemin-Cauer Best Paper Award, the 2015 Best Paper of Computer Architecture Letters, the Viterbi Fellowship, Jacobs Fellowship, ERC starting grant, the 2017 Pazy Memorial Award, the 2014 and 2017 Hershel Rich Technion Innovation Awards, the 2013 Sanford Kaplan Prize for Creative Management in High Tech, the 2010 Benin Prize, and seven Technion excellence teaching awards. His current research focuses on circuits and architectures with emerging memory technologies and design of energy-efficient architectures.

Oliver LENKE is an MSc student at the TUM Department of Electrical and Computer Engineering (EI) with a focus on embedded computing systems. He holds a BSc degree from this department, which he received in 2018. His main research interests are on the analysis and design of on-chip interconnect, as well as memory architectures for multi-core processor arrays.

Olivier MULLER received his engineering (MSc) and PhD degrees in Telecommunications and Electrical Engineering from the Ecole Nationale Supérieure des Télécommunications de Bretagne (TELECOM Bretagne), France, in 2004 and 2007, respectively. From 2004 to 2008, he was a PhD student and postdoctoral researcher with TELECOM Bretagne. Since 2008, he has been Associate Professor at TIMA/Grenoble INP. His research interests are in the areas of reconfigurable computing, methodologies and CAD tools for FPGA and FPGA virtualization.

Lars NOLTE received his BSc degree in 2018 from the Department of Electrical and Computer Engineering at the Technical University of Munich (TUM), Germany. He is currently a graduate student at TUM pursuing an MSc in the area of embedded computing systems. His special research interests are in multi-core architectures, their memory hierarchies and on FPGA-based prototyping.

Frédéric PÉTROT received his PhD degree in Computer Science in 1994, from Pierre and Marie Curie University, Paris, where he was Assistant Professor until September 2004. From 1989 to 1996, he was one of the main contributors of the open-source Alliance CAD System that implements a VHDL to layout design flow, still in use today. In 1996, he started working on electronic system-level design automation, with a specific focus on multi-processor architectures and how to simulate them efficiently. He joined the TIMA Laboratory in September 2004, and, since then, has held a professor position at Ensimag, part of the Grenoble Institute of Technology (Grenoble INP), France. His main interest is still digital system-level design and implementation, including general-purpose computing and hardware acceleration, and fast system-level simulation. He holds several administrative responsibilities at TIMA and Ensimag, and is scientific advisor at GreenSocs. He has been a member of the MPSoC community since 2004, and was twice general chair of the Forum.

Sven RHEINDT holds BSc and MSc degrees in Electrical and Computer Engineering with a focus on Integrated Systems, which he received from the TUM Department of Electrical and Computer Engineering in 2013 and 2015, respectively. Since 2016, he has been a staff member at the Chair of Integrated Systems at TUM and is working towards a doctoral degree. His research interests are on memory architecture and near-memory computing.

Frédéric ROUSSEAU received his degree in Computer Science and Electrical Engineering from Université Grenoble Alpes in 1991, and a PhD in Computer Science from the University of Evry, France, in 1997. He has held an assistant professor position at Grenoble Alpes University since October 1999 and a professor position since 2007. He is a researcher at the TIMA Lab. His research interests concern multi-processor systems-on-chip design and architecture, prototyping of hardware/software systems, including reconfigurable systems and high-level synthesis for embedded systems.

Arif SASONGKO received his Bachelor's and Master's in Electrical Engineering from the Bandung Technology Institute (ITB), Indonesia, in 1998 and 2001, respectively. He then received his PhD degree in Microelectronics from Joseph Fourier University (now part of Grenoble Alpes University, France), in 2004. Since 2006, he has been a lecturer at ITB with research interests in the areas of digital design, FPGA and cryptography. He also has a keen interest in engineering design methodology. Currently, he is chairman of the Electrical Engineering Program at ITB.

Jun SHIOMI received his BE degree in Electronic Engineering in 2014, an ME degree in Communications and Computer Engineering in 2016 and a PhD degree in Informatics in 2017, all from Kyoto University, Japan. In 2017, he joined Kyoto University, where he is currently an Assistant Professor in the Department of Communications and Computer Engineering. His research interests include modeling and computer-aided design for low-power and low-voltage system-on-chips. He is a member of the IEEE, IPSJ and IEICE.

Akshay SRIVATSA received his BEng degree in Electronics and Communication Engineering in 2013, from the Visvesvaraya Technological University, India. In 2015, he received an MSc degree in Communication Engineering from the Department of Electrical and Computer Engineering at the Technical University of Munich (TUM), Germany. Since 2016, he has been a member of the scientific staff at the Chair of Integrated Systems of TUM, pursuing a Doctor of Engineering degree. His research interests include computer architecture and micro-architecture, with a focus on memory hierarchy and cache coherence.

Yankin TANURHAN is Vice President of Engineering for DesignWare Processor Cores, IP Subsystems and Non-Volatile Memory, Security and SoC Design at Synopsys. He leads the low-power and high-performance ARC and EV embedded processor developments, ASIP tool development, IP Subsystems development and CMOS-based Non-Volatile IP development. Before joining Synopsys, Dr. Tanurhan was Vice President and General Manager of Virage Logic's Processors, SoC Infrastructure and NVM Solutions business units. Virage Logic was acquired by Synopsys in September 2010. Prior to this, Dr. Tanurhan served as Vice President of Actel's Advanced Applications and System Solutions, where he led Actel's new architecture design, IP and MPU business units, system and hardware tools and product validation departments. He was also responsible for leading Actel's embedded FPGA, embedded processor and DSP activities. Previously in his research career he served as the director of the department of electronic systems and microsystems of

FZI (Forschungszentrum Informatik), a German contract research institute attached to the University of Karlsruhe. Dr. Tanurhan has authored more than 100 papers in revered publications. He holds a BSc and MSc in Electrical and Computer Engineering from Rheinisch Westfaellische Technische Hochschule (RWTH) in Aachen, Germany, and an engineering doctorate, summa cum laude, in Electrical Engineering from the University of Karlsruhe (TH), Germany.

Pieter VAN DER WOLF is a principal product architect at Synopsys. He received his MSc and PhD degrees in Electrical Engineering from Delft University of Technology where he was an Associate Professor before joining Philips Research in 1996. In 2006, he joined NXP Semiconductors when it was spun out of Philips Electronics. In 2009, he joined VirageLogic, which was subsequently acquired by Synopsys. He has worked on a broad range of topics including (multi-)processor architectures and system design methodologies, with a focus on DSP applications.

Arthur VIANES received a Master's degree in Microelectronics from Université Grenoble Alpes, France. He is currently an industrial PhD student with Kalray and the SLS group of the TIMA Lab, working on Kalray's MPPA Coolidge architecture to turn the MPPA processor into a coprocessor for ARM.

Arief WICAKSANA obtained his Bachelor degree in Electrical Engineering from the Bandung Institute of Technology (ITB). He later received his Engineering degree and PhD degree in Computer Science from Université Grenoble Alpes in 2015 and 2018, respectively. He pursued his PhD degree in the TIMA Lab, Grenoble, from 2015 to 2018. Since then, he has been working as a research engineer at Embedded Computing Laboratory, CEA Tech. His research areas include reconfigurable computing, FPGA virtualization and virtual prototyping.

Thomas WILD received his engineer's degree in 1989 and his engineering doctorate in 2003, both from the Department of Electrical and Computer Engineering at the Technical University of Munich (TUM), Germany. He is a member of the scientific staff at the Chair of Integrated Systems (LIS) of TUM where he is responsible for activities in the area of multi-core and network processing architectures. His current research interests comprise multi-processor system-on-chip (MPSoC) architectures, networks-on-chip (NoCs) and memory hierarchies, as well as MPSoC diagnosis, system-level design methodologies and design space exploration.

Index

Printed and bound by CPI Group (UK) Ltd, Croydon, CR0 4YY

13/09/2026

14961044-0004